UNDERSTANDING LTE WITH MATLAB®

UNDERSTANDING LTE WITH MATLAB®
FROM MATHEMATICAL MODELING TO SIMULATION AND PROTOTYPING

Dr Houman Zarrinkoub
MathWorks, Massachusetts, USA

WILEY

Library of Congress Cataloging-in-Publication Data

Zarrinkoub, Houman.
  Understanding LTE with MATLAB : from mathematical foundation to simulation, performance evaluation and implementation / Houman Zarrinkoub.
       pages cm
   Includes bibliographical references and index.
   ISBN 978-1-118-44341-5 (hardback)
  1. Long-Term Evolution (Telecommunications)–Computer simulation. 2. MATLAB. I. Title.
   TK5103.48325.Z37 2014
   621.3845′6–dc23

                                                                                          2013034138

A catalogue record for this book is available from the British Library.

ISBN: 9781118443415

Typeset in 10/12pt TimesLTStd by Laserwords Private Limited, Chennai, India
Printed and bound in Malaysia by Vivar Printing Sdn Bhd

1   2014

Contents

Preface

The LTE (Long Term Evolution) and LTE-Advanced are the latest mobile communications standards developed by the Third Generation Partnership Project (3GPP). These standards represent a transformative change in the evolution of mobile technology. Within the present decade, the network infrastructures and mobile terminals have been designed and upgraded to support the LTE standards. As these systems are deployed in every corner of the globe, the LTE standards have finally realized the dream of providing a truly global broadband mobile access technology.

In this book we will examine the LTE mobile communications standard, and specifically its PHY (Physical Layer), in order to understand how and why it can achieve such a remarkable feat. We will look at it simultaneously from an academic and a pragmatic point of view. We will relate the mathematical foundation of its enabling technologies, such as Orthogonal Frequency Division Multiplexing (OFDM) and Multiple Input Multiple Output (MIMO), to its ability to achieve such a superb performance. We will also show how pragmatic engineering considerations have shaped the formulation of many of its components. As an integral part of this book, we will use MATLAB®, a technical computing language and simulation environment widely used by the scientific and engineering community, to clarify the mathematical concepts and constructs, provide algorithms, testbenches, and illustrations, and give the reader a deep understanding of the specifications through the use of simulations.

This book is written for both the academic community and the practicing professional. It focuses specifically on the LTE standard and its evolution. Unlike many titles that treat only the mathematical foundation of the standard, this book will discuss the mathematical formulation of many enabling technologies (such as OFDM and MIMO) in the context of the overall performance of the system. Furthermore, by including chapters dedicated to simulation, performance evaluation, and implementation, the book broadens its appeal to a much larger readership composed of both academicians and practitioners.

Through an intuitive and pedagogic approach, we will build up components of the LTE PHY progressively from simple to more complex using MATLAB programs. Through simulation of the MATLAB programs, the reader will feel confident that he or she has learned not only all the details necessary to fully understand the standard but also the ability to implement it.

We aim to clarify technical details related to PHY modeling of the LTE standard. Therefore, knowledge of the basics of communication theory (topics such as modulation, coding, and estimation) and digital signal processing is a prerequisite. These prerequisites are usually covered by the senior year of most electrical engineering undergraduate curricula. It also aims to teach through simulation with MATLAB. Therefore a basic knowledge of the MATLAB language is necessary to follow the text. This book is intended for professors, researchers, and students in electrical and computer engineering departments, as well as engineers, designers, and implementers of wireless systems. What they learn from both a technical and a programming point of view may be quite applicable to their everyday work. Depending on the reader's function and the need to implement or teach the LTE standard, this book may be considered introductory, intermediate, or advanced in nature.

The book is conceptually composed of two parts. The first deals with modeling the PHY of the LTE standard and with MATLAB algorithms that enable the reader to simulate and verify various components of the system. The second deals with practical issues such as simulation of the system and implementation and prototyping of its components. In the first chapter we provide a brief introduction to the standard, its genesis, and its objective, and we identify four enabling technologies (OFDM, MIMO, turbo coding, and dynamic link adaptations) as the components responsible for its remarkable performance. In Chapter 2, we provide a quick and sufficiently detailed overview of the LTE PHY specifications. Chapter 3 introduces the modeling, simulation, and implementation capabilities of MATLAB and Simulink that are used throughout this book. In Chapters 4–7 we treat each of the enabling technologies of the LTE standard (modulation and coding, OFDM, MIMO, and link adaptations) in detail and create models in MATLAB that iteratively and progressively build up LTE PHY components based on these. We wrap up the first part of the book in Chapter 8 by putting all the enabling technologies together and showing how the PHY of the LTE standard can be modeled in MATLAB based on the insight obtained in the preceding chapters.

Chapter 9 includes a discussion on how to accelerate the speed of our MATLAB programs through the use of a variety of techniques, including parallel computing, automatic C code generation, GPU processing, and more efficient algorithms. In Chapter 10 we discuss some implementation issues, such as target environments, and how they affect the programming style. We also discuss fixed-point numerical representation of data as a prerequisite for hardware implementation and its effect on the performance of the standard. Finally, in Chapter 11 we summarize what we have discussed and provide some directions for future work.

Any effort related to introducing the technical background of a complex communications system like LTE requires addressing the question of scope. We identify three conceptual elements that can combine to provide a deep understanding of the way the LTE standard works:

- The theoretical background of the enabling technologies
- Details regarding the standard specifications
- Algorithms and simulation testbenches needed to implement the design

To make the most of the time available to develop this book, we decided to strike a balance in covering each of these conceptual elements. We chose to provide a sufficient level of discussion regarding the theoretical foundations and technical specifications of the standard. To leverage our expertise in developing MATLAB applications, we decided to cover the algorithms and testbenches that implement various modes of the LTE standard in further detail. This choice was motivated by two factors:

1. There are many books that extensively cover the first two elements and do not focus on algorithms and simulations. We consider the emphasis on simulation one of the innovative characteristics of this work.

2. By providing simulation models of the LTE standard, we help the reader develop an understanding of the elements that make up a communications system and obtain a programmatic recipe for the sequence of operations that make up the PHY specifications. Algorithms and testbenches naturally reveal the dynamic nature of a system through simulation.

In this sense, the insight and understanding obtained by delving into simulation details are invaluable as they provide a better mastery of the subject matter. Even more importantly, they instill a sense of confidence in the reader that he or she can try out new ideas, propose and test new improvements, and make use of new tools and models to help graduate from a theoretical knowledge to a hands-on understanding and ultimately to the ability to innovate, design, and implement.

It is our hope that this book can provide a reliable framework for modeling and simulation of the LTE standard for the community of young researchers, students, and professionals interested in mobile communications. We hope they can apply what they learn here, introduce their own improvements and innovations, and become inspired to contribute to the research and development of the mobile communications systems of the future.

List of Abbreviations

ASIC	Application-Specific Integrated Circuit
BCH	Broadcast Channel
BER	Bit Error Rate
BPSK	Binary Phase Shift Keying
CP	Cyclic Prefix
CQI	Channel Quality Indicator
CRC	Cyclic Redundancy Check
CSI	Channel State Information
CSI-RS	Channel State Information Reference Signal
CSR	Cell-Specific Reference
CUDA	Compute Unified Device Architecture
DM-RS	Demodulation Reference Signal
DSP	Digital Signal Processor
eNodeB	enhanced Node Base station
E-UTRA	Evolved Universal Terrestrial Radio Access
FDD	Frequency Division Duplex
FPGA	Field-Programmable Gate Array
HARQ	Hybrid Automatic Repeat Request
HDL	Hardware Description Language
LTE	Long Term Evolution
MAC	Medium Access Control
MBMS	Multimedia Broadcast and Multicast Service
MBSFN	Multicast/Broadcast over Single Frequency Network
MIMO	Multiple Input Multiple Output
MMSE	Minimum Mean Square Error
MRC	Maximum Ratio Combining
MU-MIMO	Multi-User Multiple Input Multiple Output
OFDM	Orthogonal Frequency Division Multiplexing
PBCH	Physical Broadcast Channel
PCFICH	Physical Control Format Indicator Channel
PCM	Pulse Code Modulation
PDCCH	Physical Downlink Control Channel
PDSCH	Physical Downlink Shared Channel
PHICH	Physical Hybrid ARQ Indicator Channel

PHY	Physical Layer
PMCH	Physical Multicast Channel
PRACH	Physical Random Access Channel
PSS	Primary Synchronization Signal
PUCCH	Physical Uplink Control Channel
PUSCH	Physical Uplink Shared Channel
QAM	Quadrature Amplitude Modulation
QPP	Quadratic Permutation Polynomial
QPSK	Quadrature Phase Shift Keying
RLC	Radio Link Control
RMS	Root Mean Square
RRC	Radio Resource Control
RTL	Register Transfer Level
SC-FDM	Single-Carrier Frequency Division Multiplexing
SD	Sphere Decoder
SFBC	Space–Frequency Block Coding
SINR	Signal-to-Interference-plus-Noise Ratio
SNR	Signal-to-Noise Ratio
SSD	Soft-Sphere Decoder
SSS	Secondary Synchronization Signal
STBC	Space–Time Block Coding
SFBC	Space-Frequency Block Coding
SU-MIMO	Single-User MIMO
TDD	Time-Division Duplex
UE	User Equipment
ZF	Zero Forcing

1

Introduction

We live in the era of a mobile data revolution. With the mass-market expansion of smartphones, tablets, notebooks, and laptop computers, users demand services and applications from mobile communication systems that go far beyond mere voice and telephony. The growth in data-intensive mobile services and applications such as Web browsing, social networking, and music and video streaming has become a driving force for development of the next generation of wireless standards. As a result, new standards are being developed to provide the data rates and network capacity necessary to support worldwide delivery of these types of rich multimedia application.

LTE (Long Term Evolution) and LTE-Advanced have been developed to respond to the requirements of this era and to realize the goal of achieving global broadband mobile communications. The goals and objectives of this evolved system include higher radio access data rates, improved system capacity and coverage, flexible bandwidth operations, significantly improved spectral efficiency, low latency, reduced operating costs, multi-antenna support, and seamless integration with the Internet and existing mobile communication systems.

In some ways, LTE and LTE-Advanced are representatives of what is known as a fourth-generation wireless system and can be considered an organic evolution of the third-generation predecessors. On the other hand, in terms of their underlying transmission technology they represent a disruptive departure from the past and the dawn of what is to come. To put into context the evolution of mobile technology leading up to the introduction of the LTE standards, a short overview of the wireless standard history will now be presented. This overview intends to trace the origins of many enabling technologies of the LTE standards and to clarify some of their requirements, which are expressed in terms of improvements over earlier technologies.

1.1 Quick Overview of Wireless Standards

In the past two decades we have seen the introduction of various mobile standards, from 2G to 3G to the present 4G, and we expect the trend to continue (see Figure 1.1). The primary mandate of the 2G standards was the support of mobile telephony and voice applications. The 3G standards marked the beginning of the packet-based data revolution and the support of Internet

Understanding LTE with MATLAB®: From Mathematical Modeling to Simulation and Prototyping, First Edition. Houman Zarrinkoub.

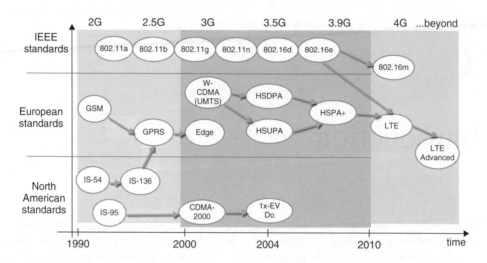

Figure 1.1 Evolution of wireless standards in the last two decades

applications such as email, Web browsing, text messaging, and other client-server services. The 4G standards will feature all-IP packet-based networks and will support the explosive demand for bandwidth-hungry applications such as mobile video-on-demand services.

Historically, standards for mobile communication have been developed by consortia of network providers and operators, separately in North America, Europe, and other regions of the world. The second-generation (2G) digital mobile communications systems were introduced in the early 1990s. The technology supporting these 2G systems were circuit-switched data communications. The GSM (Global System for Mobile Communications) in Europe and the IS-54 (Interim Standard 54) in North America were among the first 2G standards. Both were based on the Time Division Multiple Access (TDMA) technology. In TDMA, a narrowband communication channel is subdivided into a number of time slots and multiple users share the spectrum at allocated slots. In terms of data rates, for example, GSM systems support voice services up to 13 kbps and data services up to 9.6 kbps.

The GSM standard later evolved into the Generalized Packet Radio Service (GPRS), supporting a peak data rate of 171.2 kbps. The GPRS standard marked the introduction of the split-core wireless networks, in which packet-based switching technology supports data transmission and circuit-switched technology supports voice transmission. The GPRS technology further evolved into Enhanced Data Rates for Global Evolution (EDGE), which introduced a higher-rate modulation scheme (8-PSK, Phase Shift Keying) and further enhanced the peak data rate to 384 kbps.

In North America, the introduction of IS-95 marked the first commercial deployment of a Code Division Multiple Access (CDMA) technology. CDMA in IS-95 is based on a direct spread spectrum technology, where multiple users share a wider bandwidth by using orthogonal spreading codes. IS-95 employs a 1.2284 MHz bandwidth and allows for a maximum of 64 voice channels per cell, with a peak data rate of 14.4 kbps per fundamental channel. The IS-95-B revision of the standard was developed to support high-speed packet-based data transmission. With the introduction of the new supplemental code channel supporting high-speed packet data, IS-95-B supported a peak data rate of 115.2 kbps. In North America,

3GPP2 (Third Generation Partnership Project 2) was the standardization body that established technical specifications and standards for 3G mobile systems based on the evolution of CDMA technology. From 1997 to 2003, 3GPP2 developed a family of standards based on the original IS-95 that included 1xRTT, 1x-EV-DO (Evolved Voice Data Only), and EV-DV (Evolved Data and Voice). 1xRTT doubled the IS-95 capacity by adding 64 more traffic channels to achieve a peak data rate of 307 kbps. The 1x-EV-DO and 1x-EV-DV standards achieved peak data rates in the range of 2.4–3.1 Mbps by introducing a set of features including adaptive modulation and coding, hybrid automatic repeat request (HARQ), turbo coding, and faster scheduling based on smaller frame sizes.

The 3GPP (Third-Generation Partnership Project) is the standardization body that originally managed European mobile standard and later on evolved into a global standardization organization. It is responsible for establishing technical specifications for the 3G mobile systems and beyond. In 1997, 3GPP started working on a standardization effort to meet goals specified by the ITU IMT-2000 (International Telecommunications Union International Mobile Telecommunication) project. The goal of this project was the transition from a 2G TDMA-based GSM technology to a 3G wide-band CDMA-based technology called the Universal Mobile Telecommunications System (UMTS). The UMTS represented a significant change in mobile communications at the time. It was standardized in 2001 and was dubbed Release 4 of the 3GPP standards. The UMTS system can achieve a downlink peak data rate of 1.92 Mbps. As an upgrade to the UMTS system, the High-Speed Downlink Packet Access (HSDPA) was standardized in 2002 as Release 5 of the 3GPP. The peak data rates of 14.4 Mbps offered by this standard were made possible by introducing faster scheduling with shorter subframes and the use of a 16QAM (Quadrature Amplitude Modulation) modulation scheme. High-Speed Uplink Packet Access (HSUPA) was standardized in 2004 as Release 6, with a maximum rate of 5.76 Mbps. Both of these standards, together known as HSPA (High-Speed Packet Access), were then upgraded to Release 7 of the 3GPP standard known as HSPA+ or MIMO (Multiple Input Multiple Output) HSDPA. The HSPA+ standard can reach rates of up to 84 Mbps and was the first mobile standard to introduce a 2×2 MIMO technique and the use of an even higher modulation scheme (64QAM). Advanced features that were originally introduced as part of the North American 3G standards were also incorporated in HSPA and HSPA+. These features include adaptive modulation and coding, HARQ, turbo coding, and faster scheduling.

Another important wireless application that has been a driving force for higher data rates and spectral efficiency is the wireless local area network (WLAN). The main purpose of WLAN standards is to provide stationary users in buildings (homes, offices) with reliable and high-speed network connections. As the global mobile communications networks were undergoing their evolution, IEEE (Institute of Electrical and Electronics Engineers) was developing international standards for WLANs and wireless metropolitan area networks (WMANs). With the introduction of a family of WiFi standards (802.11a/b/g/n) and WiMAX standards (802.16d/e/m), IEEE established Orthogonal Frequency Division Multiplexing (OFDM) as a promising and innovative air-interface technology. For example, the IEEE 802.11a WLAN standard uses the 5 GHz frequency band to transmit OFDM signals with data rates of up to 54 Mb/s. In 2006, IEEE standardized a new WiMAX standard (IEEE 802.16m) that introduced a packet-based wireless broadband system. Among the features of WiMAX are scalable bandwidths up to 20 MHz, higher peak data rates, and better special efficiency profiles than were being offered by the UMTS and HSPA systems at the time. This advance essentially kicked off the effort by 3GPP to introduce a new wireless mobile standard that could compete with the WiMAX technology. This effort ultimately led to the standardization of the LTE standard.

Table 1.1 Peak data rates of various wireless standards introduced over the past two decades

Technology	Theoretical peak data rate (at low mobility)
GSM	9.6 kbps
IS-95	14.4 kbps
GPRS	171.2 kbps
EDGE	473 kbps
CDMA-2000 (1xRTT)	307 kbps
WCDMA (UMTS)	1.92 Mbps
HSDPA (Rel 5)	14 Mbps
CDMA-2000 (1x-EV-DO)	3.1 Mbps
HSPA+ (Rel 6)	84 Mbps
WiMAX (802.16e)	26 Mbps
LTE (Rel 8)	300 Mbps
WiMAX (802.16m)	303 Mbps
LTE-Advanced (Rel 10)	1 Gbps

1.2 Historical Profile of Data Rates

Table 1.1 summarizes the peak data rates of various wireless technologies. Looking at the maximum data rates offered by these standards, the LTE standard (3GPP release 8) is specified to provide a maximum data rate of 300 Mbps. The LTE-Advanced (3GPP version 10) features a peak data rate of 1 Gbps.

These figures represent a boosts in peak data rates of about 2000 times above what was offered by GSM/EDGE technology and 50–500 times above what was offered by the W-CDMA/UMTS systems. This remarkable boost was achieved through the development of new technologies introduced within a time span of about 10 years. One can argue that this extraordinary advancement is firmly rooted in the elegant mathematical formulation of the enabling technologies featured in the LTE standards. It is our aim in this book to clarify and explain these enabling technologies and to put into context how they combine to achieve such a performance. We also aim to gain insight into how to simulate, verify, implement, and further enhance the PHY (Physical Layer) technology of the LTE standards.

1.3 IMT-Advanced Requirements

The ITU has published a set of requirements for the design of mobile systems. The first recommendations, released in 1997, were called IMT-2000 (International Mobile Telecommunications 2000) [1]. These recommendations included a set of goals and requirements for radio interface specification. 3G mobile communications systems were developed to be compliant with these recommendations. As the 3G systems evolved, so did the IMT-2000 requirements, undergoing multiple updates over the past decade [2].

In 2007, ITU published a new set of recommendations that set the bar much higher and provided requirements for IMT-Advanced systems [3]. IMT-Advanced represents the requirements for the building of truly global broadband mobile communications systems. Such systems can provide access to a wide range of packet-based advanced mobile services, support low- to high-mobility applications and a wide range of data rates, and provide capabilities for high-quality multimedia applications. The new requirements were published to spur research and development activities that bring about a significant improvement in performance and quality of services over the existing 3G systems.

One of the prominent features of IMT-Advanced is the enhanced peak data for advanced services and applications (100 Mbps for high mobility and 1 Gbps for low mobility). These requirements were established as targets for research. The LTE-Advanced standard developed by 3GPP and the mobile WiMAX standard developed by IEEE are among the most prominent standards to meet the requirements of the IMT-Advanced specifications. In this book, we focus on the LTE standards and discuss how their PHY specification is consistent with the requirements of the IMT-Advanced.

1.4   3GPP and LTE Standardization

The LTE and LTE-Advanced are developed by the 3GPP. They inherit a lot from previous 3GPP standards (UMTS and HSPA) and in that sense can be considered an evolution of those technologies. However, to meet the IMT-Advanced requirements and to keep competitive with the WiMAX standard, the LTE standard needed to make a radical departure from the W-CDMA transmission technology employed in previous standards. LTE standardization work began in 2004 and ultimately resulted in a large-scale and ambitious re-architecture of mobile networks. After four years of deliberation, and with contributions from telecommunications companies and Internet standardization bodies all across the globe, the standardization process of LTE (3GPP Release 8) was completed in 2008. The Release 8 LTE standard later evolved to LTE Release 9 with minor modifications and then to Release 10, also known as the LTE-Advanced standard. The LTE-Advanced features improvements in spectral efficiency, peak data rates, and user experience relative to the LTE. With a maximum peak data rate of 1 Gbps, LTE-Advanced has also been approved by the ITU as an IMT-Advanced technology.

1.5   LTE Requirements

LTE requirements cover two fundamental components of the evolved UMTS system architecture: the Evolved Universal Terrestrial Radio Access Network (E-UTRAN) and the Evolved Packet Core (EPC). The goals of the overall system include the following:

- Improved system capacity and coverage
- High peak data rates
- Low latency (both user-plane and control-plane)
- Reduced operating costs
- Multi-antenna support
- Flexible bandwidth operations
- Seamless integration with existing systems (UMTS, WiFi, etc.).

As a substantial boost in mobile data rates is one of the main mandates of the LTE standards, it is useful to review some of the recent advances in communications research as well as theoretical considerations related to the maximum achievable data rates in a mobile communications link. We will now present some highlights related to this topic, inspired by an excellent discussion presented in Reference [4].

1.6   Theoretical Strategies

Shannon's fundamental work on channel capacity states that data rates are always limited by the available received signal power or the received signal-to-noise-power ratio [5]. It also relates the data rates to the transmission bandwidths. In the case of low-bandwidth utilization, where the data rate is substantially lower than the available bandwidth, any increase of the data rate will require an increase in the received signal power in a proportional manner. In the case of high-bandwidth utilization, where data rates are equal to or greater than the available bandwidth, any increase in the data rate will require a much larger relative increase in the received signal power unless the bandwidth is increased in proportion to the increase in data rate.

A rather intuitive way to increase the overall power at the receiver is to use multiple antennas at the receiver side. This is known as receive diversity. Multiple antennas can also be used at the transmitter side, in what is known as transmit diversity. A transmit diversity approach based on beamforming uses multiple transmit antennas to focus the transmitted power in the direction of the receiver. This can potentially increase the received signal power and allow for higher data rates.

However, increasing data rates by using either transmit diversity or receive diversity can only work up to a certain point. Beyond this, any boost in data rates will start to saturate. An alternative approach is to use multiple antennas at both the transmitter and the receiver. For example, a technique known as spatial multiplexing, which exploits multiple antennas at the transmitter and the receiver sides, is an important member of the class of multi-antenna techniques known as MIMO. Different types of MIMO technique, including open-loop and closed-loop spatial multiplexing, are used in the LTE standard.

Beside the received signal power, another factor directly impacting on the achievable data rates of a mobile communications system is the transmission bandwidth. The provisioning of higher data rates usually involves support for even wider transmission bandwidths. The most important challenge related to wider-band transmission is the effect of multipath fading on the radio channel. Multipath fading is the result of the propagation of multiple versions of the transmitted signals through different paths before they arrive at the receiver. These different versions exhibit varying profiles of signal power and time delays or phases. As a result, the received signal can be modeled as a filtered version of the transmitted signal that is filtered by the impulse response of the radio channel. In the frequency domain, a multipath fading channel exhibits a time-varying channel frequency response. The channel frequency response inevitably corrupts the original frequency-domain content of the transmitted signal, with an adverse effect on the achievable data rates. In order to adjust for the effects of channel frequency selectivity and to achieve a reasonable performance, we must either increase the transmit power, reduce our expectations concerning data rates, or compensate for the frequency-domain distortions with equalization.

Many channel-equalization techniques have been proposed to counter the effects of multipath fading. Simple time-domain equalization methods have been shown to provide adequate performance for transmission over transmission bandwidths of up to 5 MHz. However, for

LTE standards and other mobile systems that provision for wider bandwidths of 10, 15, or 20 MHz, or higher, the complexity of the time-domain equalizers become prohibitively large. In order to overcome the problems associated with time-domain equalization, two approaches to wider-band transmission have been proposed:

- The use of multicarrier transmission schemes, where a wider band signal is represented as the sum of several more narrowband orthogonal signals. One special case of multicarrier transmission used in the LTE standard is the OFDM transmission.
- The use of a single-carrier transmission scheme, which provides the benefits of low-complexity frequency-domain equalization offered by OFDM without introducing its high transmit power fluctuations. An example of this type of transmission is called Single-Carrier Frequency Division Multiplexing (SC-FDM), which is used in the LTE standard as the technology for uplink transmission.

Furthermore, a rather straightforward way of providing higher data rates within a given transmission bandwidth is the use of higher-order modulation schemes. Using higher-order modulation allows us to represent more bits with a single modulated symbol and directly increases bandwidth utilization. However, the higher bandwidth utilization comes at a cost: a reduced minimum distance between modulated symbols and a resultant increased sensitivity to noise and interference. Consequently, adaptive modulation and coding and other link adaptation strategies can be used to judiciously decide when to use a lower- or higher-order modulation. By applying these adaptive methods, we can substantially improve the throughput and achievable data rates in a communications link.

1.7 LTE-Enabling Technologies

The enabling technologies of the LTE and its evolution include the OFDM, MIMO, turbo coding, and dynamic link-adaptation techniques. As discussed in the last section, these technologies trace their origins to well-established areas of research in communications and together help contribute to the ability of the LTE standard to meet its requirements.

1.7.1 OFDM

As elegantly described in Reference [6], the main reasons LTE selects OFDM and its single-carrier counterpart SC-FDM as the basic transmission schemes include the following: robustness to the multipath fading channel, high spectral efficiency, low-complexity implementation, and the ability to provide flexible transmission bandwidths and support advanced features such as frequency-selective scheduling, MIMO transmission, and interference coordination.

OFDM is a multicarrier transmission scheme. The main idea behind it is to subdivide the information transmitted on a wideband channel in the frequency domain and to align data symbols with multiple narrowband orthogonal subchannels known as subcarriers. When the frequency spacing between subcarriers is sufficiently small, an OFDM transmission scheme can represent a frequency-selective fading channel as a collection of narrowband flat fading subchannels. This in turn enables OFDM to provide an intuitive and simple way of estimating the channel frequency response based on transmitting known data or reference signals. With a good estimate of the channel response at the receiver, we can then recover the best estimate of the transmitted signal using a low-complexity frequency-domain equalizer. The equalizer in a sense inverts the channel frequency response at each subcarrier.

1.7.2   SC-FDM

One of the drawbacks of OFDM multicarrier transmission is the large variations in the instantaneous transmit power. This implies a reduced efficiency in power amplifiers and results in higher mobile-terminal power consumption. In uplink transmission, the design of complex power amplifiers is especially challenging. As a result, a variant of the OFDM transmission known as SC-FDM is selected in the LTE standard for uplink transmission. SC-FDM is implemented by combining a regular OFDM system with a precoding based on Discrete Fourier Transform (DFT) [6]. By applying a DFT-based precoding, SC-FDM substantially reduces fluctuations of the transmit power. The resulting uplink transmission scheme can still feature most of the benefits associated with OFDM, such as low-complexity frequency-domain equalization and frequency-domain scheduling, with less stringent requirements on the power amplifier design.

1.7.3   MIMO

MIMO is one of the key technologies deployed in the LTE standards. With deep roots in mobile communications research, MIMO techniques bring to bear the advantages of using multiple antennas in order to meet the ambitious requirements of the LTE standard in terms of peak data rates and throughput.

  MIMO methods can improve mobile communication in two different ways: by boosting the overall data rates and by increasing the reliability of the communication link. The MIMO algorithms used in the LTE standard can be divided into four broad categories: receive diversity, transmit diversity, beamforming, and spatial multiplexing. In transmit diversity and beamforming, we transmit redundant information on different antennas. As such, these methods do not contribute to any boost in the achievable data rates but rather make the communications link more robust. In spatial multiplexing, however, the system transmits independent (nonredundant) information on different antennas. This type of MIMO scheme can substantially boost the data rate of a given link. The extent to which data rates can be improved may be linearly proportional to the number of transmit antennas. In order to accommodate this, the LTE standard provides multiple transmit configurations of up to four transmit antennas in its downlink specification. The LTE-Advanced allows the use of up to eight transmit antennas for downlink transmission.

1.7.4   Turbo Channel Coding

Turbo coding is an evolution of the convolutional coding technology used in all previous standards with impressive near-channel capacity performance [7]. Turbo coding was first introduced in 1993 and has been deployed in 3G UMTS and HSPA systems. However, in these standards it was used as an optional way of boosting the performance of the system. In the LTE standard, on the other hand, turbo coding is the only channel coding mechanism used to process the user data.

The near-optimal performance of turbo coders is well documented, as is the computational complexity associated with their implementation. The LTE turbo coders come with many improvements, aimed at making them more efficient in their implementation. For example, by appending a CRC (Cyclic Redundancy Check) checking syndrome to the input of the turbo encoder, LTE turbo decoders can take advantage of an early termination mechanism if the quality of the code is deemed acceptable. Instead of following through with a fixed number of decoding iterations, the decoding can be stopped early when the CRC check indicates that no errors are detected. This very simple solution allows the computational complexity of the LTE turbo decoders to be reduced without severely penalizing their performance.

1.7.5   Link Adaptation

Link adaptation is defined as a collection of techniques for changing and adapting the transmission parameters of a mobile communication system to better respond to the dynamic nature of the communication channel. Depending on the channel quality, we can use different modulation and coding techniques (adaptive modulation and coding), change the number of transmit or receive antennas (adaptive MIMO), and even change the transmission bandwidth (adaptive bandwidth). Closely related to link adaptation is channel-dependent scheduling in a mobile communication system. Scheduling deals with the question of how to share the radio resources between different users in order to achieve more efficient resource utilizations. Typically, we need to either minimize the amount of resources allocated to each user or match the resources to the type and priority of the user data. Channel-dependent scheduling aims to accommodate as many users as possible, while satisfying the best quality-of-service requirements that may exist based on the instantaneous channel condition.

1.8   LTE Physical Layer (PHY) Modeling

In this book we will focus on digital signal processing in the physical layer of the Radio Access networks. Almost no discussion of the LTE core networks is present here, and we will leave the discussion of higher-layer processing such as Radio Resource Control (RRC), Radio Link Control (RLC), and Medium Access Control (MAC) to another occasion.

Physical layer modeling involves all the processing performed on bits of data that are handed down from the higher layers to the PHY. It describes how various transport channels are mapped to physical channels, how signal processing is performed on each of these channels, and how data are ultimately transported to the antenna for transmission.

For example, Figure 1.2 illustrates the PHY model for the LTE downlink transmission. First, the data is multiplexed and encoded in a step known as Downlink Shared Channel processing (DLSCH). The DLSCH processing chain involves attaching a CRC code for error detection, segmenting the data into smaller chunks known as subblocks, undertaking channel-coding operations based on turbo coding for the user data, carrying out a rate-matching operation that selects the number of output bits to reflect a desired coding rate, and finally reconstructing the codeblocks into codewords. The next phase of processing is known as physical downlink

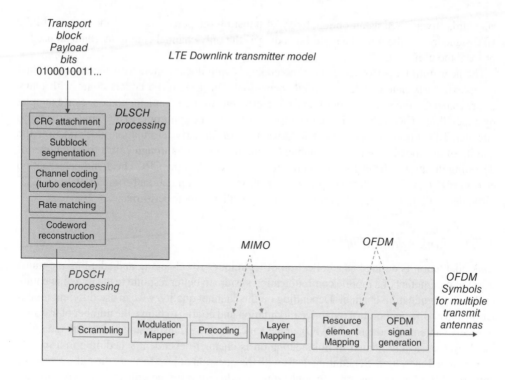

Figure 1.2   Physical layer specifications in LTE

shared channel processing. In this phase, the codewords first become subject to a scrambling operation and then undergo a modulation mapping that results in a modulated symbol stream. The next step comprises the LTE MIMO or multi-antenna processing, in which a single stream of modulated symbols is subdivided into multiple substreams destined for transmission via multiple antennas. The MIMO operations can be regarded as a combination of two steps: precoding and layer mapping. Precoding scales and organizes symbols allocated to each substream and layer mapping selects and routes data into each substream to implement one of the nine different MIMO modes specified for downlink transmission. Among the available MIMO techniques implemented in downlink transmission are transmit diversity, spatial multiplexing, and beamforming. The final step in the processing chain relates to the multicarrier transmission. In downlink, the multicarrier operations are based on the OFDM transmission scheme. The OFDM transmission involves two steps. First, the resource element mapping organizes the modulated symbols of each layer within a time–frequency resource grid. On the frequency axis of the grid, the data are aligned with subcarriers in the frequency domain. In the OFDM signal-generation step, a series of OFDM symbols are generated by applying inverse Fourier transform to compute the transmitted data in time and are transported to each antenna for transmission.

In my opinion, it is remarkable that such a straightforward and intuitive transmission structure can combine all the enabling technologies so effectively that they meet the diverse and stringent IMT-Advanced requirements set out for the LTE standardization. By focusing on PHY modeling, we aim to address challenges in understanding the development of the digital signal processing associated with the LTE standard.

1.9   LTE (Releases 8 and 9)

The introduction of the first release of the LTE standard was the culmination of about four years of work by 3GPP, starting in 2005. Following an extensive study of various technologies capable of delivering on the requirements set for the LTE standard, it was decided that the air interface transmission technology of the new standard would be based on OFDM in downlink and SC-FDM in uplink. The full specifications, including various MIMO modes, were then incorporated in the standard. The first version of the LTE standard (3GPP version 8) was released in December 2008. Release 9 came in December 2009; it included relatively minor enhancements such as Multimedia Broadcast/Multicast Services (MBMS) support, location services, and provisioning for base stations that support multiple standards [4].

1.10   LTE-Advanced (Release 10)

The LTE-Advanced was released in December 2010. LTE-Advanced is an evolution of the original LTE standard and does not represent a new technology. Among the technologies added to the LTE standard to result in the LTE-Advanced were carrier aggregation, enhanced downlink MIMO, uplink MIMO, and relays [4].

1.11   MATLAB® and Wireless System Design

In this book, we use MATLAB to model the PHY of the LTE standard and to obtain insight into its simulation and implementation requirements. MATLAB is a widely used language and a high-level development environment for mathematical modeling and numerical computations. We also use Simulink, a graphical design environment for system simulations and model-based design, as well as various MATLAB toolboxes – application-specific libraries of components that make the task of modeling applications in MATLAB easier. For example, in order to model communications systems we use functionalities from the Communication System Toolbox. The toolbox provides tools for the design, prototyping, simulation, and verification of communications systems, including wireless standards in both MATLAB and Simulink.

Among the functionalities in MATLAB that are introduced in this book are the new System objects. System objects are a set of algorithmic building blocks suitable for system design available in various MATLAB toolboxes. They are self-documented algorithms that make the task of developing MATLAB testbenches to perform system simulations easier. By covering a wide range of algorithms, they also eliminate the need to recreate the basic building blocks of communications systems in MATLAB, C, or any other programming language. System objects are designed not only for modeling and simulation but also to provide support for implementation. For example, they have favorable characteristics that help accelerate simulation speeds and support C/C++ code generation and fixed-point numeric, and a few of them support automatic HDL (Hardware Description Language) code generation.

1.12   Organization of This Book

The thesis of this book is that by understanding its four enabling technologies (OFDMA, MIMO, turbo coding, and link adaptation) the reader can obtain an adequate understanding of the PHY model of the LTE standard. Chapter 2 provides a short overview of the technical specifications of the LTE standard. Chapter 3 provides an introduction to the tools and features in MATLAB that are useful for the modeling and simulation of mobile communications systems. In Chapters 4–7, we treat each of the OFDM, MIMO, modulation, and coding and link adaptation techniques in detail. In each chapter, we create models in MATLAB that iteratively and progressively build up components of the LTE PHY based on these techniques. Chapter 8, on system-level specifications and performance evaluation, discusses various channel models specified in the standard and ways of performing system-level qualitative and quantitative performance analysis in MATLAB and Simulink. It also wraps up the first part of the book by putting together a system model and showing how the PHY of the LTE standard can be modeled in MATLAB based on the insight obtained in the preceding chapters.

The second part deals with practical issues such as simulation of the system and implementation of its components. Chapter 9 includes discussion on how to accelerate the speed of our MATLAB programs using a variety of techniques, including parallel computing, automatic C code generation, GPU (Graphics Progressing Unit) processing, and the use of more efficient algorithms. In Chapter 10, we discuss related implementation issues such as automatic C/C++ code generation from the MATLAB code, target environments, and code optimizations, and how these affect the programming style. We also discuss fixed-point numerical representation of data as a prerequisite for hardware implementation and its effect on the performance of various modeling components. Finally, in Chapter 11, we summarize our discussions and highlight directions for future work.

References

[1] ITU-R (1997) International Mobile Telecommunications-2000 (IMT-2000). Recommendation ITU-R M. 687-2, February 1997.
[2] ITU-R (2010) Detailed specifications of the radio interfaces of international mobile telecommunications-2000 (IMT-2000). Recommendation ITU-R M.1457-9, May 2010.
[3] ITU-R (2007) Principles for the Process of Development of IMT-Advanced. Resolution ITU-R 57, October 2007.
[4] Dahlman, E., Parkvall, S. and Sköld, J. (2011) *4G LTE/LTE-Advanced for Mobile Broadband*, Elsevier.
[5] Shannon, C.E. (1948) A mathematical theory of communication. *Bell System Technical Journal*, 379–423, 623–656.
[6] Ghosh, A. and Ratasuk, R. (2011) *Essentials of LTE and LTE-A*, Cambridge University Press, Cambridge.
[7] Proakis, J.G. (2001) *Digital Communications*, McGraw-Hill, New York.

2

Overview of the LTE Physical Layer

The focus of this book is the LTE (Long Term Evolution) radio access technology and particularly its PHY (Physical Layer). Here, we will highlight the major concepts related to understanding the technology choices made in the design of the LTE PHY radio interface. Focusing on this topic will best explain the remarkable data rates achievable by LTE and LTE-Advanced standards.

LTE specifies data communications protocols for both the uplink (mobile to base station) and downlink (base station to mobile) communications. In the 3GPP (Third Generation Partnership Project) nomenclature, the base station is referred to as eNodeB (enhanced Node Base station) and the mobile unit is referred to as UE (User Equipment).

In this chapter, we will cover topics related to PHY data communication and the transmission protocols of the LTE standards. We will first provide an overview of frequency bands, FDD (Frequency Division Duplex) and TDD (Time Division Duplex) duplex methodologies, flexible bandwidth allocation, time framing, and the time–frequency resource representation of the LTE standard. We will then study in detail both the downlink and uplink processing stacks, which include multicarrier transmission schemes, multi-antenna protocols, adaptive modulation, and coding schemes and channel-dependent link adaptations.

In each case, we will first describe the various channels that connect different layers of the communication stacks and then describe in detail the signal processing in the PHY applied on each of the downlink and uplink physical channels. The amount of detail presented will be sufficient to enables us to model the downlink PHY processing as MATLAB® programs. In the subsequent four chapters we will iteratively and progressively derive a system model from simpler algorithms in MATLAB.

2.1 Air Interface

The LTE air interface is based on OFDM (Orthogonal Frequency Division Multiplexing) multiple-access technology in the downlink and a closely related technology known as

Understanding LTE with MATLAB®: From Mathematical Modeling to Simulation and Prototyping, First Edition. Houman Zarrinkoub.
© 2014 John Wiley & Sons, Ltd. Published 2014 by John Wiley & Sons, Ltd.

Single-Carrier Frequency Division Multiplexing (SC-FDM) in the uplink. The use of OFDM provides significant advantages over alternative multiple-access technologies and signals a sharp departure from the past. Among the advantages are high spectral efficiency and adaptability for broadband data transmission, resistance to intersymbol interference caused by multipath fading, a natural support for MIMO (Multiple Input Multiple Output) schemes, and support for frequency-domain techniques such as frequency-selective scheduling [1].

The time–frequency representation of OFDM is designed to provide high levels of flexibility in allocating both spectra and the time frames for transmission. The spectrum flexibility in LTE provides not only a variety of frequency bands but also a scalable set of bandwidths. LTE also provides a short frame size of 10 ms in order to minimize latency. By specifying short frame sizes, LTE allows better channel estimation to be performed in the mobile, allowing timely feedbacks necessary for link adaptations to be provided to the base station.

2.2 Frequency Bands

The LTE standards specify the available radio spectra in different frequency bands. One of the goals of the LTE standards is seamless integration with previous mobile systems. As such, the frequency bands already defined for previous 3GPP standards are available for LTE deployment. In addition to these common bands, a few new frequency bands are also introduced for the first time in the LTE specification. The regulations governing these frequency bands vary between different countries. Therefore, it is conceivable that not just one but many of the frequency bands could be deployed by any given service provider to make the global roaming mechanism much easier to manage.

As was the case with previous 3GPP standards, LTE supports both FDD and TDD modes, with frequency bands specified as paired and unpaired spectra, respectively. FDD frequency bands are paired, which enables simultaneous transmission on two frequencies: one for the downlink and one for the uplink. The paired bands are also specified with sufficient separations for improved receiver performance. TDD frequency bands are unpaired, as uplink and downlink transmissions share the same channel and carrier frequency. The transmissions in uplink and downlink directions are time-multiplexed.

Release 11 of the 3GPP specifications for LTE shows the comprehensive list of ITU IMT-Advanced (International Telecommunications Union International Mobile Telecommunication) frequency bands [2]. It includes 25 frequency bands for FDD and 11 for TDD. As shown in Table 2.1, the paired bands used in FDD duplex mode are numbered from 1 to 25; the unpaired bands used in TDD mode are numbered from 33 to 43, as illustrated in Table 2.2. The band number 6 is not applicable to LTE and bands 15 and 16 are dedicated to ITU Region 1.

2.3 Unicast and Multicast Services

In mobile communications, the normal mode of transmission is known as a unicast transmission, where the transmitted data are intended for a single user. In addition to unicast services, the LTE standards support a mode of transmission known as Multimedia Broadcast/Multicast Services (MBMS). MBMS delivers high-data-rate multimedia services such as TV and radio broadcasting and audio and video streaming [1].

Table 2.1   Paired frequency bands defined for E-UTRA

Operating band index	Uplink (UL) operating band frequency range (MHz)	Downlink (DL) operating band frequency range (MHz)	Duplex mode
1	1920–1980	2110–2170	FDD
2	1850–1910	1930–1990	FDD
3	1710–1785	1805–1880	FDD
4	1710–1755	2110–2155	FDD
5	824–849	869–894	FDD
6	830–840	875–885	FDD
7	2500–2570	2620–2690	FDD
8	880–915	925–960	FDD
9	1749.9–1784.9	1844.9–1879.9	FDD
10	1710–1770	2110–2170	FDD
11	1427.9–1447.9	1475.9–1495.9	FDD
12	699–716	729–746	FDD
13	777–787	746–756	FDD
14	788–798	758–768	FDD
15	Reserved	Reserved	FDD
16	Reserved	Reserved	FDD
17	704–716	734–746	FDD
18	815–830	860–875	FDD
19	830–845	875–890	FDD
20	832–862	791–821	FDD
21	1447.9–1462.9	1495.9–1510.9	FDD
22	3410–3490	3510–3590	FDD
23	2000–2020	2180–2200	FDD
24	1626.5–1660.5	1525–1559	FDD
25	1850–1915	1930–1995	FDD

MBMS has its own set of dedicated traffic and control channels and is based on a multicell transmission scheme forming a Multimedia Broadcast Single-Frequency Network (MBSFN) service area. A multimedia signal is transmitted from multiple adjacent cells belonging to a given MBSFN service area. When the content of a single Multicast Channel (MCH) is transmitted from different cells, the signals on the same subcarrier are coherently combined at the UE. This results in a substantial improvement in the SNR (signal-to-noise ratio) and significantly improves the maximum allowable data rates for the multimedia transmission. Being in either a unicast or a multicast/broadcast mode of transmission affects many parameters and components of the system operation. As we describe various components of the LTE technology, we will highlight how different channels, transmission modes, and physical signals and parameters are used in the unicast and multicast modes of operations. The focus throughout this book will be on unicast services and data transmission.

Table 2.2   Unpaired frequency bands defined
for E-UTRA

Operating band index	Uplink and downlink operating band frequency range (MHz)	Duplex mode
33	1900–1920	TDD
34	2010–2025	TDD
35	1850–1910	TDD
36	1930–1990	TDD
37	1910–1930	TDD
38	2570–2620	TDD
39	1880–1920	TDD
40	2300–2400	TDD
41	2496–2690	TDD
42	3400–3600	TDD
43	3600–3800	TDD

2.4   Allocation of Bandwidth

The IMT-Advanced guidelines require spectrum flexibility in the LTE standard. This leads to scalability in the frequency domain, which is manifested by a list of spectrum allocations ranging from 1.4 to 20 MHz. The frequency spectra in LTE are formed as concatenations of resource blocks consisting of 12 subcarriers. Since subcarriers are separated by 15 kHz, the total bandwidth of a resource block is 180 kHz. This enables transmission bandwidth configurations of from 6 to 110 resource blocks over a single frequency carrier, which explains how the multicarrier transmission nature of the LTE standard allows for channel bandwidths ranging from 1.4 to 20.0 MHz in steps of 180 kHz, allowing the required spectrum flexibility to be achieved.

Table 2.3 illustrates the relationship between the channel bandwidth and the number of resource blocks transmitted over an LTE RF carrier. For bandwidths of 3–20 MHz, the totality of resource blocks in the transmission bandwidth occupies around 90% of the channel

Table 2.3   Channel bandwidths
specified in LTE

Channel bandwidth (MHz)	Number of resource blocks
1.4	6
3	15
5	25
10	50
15	75
20	100

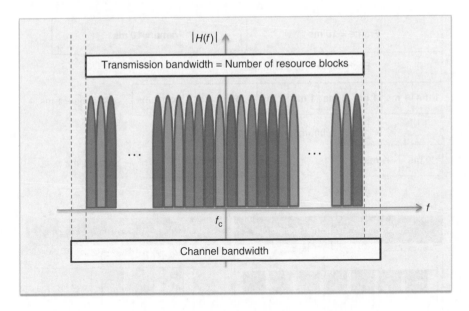

Figure 2.1   Relationship between channel bandwidth and number of resource blocks

bandwidth. In the case of 1.4 kHz, the percentage drops to around 77%. This helps reduce unwanted emissions outside the bandwidth, as illustrated in Figure 2.1. A formal definition of the time–frequency representation of the spectrum, the resource grid, and the blocks will be presented shortly.

2.5   Time Framing

The time-domain structure of the LTE is illustrated in Figure 2.2. Understanding of LTE transmission relies on a clear understanding of the time–frequency representation of data, how it maps to what is known as the resource grid, and how the resource grid is finally transformed into OFDM symbols for transmission.

In the time domain, LTE organizes the transmission as a sequence of radio frames of length 10 ms. Each frame is then subdivided into 10 subframes of length 1 ms. Each subframe is composed of two slots of length 0.5 ms each. Finally, each slot consists of a number of OFDM symbols, either seven or six depending on whether a normal or an extended cyclic prefix is used. Next, we will focus on the time–frequency representation of the OFDM transmission.

2.6   Time–Frequency Representation

One of the most attractive features of OFDM is that it maps explicitly to a time–frequency representation for the transmitted signal. After coding and modulation, a transformed version of the complex-valued modulated signal, the physical resource element, is mapped on to a time-frequency coordinate system, the resource grid. The resource grid has time on the x-axis and frequency on the y-axis. The x-coordinate of a resource element indicates the OFDM

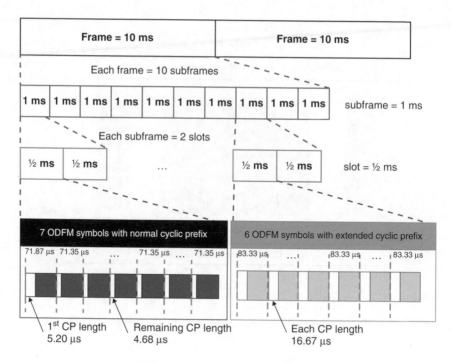

Figure 2.2   LTE time-domain structure

symbol to which it belongs in time. The y-coordinate signifies the OFDM subcarrier to which it belongs in frequency.

Figure 2.3 illustrates the LTE downlink resource grid when a normal cyclic prefix is used. A resource element is placed at the intersection of an OFDM symbol and a subcarrier. The subcarrier spacing is 15 kHz and, in the case of normal cyclic prefix, there are 14 OFDM symbols per subframe or seven symbols per slot. A resource block is defined as a group of resource elements corresponding to 12 subcarriers or 180 kHz in the frequency domain and one 0.5 ms slot in the time domain. In the case of a normal cyclic prefix with seven OFDM symbols per slot, each resource block consists of 84 resource elements. In the case of an extended cyclic prefix with six OFDM symbols per slot, the resource block contains 72 resource elements. The definition of a resource block is important because it represents the smallest unit of transmission that is subject to frequency-domain scheduling.

As we discussed earlier, the LTE PHY specification allows an RF carrier to consist of any number of resource blocks in the frequency domain, ranging from a minimum of six resource blocks up to a maximum of 110 resource blocks. This corresponds to transmission bandwidths ranging from 1.4 to 20.0 MHz, with a granularity of 15 kHz, and allows for a very high degree of LTE bandwidth flexibility. The resource-block definition applies equally to both the downlink and the uplink transmissions. There is a minor difference between the downlink and the uplink regarding the location of the carrier center frequency relative to the subcarriers.

In the uplink, as illustrated in Figure 2.4, no unused DC subcarrier is defined and the center frequency of an uplink carrier is located between two uplink subcarriers. In the downlink, the subcarrier that coincides with the carrier-center frequency is left unused. This is shown in Figure 2.5. The reason why the DC subcarrier is not used for downlink transmission is the possibility of disproportionately high interference.

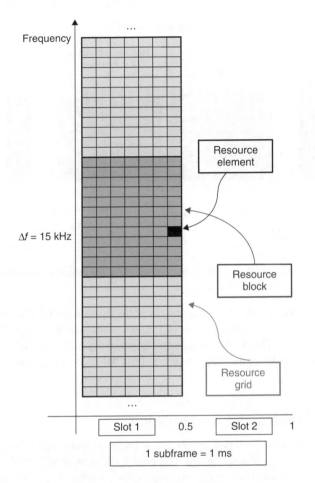

Figure 2.3   Resource elements, blocks, and grid

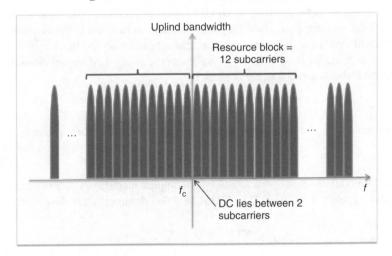

Figure 2.4   Resource blocks and DC components of the frequency in uplink transmission

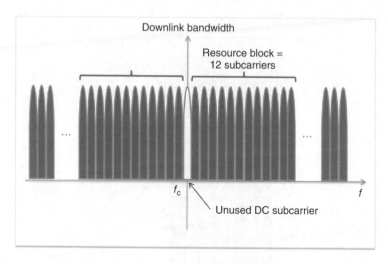

Figure 2.5   Resource blocks and DC components of the frequency in downlink transmission

The choice of 15 kHz as subcarrier spacing fits perfectly with the OFDM mandate that turns a frequency-selective channel into a series of frequency-flat subchannels with fine resolution. This is turn helps the OFDM to efficiently combat frequency-selective fading by using a bank of low-complexity equalizers that apply to each of the flat-faded subchannels in the frequency domain.

2.7   OFDM Multicarrier Transmission

In the LTE standard, the downlink transmission is based on an OFDM scheme and the uplink transmission is based on a closely related methodology known as SC-FDM. OFDM is a multicarrier transmission methodology that represents the broadband transmission bandwidth as a collection of many narrowband subchannels.

There are multiple steps involved in OFDM signal generation. First, modulated data are mapped on to the resource grid, where they are organized and aligned in the frequency domain. Each modulated symbol a_k is assigned to a single subcarrier on the frequency axis. With N subcarriers occupying the bandwidth with a subcarrier spacing of Δf, the relationship between the bandwidth and subcarrier spacing is given by:

$$BW = N_{rb}\Delta f \tag{2.1}$$

Each subcarrier f_k can be considered an integer multiple of subcarrier spacing:

$$f_k = k\Delta f \tag{2.2}$$

The OFDM modulator consists of a bank of N complex modulators, where each modulator corresponds to a single subcarrier. The OFDM modulated output $x(t)$ is thus expressed as:

$$x(t) = \sum_{k=1}^{N} a_k e^{j2\pi f_k t} = \sum_{k=1}^{N} a_k e^{j2\pi k\Delta f t} \tag{2.3}$$

Assuming that the channel sample rate is Fs and the channel sample time is $Ts = 1/Fs$, the discrete-time representation of the OFDM modulator can be expressed as:

$$x(n) = \sum_{k=1}^{N} a_k e^{j2\pi k \Delta fn/N} \qquad (2.4)$$

The OFDM modulation lends itself naturally to an efficient implementation based on Inverse Fast Fourier Transform (IFFT). After the OFDM modulation, an OFDM symbol is generated and a cyclic prefix is added to the modulated signal. Insertion of a cyclic prefix is essentially copying of the last part of the OFDM symbol to its beginning.

2.7.1 Cyclic Prefix

Cyclic prefix insertion is an important function during OFDM signal generation. A cyclic prefix is necessary to prevent interference from previously transmitted OFDM symbols. The intersymbol interference can be viewed as a direct result of multipath propagation. At first glance, cyclic prefix insertion may be regarded as a useless operation since it is merely repeats a copy of the existing data in the OFDM symbol and does not add any new information. However, it is instrumental for multiple reasons. First, it helps maintain orthogonality between subcarriers in the receiver, which is one of the foundations of an orthogonal frequency division transmission. It also provides a periodic extension to the OFDM signal through which the "linear convolution" operation performed on the transmitted signal by the channel can be approximated by a "circular convolution" operation. Mimicking a circular convolution with a cyclic prefix is quite important if you want OFDM to represent the modulated signal in the frequency domain. The validity of the frequency-domain equalization performed in the receiver is only ensured if channel response can be viewed as circular convolution, something that cyclic prefix insertion can ensure [2].

The length of the cyclic prefix is an important design parameter for a multicarrier transmission system. On one hand, the length of the cyclic prefix must be sufficient to cover typical delay spreads encountered in most propagation scenarios within a cellular environment. On the other hand, the cyclic prefix represents redundant data and a necessary overhead. As the name "prefix" implies, the first portion of the received OFDM signal is discarded at the receiver. Therefore, LTE must specify as small a cyclic prefix as possible in order to minimize the overhead and maximize the spectral efficiency. To resolve this tradeoff, LTE specifies the cyclic prefix length as the expected delay spread of the propagation channel and provides a margin for error to account for imperfect timing alignment.

As shown in Table 2.4, the LTE standard specifies three different cyclic prefix values: (i) normal (4.7 μs) and (ii) extended (16.6 μs) for subcarrier spacing of 15 kHz and (iii)

Table 2.4 Normal and extended cyclic prefix specifications

Configuration	Subcarrier spacing (Δf) (kHz)	Number of subcarriers per resource block	Number of OFDM symbols per resource block
Normal cyclic prefix	15	12	7
Extended cyclic prefix	15	12	6
	7.5	24	3

extended (33 µs) for subcarrier spacing of 7.5 kHz. Note that the subcarrier spacing 7.5 kHz can only be used in a multicast/broadcast context. The normal cyclic prefix length of 4.7 µs is appropriate for transmissions over most urban and suburban environments and reflects typical delay spread values for those environments. Given that the time occupied by each OFDM modulated symbol is about 66.7 µs, the cyclic prefix in normal mode accounts for an overhead of about 7%. The overhead associated with an extended cyclic prefix of length 16.7 µs is 25%. This rather excessive overhead is necessary for transmissions over rural environments with longer delay spread and for broadcast services.

2.7.2   Subcarrier Spacing

Small subcarrier spacing ensures that the fading on each subcarrier is frequency nonselective. However, subcarrier spacing cannot be arbitrarily small. Performance degrades as subcarrier spacing decreases beyond a certain limit as a result of Doppler shift and phase noise [1]. Doppler shift is caused when a mobile terminal moves, and it increases with higher velocity. Doppler shift causes intercarrier interference and the resulting degradations get amplified with small subcarrier spacing. Phase noise or jitter results from fluctuations in the frequency of the local oscillator and will cause intercarrier interference. To minimize the degradations caused by phase noise and Doppler shift, a subcarrier spacing of 15 kHz is specified in the LTE standard.

2.7.3   Resource Block Size

In LTE, a block of resource elements, known as a resource block, forms the unit of resource scheduling. Several factors must be considered in the selection of the resource block size. First, it should be small enough that the gain in frequency-selective scheduling (i.e., scheduling of data transmission on good-frequency subcarriers) is large. Small resource block size ensures that the frequency response within each resource block is similar, thereby enabling the scheduler to assign only good resource blocks. However, since the eNodeB does not know which resource blocks are experiencing good channel conditions, the UE must report this information back to the eNodeB. Thus, the resource block size must be sufficiently large to avoid excessive feedback overhead. Since in LTE a subframe size of 1 ms is used to ensure low latency, the resource block size in frequency should be small, so that small data packets can be efficiently supported. As a result, 180 kHz (12 subcarriers) was chosen as the resource block bandwidth.

2.7.4   Frequency-Domain Scheduling

LTE supports different system bandwidths. OFDM and SC-FDM generate the transmitted signal with an IFFT operation. We can thus accommodate different bandwidths by choosing different FFT lengths. Regardless of the bandwidth used, LTE keeps the OFDM symbol duration constant at a fixed value of 66.7 µs. This enables the use of the same subcarrier of 15 kHz for all bandwidths. These design choices ensure that the same frequency-domain equalization techniques can be applied across multiple bandwidths. Having constant symbol durations also means having the same subframe length in different bandwidths, a feature that greatly

Table 2.5 Resource blocks, FFT, and cyclic prefix sizes for each LTE bandwidth

OFDM parameters for downlink transmission subframe duration (1 ms) subcarrier spacing (15 kHz)						
Bandwidth (MHz)	1.4	3	5	10	15	20
Sampling frequency (MHz)	1.92	3.84	7.68	15.36	23.04	30.72
FFT size	128	256	512	1024	1536	2048
Number of resource blocks	6	15	25	50	75	100
OFDM symbols per slot			14/12		(Normal/extended)	
CP length			4.7/5.6		(Normal/extended)	

simplifies the time framing of the transmissions model. Although the actual FFT size used in each bandwidth is not specified by the standard, an FFT size of 2048 is usually associated with 20 MHz. The FFT sizes for other bandwidths are usually the scaled-down versions of this value, as shown in Table 2.5.

2.7.5 Typical Receiver Operations

In the receiver, we perform the inverses of the transmitter operations. Although the LTE standard, like many other requirement-based standards, does not specify the way receiver-side operations are performed, discussing typical receiver operations is useful in understanding the motivations behind specific transmitter-side operations defined in the standard.

The OFDM receiver reverses the operations of OFDM signal generation and transmission. First, we delete the cyclic prefix samples from the beginning of the received OFDM symbol. Then, by performing an FFT operation, we compute the received resource grid elements of a particular OFDM symbol. At this stage we need to perform an equalization operation on the received resource elements in order to undo the effects of channel and intersymbol interference in order to recover the best estimate of the transmitted resource elements.

In order to perform equalization, we first need to estimate the channel frequency response for the entire bandwidth; that is, for all resource elements. This is where the importance of pilots or cell-specific reference (CSR) signals becomes evident. By transmitting known signal values as pilots at various known points in the resource grid, we can estimate the actual channel response at the corresponding subcarriers easily. These channel responses can be computed in multiple ways, including via a simple ratio of received signal to transmitted signal. Now that we have the channel responses at some regular points within the resource grid, we can employ various averaging or interpolation operations to estimate the channel response for the entire resource grid. After estimating the channel response for the grid, we recover the best estimates of the transmitted resource elements through multiplication of the resource elements received by the reciprocal values of the estimated channel responses.

2.8 Single-Carrier Frequency Division Multiplexing

The LTE uplink is based on a variant of the OFDM transmission scheme known as SC-FDM. SC-FDM reduces the instantaneous power fluctuations observed in OFDM transmission.

Therefore, it is a better choice for the design of low-power amplifiers suitable for user terminals (UE). The way SC-FDM is implemented in the LTE standard is by essentially preceding the OFDM modulator with a DFT (Discrete Fourier Transform) precoder. This technique is known as Discrete Fourier Transform-Spread Orthogonal Frequency Division Multiplexing (DFTS-OFDM).

The distinguishing feature of single-carrier transmission is that each data symbol is essentially spread over the entire allocated bandwidth. This is in contrast to OFDM, where each data symbol is assigned to one subcarrier. By spreading the data power over the bandwidth, SC-FDM reduces the mean transmission power and guarantees that the dynamic range of the transmitted signal stays within the linear region of the power amplifier. SC-FDM is capable of providing the same advantages offered by OFDM, including (i) maintaining orthogonality among multiple uplink users, (ii) recovering data using a frequency-domain equalization, and (iii) combating multipath fading. However, the performance of SC-FDM transmission is usually inferior to that of OFDM when the same receiver is used [1]. DFTS-OFDM is discussed in more detail later in this chapter.

2.9 Resource Grid Content

The LTE transmission scheme provides a time resolution of 12 or 14 OFDM symbols for each subframe of 1 ms, depending on the length of the OFDM cyclic prefix. Regarding the frequency resolution, it provides for a number of resource blocks ranging from 6 to 100, depending on the bandwidth, each containing 12 subcarriers with 15 kHz spacing. The next question is what type of data occupies the resource elements that make up the resource grid. To answer this, we must describe the various physical channels and signals that constitute the content of the resource grid.

There are essentially three types of information contained in the physical resource grid. Each resource element contains the modulated symbol of either user data or a reference or synchronization signal or control information originating from various higher-layer channels. Figure 2.6 shows the relative locations of the user data, control information, and reference signal in a resource grid as defined for a unicast mode of operation.

In unicast mode, the user data carry the information that each user wants to communicate and are delivered from the MAC (Medium Access Control) layer to the PHY as a transport block. Various types of reference and synchronization signal are generated in a predictable manner by the base station and the mobile set. These signals are used for such purposes as channel estimation, channel measurement, and synchronization. Finally we have various types of control information, which are obtained via the control channels and carry information that the receiver requires in order to correctly decode the signal.

Next, we will describe the physical channels used in downlink and uplink transmission and their relationships to higher-layer channels; that is, transport channels and logical channels. Compared with UMTS (Universal Mobile Telecommunications System) and other 3GPP standards, LTE has substantially reduced its use of dedicated channels and relies more on shared channels. This explains the convergence of many different types of logical and transport channel on the shared physical channels. Beside physical channels, two types of physical signals – reference signals and synchronization signals – are also transmitted within the shared physical channel. The details of LTE channels and signals are presented in the following sections.

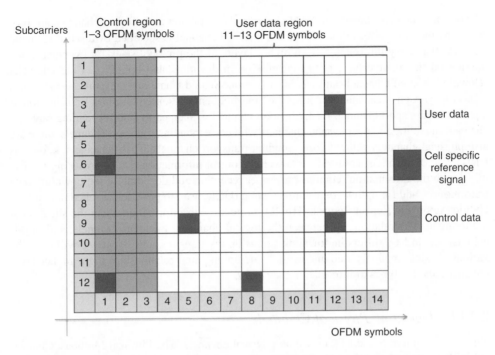

Figure 2.6   Physical channel and signal content of LTE downlink subframe in unicast mode

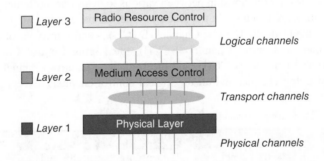

Figure 2.7   Layer architecture in a LTE radio access network

2.10   Physical Channels

Among the objectives of the LTE standard is to create a more efficient and streamlined protocol stack and architecture. Many dedicated channels specified in previous 3GPP standards have been replaced by shared channels and the total number of physical channels has been reduced. Figure 2.7 shows the protocol stack of the radio access network and its layer architecture.

Logical channels represent the data transfers and connections between the radio link control (RLC) layer and the MAC layer. LTE defines two types of logical channel: a traffic channel and a control channel. While the traffic logical channel transfers user-plane data, the control logical channels transfer the control-plane information.

Transport channels connect the MAC layer to the PHY and the physical channels are processed by the transceiver at the PHY. Each physical channel is specified by a set of resource elements that carry information from higher layers of the protocol stack for eventual transmission on the air interface. Data transmission in downlink and uplink uses the DL-SCH (Downlink Shared Channel) and UL-SCH (Uplink Shared Channel) transport channel types respectively. A physical channel carries the time-frequency resources used for transmission of a particular transport channel. Each transport channel is mapped to a corresponding physical channel. In addition to the physical channels with corresponding transport channels, there are also physical channels without corresponding transport channels. These channels, known as L1/L2 control channels, are used for downlink control information (DCI), providing the terminal with the information required for proper reception and decoding of the downlink data transmission, and for uplink control information (UCI), used to provide the scheduler and the Hybrid Automatic Repeat Request (HARQ) protocol with information about the situation at the terminal. The relationship between the logical channels, transport channels, and physical channels in LTE differs in downlink versus uplink transmissions. Next, we will describe various physical channels used in downlink and uplink, their relationships to the higher-layer channels, and the types of information they carry.

2.10.1   Downlink Physical Channels

Table 2.6 summarizes the LTE downlink physical channels. The Physical Multicast Channel (PMCH) is used for the purpose of MBMS. The rest of the physical channels are used in the traditional unicast mode of transmission.

Figure 2.8 illustrates the relationship between various logical, transport, and physical channels in LTE downlink architecture. In the unicast mode, we have only a single type of traffic logical channel – the Dedicated Traffic Channel (DTCH) – and four types of control logical channel: the Broadcast Control Channel (BCCH), the Paging Control Channel (PCCH), the Common Control Channel (CCCH), and the Dedicated Control Channel (DCCH). The dedicated logical traffic channel and all the logical control channels, except for PCCH, are

Table 2.6   LTE downlink physical channels

Downlink physical channel	Function
Physical Downlink Shared Channel (PDSCH)	Unicast user data traffic and paging information
Physical Downlink Control Channel (PDCCH)	Downlink Control Information (DCI)
Physical Hybrid-ARQ Indicator Channel (PHICH)	HARQ Indicator (HI) and ACK/NACKs for the uplink packets
Physical Control Format Indicator Channel (PCFICH)	Control Format Information (CFI) containing information necessary to decode PDCCH information
Physical Multicast Channel (PMCH)	Multimedia Broadcast Single-Frequency Network (MBSFN) operation
Physical Broadcast Channel (PBCH)	System information required by the terminal in order to access the network during cell search

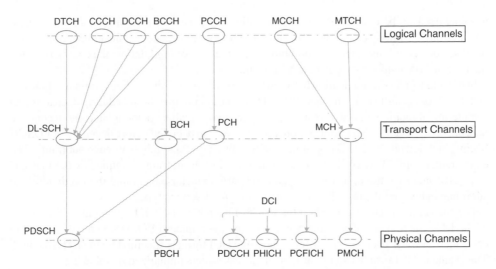

Figure 2.8   Mapping LTE downlink logical, transport, and physical channels

multiplexed to form a transport channel known as the Downlink Shared Channel. The Paging Control Channel (PCCH) is mapped to the Paging Channel (PCH) and combined with the DLSCH to form the Physical Downlink Shared Channel (PDSCH). The PDSCH and four other physical channels (PDCCH, Physical Downlink Control Channel; PHICH, Physical Hybrid Automatic Repeat Request Indicator Channel; PCFICH, Physical Control Format Indicator Channel; and PBCH, Physical Broadcast Channel) provide all the user data, control information, and system information needed in the unicast mode, which are delivered from higher layers.

In the multicast/broadcast mode, we have a traffic logical channel known as the Multicast Traffic Channel (MTCH) and a control logical channel known as the Multicast Control Channel (MCCH). These are combined to form the transport channel known as the Multicast Channel (MCH). Finally, the PMCH is formed as the physical channel for the MBMS mode.

2.10.2   Function of Downlink Channels

The PDSCH carries downlink user data as transport blocks that are handed down from the MAC layer to the PHY. Usually, transport blocks are transmitted one at a time in each subframe, except in a particular case of MIMO known as spatial multiplexing, where one or two transport blocks can be transmitted per given subframe. Following adaptive modulation and coding, the modulated symbols are mapped on to multiple time–frequency resource grids, which are eventually mapped to multiple transmit antennas for transmission. The type of multi-antenna technique used in each subframe is also subject to adaptation based on channel conditions.

The use of adaptive modulation, coding, and MIMO in the LTE standard implies that in each subframe, depending on the channel quality observed at the mobile terminal, the base station needs to make decisions about the type of modulation scheme, coding rate, and MIMO mode. The measurements made in the terminal must feed back to the base station in order to help the scheduling decisions made there for the ensuing transmissions. At each subframe, the mobile terminal needs to be notified about the scheduling from the base station for each transmitted resource block. Among the information that must be communicated are the number of resource blocks allocated to a user, the transport block size, the type of modulation, the coding rate, and the type of MIMO mode used per each subframe.

In order to foster communication between the base station and the mobile terminal, a PDCCH is defined for each PDSCH channel. PDCCH primarily contains the scheduling decisions that each terminal requires in order to successfully receive, equalize, demodulate, and decode the data packets. Since PDCCH information must be read and decoded before decoding of PDSCH begins, in a downlink PDCCH occupies the first few OFDM symbols of each subframe. The exact number of OFDM symbols at the beginning of each subframe occupied by the PDCCH (typically one, two, three, or four) depends on various factors, including the bandwidth, the subframe index, and the use of unicast versus multicast service type.

The control information carried on the PDCCH is known as DCI. Depending on the format of the DCI, the number of resource elements (i.e., the number of OFDM symbols needed to carry them) varies. There are 10 different possible DCI formats specified by the LTE standard. The available DCI formats and their typical use cases are summarized in Table 2.7.

Each DCI format contains the following types of control information: resource allocation information, such as resource block size and resource assignment duration; transport information, such as multi-antenna configuration, modulation type, coding rate, and transport block payload size; and finally information related to the HARQ, including its process number, the redundancy version, and the indicator signaling availability of new data. For example, the content fields of DCI format 1 are summarized in Table 2.8.

Table 2.7 LTE Downlink Control Information (DCI) formats and their use cases

DCI format	Use case
0	Uplink scheduling assignment
1	Downlink scheduling for one PDSCH codeword in SISO and SIMO modes
1A	Compact version of format 1 scheduling for one PDSCH codeword or dedicated preamble assignment to iniate random access
1B	Very compact downlink scheduling for one PDSCH codeword used in MIMO mode number 6
1C	Very compact downlink scheduling for paging or system information
1D	Compact downlink scheduling for one PDSCH codeword with MIMO precoding and power offset information necessary for multi-user MIMO
2	Downlink scheduling assignment for MIMO with closed-loop spatial multiplexing
2A	Downlink scheduling assignment for MIMO with open-loop spatial multiplexing
3	Transmit Power Control (TPC) information for PUCCH and PUSCH with 2 bit power adjustment
3A	Transmit power control (TPC) information for PUCCH and PUSCH with 1 bit power adjustment

Table 2.8   Content of the DCI format 1

Field	Number of bits on PDCCH	Description
Resource allocation header	1	Indicates the selected resource allocation of either type 0 or type 1
Resource block assignment	Depends on resource allocation type	Indicates resource blocks on PDSCH to be assigned to the terminal
Modulation and Coding Scheme (MCS)	5	Indicates the type of modulation and coding used, together with the transport block size and the number of resource blocks allocated
HARQ process number	3 (FDD) 4 (TDD)	Indicates the HARQ ID used in asynchronous stop-and-wait protocol
New data indicator	1	Indicates whether the current packet is a new transmission or a retransmission
Redundancy version	2	Indicates the incremental redundancy state of the HARQ process
PUCCH TPC command	2	Indicates the transmit power control command for adaptation of transmit power on PUCCH
Downlink assignment index	2	(Only for TDD mode) Indicates the number of downlink subframes used for uplink ACK/NACK bundling

The PCFICH is used to define the number of OFDM symbols that the DCI occupies in a subframe. The PCFICH information is mapped to specific resource elements belonging to the first OFDM symbol in each subframe. The possible values for PCFICH (one, two, three, or four) depend on the bandwidth, frame structure, and subframe index. For bandwidths larger than 1.4 MHz, PCFICH code can take up to three OFDM symbols. For 1.4 MHz bandwidth, since the number of resource blocks is quite small, PCFICH may need up to four symbols for control signaling.

Besides the PDCCH and PCFICH control channels, LTE defines another control channel known as the Physical HARQ Indicator Channel (PHICH). The PHICH contains information regarding the acknowledgment response for received packets in the uplink. Following the transmission of an uplink packet, the UE will receive an acknowledgment for that packet on a PHICH resource after a predetermined time delay. The duration of the PHICH is determined by higher layers. In the case of a normal duration, the PHICH is only found in the first OFDM symbol of a subframe; with extended duration, it is found in the first three subframes.

The PBCH carries the Master Information Block (MIB), which contains the basic PHY system information and cell-specific information during the cell search. After the mobile terminal correctly acquires the MIB, it can then read the downlink control and data channels and perform necessary operations to access the system. The MIB is transmitted on the PBCH over 40 ms periods, corresponding to four radio frames, with portions transmitted in the first subframe of every frame. The MIB contains four fields of information. The first two fields hold information regarding downlink system bandwidth and PHICH configuration. The downlink system bandwidth is communicated as one of six values for the number of resource blocks in downlink (6, 15, 25, 50, 75, or 100). As discussed earlier, these values for the number of resource blocks map directly to bandwidths of 1.4, 3, 5, 10, 15, and 20 MHz, respectively. The PHICH configuration field of the MIB specifies the duration and amount of the PHICH, as described earlier. The PBCH is always confined to the first four OFDM symbols found in the first slot of the first subframe of every radio frame. In frequency, the PBCH occupies 72 subcarriers centered on the DC subcarrier. Following a description of the physical signals, we can completely describe the content of the frame structures in the LTE standard.

2.10.3   Uplink Physical Channels

Table 2.9 summarizes the LTE uplink physical channels. The Physical Uplink Shared Channel (PUSCH) carries the user data transmitted from the user terminal. The Physical Random Access Channel (PRACH) is used for initial access of a UE to the network through transmission of random access preambles. The Physical Uplink Control Channel (PUCCH) carries the UCI, including scheduling requests (SRs), acknowledgments of transmission success or failure (ACKs/NACKs), and reports of downlink channel measurements including the Channel Quality Indicator (CQI), Precoding Matrix Information (PMI), and Rank Indication (RI).

Figure 2.9 illustrates the relationship between logical, transport, and physical channels in the LTE uplink architecture. Starting with logical channels, we have a Dedicated Traffic Channel (DTCH) and two logical control channels, a Common Control Channel (CCCH), and a Dedicated Control Channel (DCCH). These three channels are combined to form the transport channel known as the Uplink Shared Channel (UL-SCH). Finally, the Physical Uplink Shared Channel (PUSCH) and the Physical Uplink Control Channel (PUCCH) are formed as the physical channels. The transport channel known as the Random Access Channel (RACH) is also mapped to the Physical Random Access Channel (PRACH).

2.10.4   Function of Uplink Channels

The PUCCH carries three types of control signaling information: ACK/NACK signals for downlink transmission, scheduling requests (SR) indicators, and feedback from the downlink channel information, including the CQI, the PMI, and the RI.

The feedback of the downlink channel information relates to MIMO modes in downlink. In order to ensure that the MIMO transmission schemes work correctly in downlink, each terminal must perform measurements on the quality of the radio link and report the channel characteristic to the base station. This essentially describes the channel quality functions of the UCI as contained in the PUCCH.

Table 2.9   LTE uplink physical channels

Uplink physical channel	Function
Physical Uplink Shared Channel (PUSCH)	Uplink user data traffic
Physical Uplink Control Channel (PUCCH)	Uplink Control Information (UCI)
Physical Random Access Channel (PRACH)	Initial access to network through random access preambles

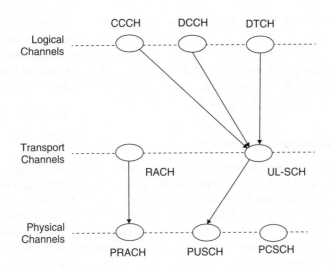

Figure 2.9   Mapping LTE uplink logical, transport, and physical channels

The CQI is an indicator of downlink mobile radio channel quality measures as taken by the UE and transmitted to the base station for use in subsequent scheduling. It allows the UE to propose to the base station a set of optimal modulation schemes and coding rates matched to the present radio link quality. There are 16 combinations of the modulation schemes and coding rates that are transmitted as CQI information. Higher CQI values stand for higher modulation orders and higher coding rates. Either a wideband CQI is used, which applies to all resource blocks forming the bandwidth, or else a subband CQI is used, which assigns a given CQI value to a certain number of resource blocks. The higher-layer configurations determine the rate, periodicity, or frequency of CQI measurements in the terminal.

The PMI is an indication of a preferred precoding matrix to be used in a base station for a given radio link. The PMI values represent precoding table indices for a two, four, or eight transmit antenna configuration. The RI signals the number of useful transmit antennas, esti- mated based on the channel quality and its effect on the correlations observed between adjacent receive antennas. In the following sections, we will describe MIMO modes of transmission in the LTE standard. From this, the roles of the CQI, PMI and RI indicators will become clear.

2.11   Physical Signals

A variety of physical signals, including reference and synchronization signals, are transmitted within the shared physical channel. Physical signals map to a specific resource element used by the PHY but do not carry information originating from higher layers. The details of LTE signals are presented next.

2.11.1   Reference Signals

Channel-dependent scheduling in the frequency domain is one of the most attractive features of the LTE standard. For example, in order to perform downlink scheduling that is aware of the actual channel quality, the mobile terminal must provide the base station with the Channel-State Information (CSI). The CSI may be obtained by measuring reference signals transmitted in the downlink. Reference signals are transmitted signals that are generated with synchronized sequence generators in the transmitter and the receiver. These signals are placed in specific resource elements in the time-frequency grid. LTE specifies several types of downlink and uplink reference signal, which are described next.

2.11.1.1 Downlink Reference Signals

Downlink reference signals support the channel estimation functionality needed to equalize and demodulate the control and data information. They are also instrumental in CSI measurements (such as RI, CQI, and PMI) needed for channel quality feedback. LTE specifies five types of reference signal for downlink transmission: Cell-Specific Reference Signals (CSR), Demodulation Reference Signal (DM-RS, otherwise known as UE-specific reference signal), Channel-State Information Reference Signal (CSI-RS), MBSFN reference signals, and positioning reference signals.

CSRs are common to all users in a cell and are transmitted in every downlink subframe. DM-RSs are used in downlink multi-user transmission modes 7, 8, or 9. As the name implies, they are intended for channel estimation performed by each individual mobile terminal in a cell. CSI-RSs were first introduced in LTE Release 10. Their main function is to alleviate the density problem associated with using CSRs for CSI measurements when more than eight antennas are in use. Therefore, the use of CSI-RSs is limited to the multi-user downlink transmission mode 9. MBSFN reference signals are used in the coherent demodulation employed in multicast/broadcast services. Finally, positioning reference signals, first introduced in LTE Release 9, help support measurements on multiple cells in order to estimate the position of a given terminal. In this section, we provide more detail on the first three types of reference signal enumerated here.

Cell-Specific Reference Signals
CRSs are transmitted in every downlink subframe and in every resource block in the frequency domain, and thus cover the entire cell bandwidth. The CRSs can be used by the terminal for channel estimation for coherent demodulation of any downlink physical channel except PMCH and PDSCH in the case of transmission modes 7, 8, or 9, corresponding to non-codebook-based precoding.

The CRSs can also be used by the terminal to acquire CSI. Finally, terminal measurements such as CQI, RI, and PMI performed on CRSs are used as the basis for cell selection and handover decisions.

UE-Specific Reference Signals
DM-RSs, or UE-specific reference signals, are only used in downlink transmission modes 7, 8, or 9, where CSRs are not used for channel estimation. DM-RSs were first introduced in LTE Release 8 in order to support a single layer. In LTE Release 9, up to two layers were supported. The extended specification introduced in Release 10 aimed to support up to eight simultaneous reference signals.

When only one DM-RS is used, we have 12 reference symbols within a pair of resource blocks. As will be discussed shortly, CSRs require spectral nulls or unused resource elements on all other antenna ports when a resource element on any given antenna is transmitting a reference signal. This is a major difference between CSR and DM-RS. When two DM-RSs are used on two antennas, all 12 reference symbols are transmitted on both antenna ports. The interference between the reference signals is mitigated by generating mutually orthogonal patterns for each pair of consecutive reference symbols.

CSI Reference Signals

CSI-RSs are designed for cases where we have between four and eight antennas. CSI-RSs were first introduced in LTE Release 10. They are designed to perform a complementary function to the DM-RS in LTE transmission mode 9. While the DM-RS supports channel estimation functionality, a CSI-RS acquires CSI. To reduce the overhead resulting from having two types of reference signal within the resource grid, the temporal resolution of CSI-RSs is reduced. This makes the system incapable of tracking rapid changes in the channel condition. Since CSI-RSs are only used with four to eight MIMO antenna configurations, and this configuration is only active with low mobility, the low temporal resolution of CSI-RSs does not pose a problem.

2.11.1.2 Uplink Reference Signals

There are two kinds of uplink reference signal in the LTE standard: the DM-RS and the Sounding Reference Signal (SRS). Both uplink reference signals are based on Zadoff–Chu sequences. Zadoff–Chu sequences are also used in generating downlink Primary Synchronization Signals (PSSs) and uplink preambles. Reference signals for different UEs are derived from different cyclic shift parameters of the base sequence.

Demodulation Reference Signals

DM-RSs are transmitted by UE as part of the uplink resource grid. They are used by the base station receiver to equalize and demodulate the uplink control (PUCCH) and data (PUSCH) information. In the case of PUSCH, when a normal cyclic prefix is used DSR signals are located on the fourth OFDM symbol in each 0.5 ms slot and extend across all the resource blocks. In the case of PUCCH, the location of DSR will depend on the format of the control channel.

Sounding Reference Signals

SRSs are transmitted on the uplink in order to enable the base station to estimate the uplink channel response at different frequencies. These channel-state estimates may be used for uplink channel-dependent scheduling. This means the scheduler can allocate user data to portions of the uplink bandwidth where the channel responses are favorable. SRS transmissions have other applications, such as timing estimation and control of downlink channel conditions when downlink and uplink channels are reciprocal or identical, as is the case in the TDD mode.

2.11.2 Synchronization Signals

In addition to reference signals, LTE also defines synchronization signals. Downlink synchronization signals are used in a variety of procedures, including the detection of frame boundaries, determination of the number of antennas, initial cell search, neighbor cell search, and handover. Two synchronization signals are available in the LTE: the Primary Synchronization Signal (PSS) and the Secondary Synchronization Signal (SSS).

Both the PSS and the SSS are transmitted as 72 subcarriers located around the DC subcarrier. However, their placement in FDD mode differs from that in TDD mode. In an FDD frame, they are positioned in subframes 0 and 5, next to each other. In a TDD frame, they are not placed close together. The SSS is placed in the last symbols of subframes 0 and 5 and the PSS is placed as the first OFDM symbols of the ensuing special subframe.

Synchronization signals are related to the PHY cell identity. There are 504 cell identities defined in the LTE, organized into 168 groups, each of which contains three unique identities. The PSS carries the unique identities 0, 1, or 2, whereas the SSS carries the group identity with values 0–167.

2.12 Downlink Frame Structures

LTE specifies two downlink frame structures. A type 1 frame applies to an FDD deployment and a type 2 frame is used for a TDD deployment. Each frame is composed of 10 subframes and each subframe is characterized by the time–frequency resource grid. We have identified the three components of a resource grid: user data, control channels and reference, and synchronization signals. Now we can explain how and where each of these components is placed as the LTE resource grid is populated per subframe before OFDM symbols are generated and transmitted. Without a loss in generality, in this book we focus on FDD frame structures and type 1 frames.

Figure 2.10 shows the type 1 radio frame structure. The duration of each frame is 10 ms, composed of ten 1 ms subframes denoted by indices ranging from 0 to 9. Each subframe is subdivided into two slots of 0.5 ms duration. Each slot is composed of seven or six OFDM, depending on whether a normal or an extended cyclic prefix is used. The DCI is placed within the first slot of each subframe. The DCI carries the content of the PDCCH, PCFICH, and PHICH, and together they occupy up to the first three OFDM symbols in each subframe. This region is also known as the L1/L2 control region, since it contains information that is transferred to layer 1 (PHY) from layer 2 (the MAC layer).

The PBCH containing the MIB is located within subframe 0 and the PSS and SSS are located within subframes 0 and 5. The PBCH channel and both the PSS and SSS signals are placed

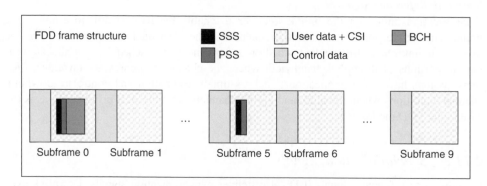

Figure 2.10 Downlink FDD subframe structure

within the six resource blocks centered on the DC subcarrier. In addition, CSRs are placed throughout each resource block in each subframe with a specific pattern of time and frequency separations. The pattern of placement for the CSR signals depends on the MIMO mode and the number of antennas in use, as will be discussed shortly. The rest of the resource elements in each subframe are allocated to user traffic data.

2.13   Uplink Frame Structures

The uplink subframe structure is in some ways similar to that for the downlink. It is composed of 1 ms subframes divided into two 0.5 ms slots. Each slot is composed of either seven or six SC-FDM symbols, depending on whether a normal or an extended cyclic prefix is used. The inner-band resource blocks are reserved for data resource elements (PUSCH) in order to reduce out-of-band emissions. Different users are assigned different resource blocks, a fact that ensures orthogonality among users in the same cell. Data transmission can hop at the slot boundary to provide frequency diversity. Control resources (PUCCH) are then placed at the edge of the carrier band, with interslot hopping providing frequency diversity. The reference signals necessary for data demodulation are interspersed throughout the data and control channels. Figure 2.11 illustrates an uplink frame structure.

2.14   MIMO

The LTE and LTE-Advanced standards achieve their maximum data rates in part due to their incorporation of many multi-antenna or MIMO techniques. LTE standards perfectly combine

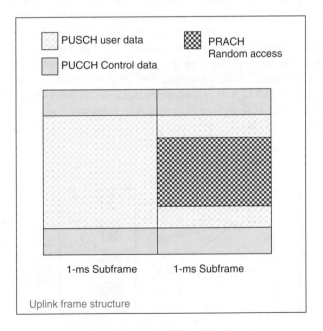

Figure 2.11   Uplink frame structure

the OFDM transmission structure with various MIMO methodologies. As such, LTE standards represent a MIMO-OFDM system. As we saw earlier, the OFDM transmission scheme in each antenna constructs the resource grid, generates the OFDM symbols, and transmits. In a MIMO-OFDM system, this process is repeated for multiple transmit antennas. Following transmission of OFDM symbols associated with multiple resource grids on multiple transmit antennas, at each receive antenna the OFDM symbols of all transmitted antennas are combined. The objective of a MIMO receiver is thus to separate the combined signals, and based on the received estimates of resource elements, to resolve each resource element transmitted on each of the transmit antennas.

Multi-antenna techniques rely on transmission by more than one antenna at the receiver or the transmitter, in combination with advanced signal processing. Although multi-antenna techniques raise the computational complexity of the implementation, they can be used to achieve improved system performance, including improved system capacity (in other words, more users per cell), and improved coverage or the possibility of transmitting over larger cells. The availability of multiple antennas at the transmitter or the receiver can be utilized in different ways to achieve different aims.

2.14.1 Receive Diversity

The simplest and most common multi-antenna configuration is the use of multiple antennas at the receiver side (Figure 2.12). This is often referred to as receive diversity. The most important algorithm used in receive diversity is known as Maximum-Ratio Combining (MRC). It is used within mode 1 of transmission in the LTE standard, which is based on single-antenna transmission. This mode is also known as SISO (Single Input Single Output) where only one receiver antenna is deployed or SIMO (Single Input Multiple Output) where multiple receive antennas are used. Two types of combining method can be used at the receiver: MRC and Selection Combining (SC) [2]. In MRC, we combine the multiple received signals (usually by averaging them) to find the most likely estimate of the transmitted signal. In SC, only the received signal with the highest SNR is used to estimate the transmitted signal.

MRC is a particularly good MIMO technique when, in a fading channel, the number of interfering signals is large and all signals exhibit rather equal strengths. As such, MRC works best in transmission over a flat-fading channel. In practice, most wideband channels, as specified in LTE, are subject to time dispersion, resulting in a frequency-selective fading response. To

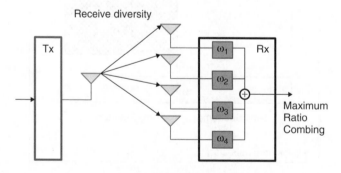

Figure 2.12 MIMO receive diversity

counteract the effects of frequency-selective coding, we must perform linear equalization, and in order to make this more efficient it should be done in the frequency domain. The MIMO techniques that handle these types of degradation best are discussed next.

2.14.2   Transmit Diversity

Transmit diversity exploits multiple antennas at the transmitter side to introduce diversity by transmitting redundant versions of the same signal on multiple antennas. This type of MIMO technique is usually referred to as Space–Time Block Coding (STBC). In STBC modulation, symbols are mapped in the time and space (transmit antenna) domains to capture the diversity offered by the use of multiple transmit antennas.

Space–Frequency Block Coding (SFBC) is a technique closely related to STBC that is selected as the transmit diversity technique in the LTE standard. The main difference between the two techniques is that in SFBC the encoding is done in the antenna (space) and frequency domains rather than in the antenna (space) and time domains, as is the case for STBC. A block diagram of SFBC is given in Figure 2.13.

In LTE, the second transmission mode is based on transmit diversity. SFBC and Frequency-Switched Transmit Diversity (FSTD) are used for two- and four-antenna transmission, respectively. Transmit diversity does not help with any boost in data rate; it only contributes to the increased robustness against channel fading and improves the link quality. Other MIMO modes – specifically, spatial multiplexing – contribute directly to the increased data rate in the LTE standard.

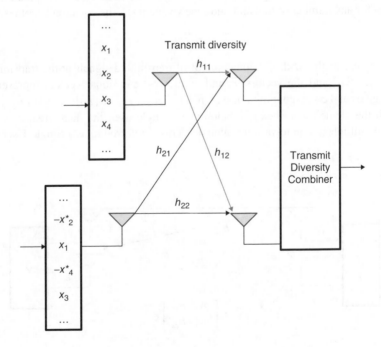

Figure 2.13   MIMO Space–Frequency Block Coding

2.14.3 Spatial Multiplexing

In spatial multiplexing, completely independent streams of data are transmitted simultaneously over each transmit antenna. The use of spatial multiplexing enables a system to increase its data proportionally to the number of transmit antenna ports. At the same time, and at the same sub-carrier in frequency, different modulated symbols are transmitted over different antennas. This means spatial multiplexing can directly increase the bandwidth efficiency and result in a system with high bandwidth utilization. The benefits of spatial multiplexing can be realized only if transmissions over different antennas are not correlated. This is where the multipath fading nature of a communication link actually helps the performance. Since multipath fading can decorrelate the received signals at each receive antenna port, spatial multiplexing transmitted over a multipath fading channel can actually enhance the performance.

All the benefits of spatial multiplexing can be realized only if a system of linear equations describing the relationship between transmit and receive antennas can be solved. Figure 2.14 illustrates the spatial multiplexing for a 2×2 antenna configuration. At each subcarrier, the symbols $s1$ and $s2$ are transmitted over two transmit antennas. The received symbols at the same subcarrier $r1$ and $r2$ may be considered the result of a linear combination of $s1$ and $s2$ weighted by the channel matrix H with the addition of AWGN (Additive White Gaussian Noise) $n1$ and $n2$. The resulting MIMO equation can be expressed as:

$$\begin{bmatrix} r_1 \\ r_2 \end{bmatrix} = \begin{bmatrix} H_{11} & H_{12} \\ H_{21} & H_{22} \end{bmatrix} \begin{bmatrix} s_1 \\ s_2 \end{bmatrix} + \begin{bmatrix} n_1 \\ n_2 \end{bmatrix} \tag{2.5}$$

where the MIMO channel matrix H contains the channel frequency responses at each subcarrier H_{ij} for any combination of transmit antenna i and receive antenna j. In a matrix notation generalized for any number of transmit and receive antennas, the equation becomes:

$$\vec{r} = H\vec{s} + \vec{n} \tag{2.6}$$

where $\vec{s}$ represents the M-dimensional vector of transmitted signals at the transmitter side: $\vec{s} = [s_1, s_2, \ldots, s_M]$ and the vectors $\vec{r}$ and $\vec{n}$ are N-dimensional vectors representing the received signals and corresponding noise signals: $\vec{r} = [r_1, r_2, \ldots, r_M]$; $\vec{n} = [n_1, n_2, \ldots, n_M]$.

When all the elements of vector $\vec{s}$ belong to a single user, the data streams of this single user are multiplexed on to various antennas. This is referred to as a Single-User Multiple

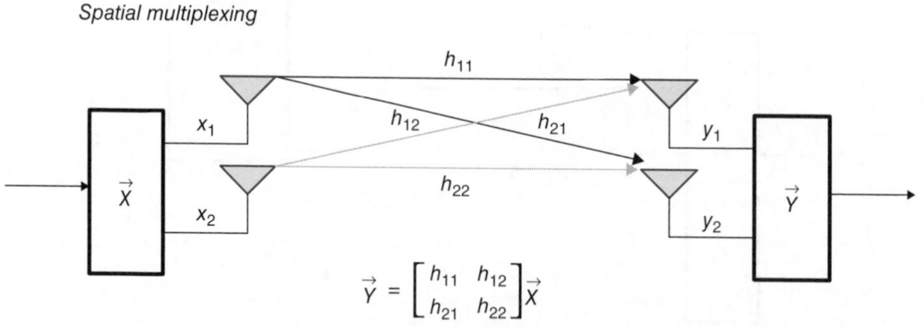

Figure 2.14   MIMO spatial multiplexing

Input Multiple Output (SU-MIMO) system. When data streams of different users are multi-plexed on to different antennas, the resulting system is known as a Multi-User Multiple Input Multiple Output (MU-MIMO) system. SU-MIMO systems substantially increase the data rate for a given user and MU-MIMO systems increase the overall capacity of a cell to handle multiple calls.

One of the most fundamental questions regarding the operation of a spatial multiplexing system is whether or not the corresponding MIMO equation can be solved and whether it has a unique solution. This question relates to the singularity of the corresponding MIMO chan-nel matrix and whether or not it can be inverted. When the received signals on many receive antennas are correlated, the channel matrix H may have rows or columns that are linearly dependent. In that case, the resulting channel matrix will have a rank less than its dimension and the matrix will be deemed non-invertible. Therefore, rank estimation is necessary for the spatial multiplexing since it determines whether it is possible to perform the spatial multiplex-ing operations under any given channel condition. The actual value of the rank of the matrix indicates the maximum number of transmit antennas that can be successfully multiplexed. In LTE terminology, the rank is known as the number of layers in the spatial multiplexing modes of MIMO.

In closed-loop MIMO operations, the rank of the channel matrix is computed by the mobile and transmitted to the base station via the uplink control channels. If the channel is deemed to have less than a full rank, only a reduced number of independent data streams can take part in spatial multiplexing in the upcoming downlink transmissions. This feature, known as rank adaptation, is part of the adaptive MIMO schemes and complements other adaptive features of the LTE standard.

2.14.4   Beam Forming

In beam forming, multiple transmit antennas can be used to shape the overall antenna radi-ation pattern (or the beam) in order to maximize the overall antenna gain in the direction of the mobile terminal. This type of beam forming provides the basis of downlink MIMO transmission mode 7.

The use of beam-forming techniques can lead to an increase in the signal power at the receiver proportional to the number of transmit antennas. Typically, beam forming relies on the use of an antenna array of at least eight antenna elements [3]. Beam forming is then imple-mented by applying different complex-valued gains (otherwise known as weights) to different elements of the antenna array. The overall transmission beam can then be steered in differ-ent directions by applying different phase shifts to the signals on the different antennas, as illustrated in Figure 2.15.

The LTE standard specifies neither the number of antennas in the antenna array nor the algorithms that are to be used in adjusting the complex-valued gains applied to each array ele-ment. The LTE specification refers to an antenna port 5, which represents the virtual antenna port created by the use of beam-forming techniques. UE-specific reference signals are used for channel estimation in beam forming MIMO mode 7. Higher layers call the use of UE-specific reference signals to the mobile terminal. Since mutually orthogonal reference sig-nals are generated scheduled on the same pairs of resource blocks, different UEs (mobile terminals) can resolve their allocated reference signals and use them for equalization and demodulation.

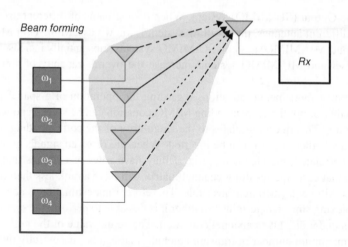

Figure 2.15 MIMO beam forming

2.14.5 *Cyclic Delay Diversity*

Cyclic Delay Diversity (CDD) is another form of diversity that is used in the LTE standard in conjunction with open-loop spatial multiplexing. CDD applies cyclic shifts to vectors or blocks of signal transmitted at any given time on different antennas. This is an effect analogous to the application of a known precoder. As such, CDD fits very well with block-based transmission schemes such as OFDM and SC-FDM. In the case of OFDM transmission, for example, a cyclic shift of the time domain corresponds to a frequency-dependent phase shift in the frequency domain. Since the phase shift in frequency – that is, the precoder matrix – is known and predictable, CDD is used in open-loop spatial multiplexing and in high-mobility scenarios where closed-loop feedback of an optimal precoder matrix is not desirable. The net effect of applying CDD is the introduction of artificial frequency diversity as experienced by the receiver. We can easily extend CDD to more than two transmit antennas, with different cyclic shifts for each.

2.15 MIMO Modes

Table 2.10 summarizes the LTE transmission modes and the associated multi-antenna transmission schemes. Mode 1 uses receive diversity and mode 2 is based on transmit diversity. Modes 3 and 4 are single-user implementations of spatial multiplexing based on open-loop and closed-loop precoding, respectively. Mode 3 also uses CDD (discussed earlier).

LTE mode 5 specifies a very simple implementation of multi-user MIMO based on mode 4 with the maximum number of layers set to one. Mode 6 features beam forming and a special case of mode 4 where the number of layers is set to two. LTE modes 7–9 implement versions of spatial multiplexing without the use of codebooks, with a number of layers of 1, up to 2, and 4–8, respectively. The LTE-Advanced (Release 10) introduced major enhancements to downlink MU-MIMO by introducing modes 8 and 9. For example, mode 9 supports eight

Table 2.10   LTE transmission modes and their associated multi-antenna transmission schemes

LTE transmission modes	
Mode 1	Single-antenna transmission
Mode 2	Transmit diversity
Mode 3	Open-loop codebook-based precoding
Mode 4	Closed-loop codebook-based precoding
Mode 5	Multi-user MIMO version of transmission mode 4
Mode 6	Single-layer special case of closed-loop codebook-based precoding
Mode 7	Release 8 non-codebook-based precoding supporting only a single layer, based on beam forming
Mode 8	Release 9 non-codebook-based precoding supporting up to two layers
Mode 9	Release 10 non-codebook-based precoding supporting up to eight layers

transmit antennas for transmissions of up to eight layers. These advances result directly from the introduction of new reference signals (CSI-RS and DM-RS), enabling a non-codebook-based precoding and thus adopting a lower-overhead double-codebook structure [4].

2.16   PHY Processing

In order to understand the LTE PHY, we have to specify the following sequence of operations. First, describe channel coding, scrambling, and modulation resulting in modulated symbols, then describe the steps in mapping the modulated signals to the resource grid, including mapping the user data, the reference signals, and the control data. Then, specify the MIMO modes that enable multiple antenna transmissions. The different MIMO algorithms involve specifying layer mapping, which describes how many transmit antennas are used in every frame and what precoding transformation is applied to the modulated bits before they are mapped to the resource grids of all transmit antennas.

2.17   Downlink Processing

The chain of signal processing operations performed in the transmitter can be summarized as the combination of transport block processing and physical channel processing. The processing stack is completely specified in 3GPP documents describing the multiplexing and channel coding [5] and physical channels and modulation [3]. The baseband signal processing chain applied to the combination of DLSCH and PDSCH can be summarized as follows:

- Transport-block CRC (Cyclic Redundancy Check) attachment
- Code-block segmentation and code-block CRC attachment
- Turbo coding based on a one-third rate
- Rate matching to handle any requested coding rates
- Code-block concatenation to generate codewords

- Scrambling of coded bits in each of the codewords to be transmitted on a physical channel
- Modulation of scrambled bits to generate complex-valued modulation symbols
- Mapping of the complex-valued modulation symbols on to one or several transmission layers
- Precoding of the complex-valued modulation symbols on each layer for transmission on the antenna ports
- Mapping of complex-valued modulation symbols for each antenna port to resource elements
- Generation of complex-valued time-domain OFDM signal for each antenna port.

Figure 2.16 illustrates the combination of the signal processing applied to transport blocks delivered to the PHY from the MAC layer until the OFDM signal is transferred to antennas for transmission.

Each of the components of LTE downlink transmission is described in detail in Chapters 4–7. In Chapter 4, we will elaborate on DLSCH processing and on scrambling and modulation mapper functionality. In Chapter 5, we will detail the OFDM multicarrier transmission scheme used in downlink. In Chapter 6, we will review details regarding various MIMO implementations of the standard. In Chapter 7, we will describe the link adaptation functionalities that use various control channels for dynamic scheduling of resources according to the channel conditions.

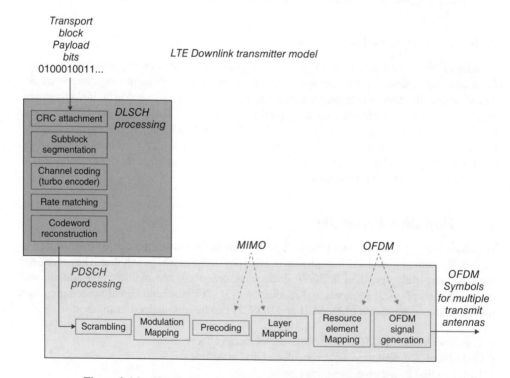

Figure 2.16   Signal processing chain of downlink DLSCH and PDSCH

2.18   Uplink Processing

The chain of signal processing operations applied to the combination of ULSCH and PUSCH is summarized as follows:

- Transport-block CRC attachment
- Code-block segmentation and code-block CRC attachment
- Turbo coding based on a one-third rate
- Rate matching to handle any requested coding rates
- Code-block concatenation to generate codewords
- Scrambling
- Modulation of scrambled bits to generate complex-valued symbols
- Mapping of modulation symbols on to one or several transmission layers
- DCT transform precoding to generate complex-valued symbols
- Precoding of the complex-valued symbols
- Mapping of precoded symbols to resource elements
- Generation of a time-domain SC-FDM signal for each antenna port.

Figure 2.17 illustrates the combination of the signal processing applied to transport blocks delivered to the PHY until the SC-FDM signal is transferred to antennas for transmission. The

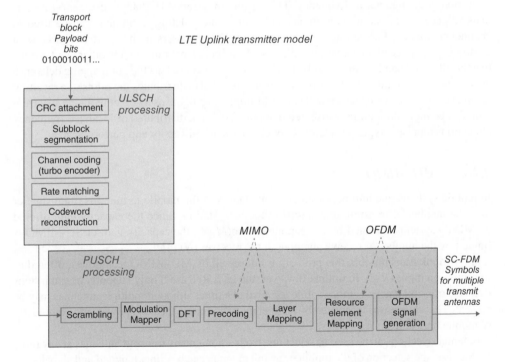

Figure 2.17   Signal processing chain of downlink ULSCH and PUSCH

processing stack is also fully specified in 3GPP documents describing the multiplexing and channel coding [5] and physical channels and modulation [3].

In this section, we will describe two distinguishing components of uplink transmission: SC-FDM based on DFT-precoded OFDM and MU-MIMO.

2.18.1   SC-FDM

In LTE, a special precoding, based on the application of DFT to modulated symbols, is used to generate the SC-FDM signal in the frequency domain. Note that SC-FDM signal generation is almost identical to that of OFDM, with the exception that an additional M-point DFT is introduced. Usually, computing the DFT is less computationally efficient than computing the FFT. However, we can find efficient implementations for certain DFTs whose sizes are of prime lengths. This is the reason why LTE specifies the M-point DFT sizes as multiples of two, three, or five (all prime numbers).

In uplink transmission, following coding, scrambling, and modulation and prior to resource element mapping, a DFT-based precoder is applied to the modulated symbols of each layer. The DFT-transformed symbols are then mapped to frequency subcarriers prior to the IFFT operation and cyclic prefix insertion, which finally leads to SC-FDM signal generation. The data symbols of any individual user transmitted as a SC-FDM symbol must be either contiguous or evenly spaced in the resource grid.

Localized mapping of DF-precoded symbols within the resource grid means that the entire allocation is contiguous in frequency. This results in acceptable channel-estimation performance, since the pilots are contiguous and simple interpolating techniques can be used in channel estimation. Furthermore, multiplexing of different users in the spectrum based on a contiguous resource block pattern is quite easy. Distributed mapping, on the other hand, means that the allocated bandwidth is evenly distributed in frequency. This type of mapping delivers a good measure of frequency diversity. However, since it also distributes the pilots, the resulting channel estimation performance will suffer. Multiplexing all the users together in the spectrum will also be more difficult in distributed mapping. As such, distributed or localized frequency allocations represent typical tradeoffs between frequency diversity and performance.

2.18.2   MU-MIMO

In mobile systems, the number of receive antennas N at the mobile terminal is often smaller than the number of transmit antennas M at the base station. Since the capacity gain offered by MIMO systems is scaled by the parameter $\min(M, N)$, the capacity gain of SU-MIMO is limited by the number of receive antennas at the receiver (N) [4].

In downlink transmission, this problem is addressed by MU-MIMO techniques, offered as transmission modes 7–9. In uplink, however, the LTE Release 8 only supports transmissions over one transmit antenna at the mobile terminal at a time, although multiple antennas may be present. This choice was motivated by an attempt to minimize the cost, power, and complexity of mobile hardware.

Antenna selection can be used to select one from among many transmit antennas at any time. In this case, the selection of the mobile transmit antenna can be either handled and signaled by

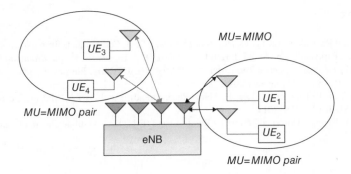

Figure 2.18   Uplink MU-MIMO

the base station or locally managed by the mobile terminal. Uplink MU-MIMO can be viewed as a MIMO system where, different users transmit their streams on the same resource blocks while each transmitting on a single antenna in their mobile units.

Figure 2.18 shows a block diagram of such an uplink MU-MIMO scenario. In this example, we form a bunch of MU-MIMO pairs by pairing together transmissions from couples of mobile units. The base station schedules the uplink transmission for each UE within the MU-MIMO pair in the same subframe and on the same resource blocks. Depending on the number of resource blocks available in the system bandwidth, we can schedule multiple MU-MIMO pairs simultaneously. The pairing can change in time based on such considerations as power control, individual channel quality, and interference profile. Although in our example we showed two paired users, the combination of DM-RS and CSI-RS reference signals in LTE-Advanced allows us to share up to eight mobile terminals in MU-MIMO that share the same resource blocks. For more information on MU-MIMO, the reader is referred to [4].

2.19   Chapter Summary

In this chapter we studied the PHY specifications of the LTE standards. We focused on identifying an adequate set of elements of the PHY model necessary for a deeper understanding of this subject. First, we examined the air interface of the standard, detailing its frequency bands, bandwidths, time framing, and time–frequency structure. We then elaborated on the multicarrier schemes of the standard: OFDM for downlink transmission and SC-FDM for uplink transmission. We identified the constituents of the OFDM resource grid, which is fundamental to understanding the PHY modeling. We also discussed the frame structures in uplink and downlink.

We then covered the physical channels and physical signals used in both uplink and downlink transmissions. We also provided an introduction to the MIMO schemes used in the standard, which completely specify various transmission modes. Finally, we summarized the sequence of operations performed in downlink and uplink transmissions. We have left the details regarding modeling of the processing chain in MATLAB for Chapters 4–7.

References

[1] Ghosh, A. and Ratasuk, R. (2011) *Essentials of LTE and LTE-A*, Cambridge University Press, Cambridge.

[2] Dahlman, E., Parkvall, S. and Sköld, J. (2011) *4G LTE/LTE-Advanced for Mobile Broadband*, Elsevier.

[3] 3GPP (2011) Evolved Universal Terrestrial Radio Access (E-UTRA), , Physical Channels and Modulation Version 10.0.0. TS 36.211, January 2011.

[4] C. Lim, T. Yoo, B. Clerckx, B. Lee, B. Shim, Recent trend of multiuser MIMO in LTE-advanced, *IEEE Magazine*, 51, 3, 127–136, 2013.

[5] 3GPP (2011) Evolved Universal Terrestrial Radio Access (E-UTRA), Multiplexing and Channel Coding. TS 36.212.

3

MATLAB® for Communications System Design

In this chapter, we introduce some of the capabilities in MATLAB related to the analysis, design, modeling, simulation, implementation, and verification of communications systems. We attempt to answer the following question: How can MATLAB, a high-level programming language and a design and simulation environment with an extensive library of software toolboxes, help academics and practitioners in the development of mobile and wireless systems?

3.1 System Development Workflow

To answer this question, we review multiple stages of development: from early research and algorithm design to integration of individual algorithms into a prototype system model, to verification using simulations that the system works as intended, to checking whether the system is realizable, to assessing its resource consumption, memory, complexity, and so on, to coding the design as a software or hardware implementation. The step before implementation – that is, system-level resource assessment – requires some form of software coding for system-level simulation. It also involves integrating real constraints such as data types and memory with complexity trade-offs. This system-level code can be used as the basis for hardware implementation, with the aim being to integrate sufficient detail that the task of the implementer becomes the creation of a bit-accurate model of the software simulation as either assembly code for implementation on Digital Signal Processors (DSPs) or as Hardware Description Language (HDL) code for implementation on a Field-Programmable Gate Array (FPGA) or an Application-Specific Integrated Circuit (ASIC). Throughout this process we must continuously monitor new details as they are added to the model in order to ensure that the elaborated design still meets the requirements set out at the research and development level.

Understanding LTE with MATLAB®: From Mathematical Modeling to Simulation and Prototyping, First Edition. Houman Zarrinkoub.
© 2014 John Wiley & Sons, Ltd. Published 2014 by John Wiley & Sons, Ltd.

3.2   Challenges and Capabilities

We face a number of challenges when we start from a typical standards specification until we implement the design. These challenges include:

- Translation from a specification based on text-based explanations to a software model that can act as a blueprint for implementation
- Introduction of innovative proprietary algorithms specially for the receiver operations where standards provide flexibility
- Execution of the software model in order to perform a dynamic system-level performance evaluation
- Acceleration of simulation for the handling of large data sets
- Resolving gaps in the implementation workflow.

MATLAB and its toolboxes can help address some of these challenges.

- Digital signal processing and advanced linear algebra, as foundations of the LTE (Long Term Evolution) standard, form the core competency of the MATLAB language. The make-up of the standard can gradually and intuitively be synthesized with a series of MATLAB programs
- The Communications System Toolbox provides ready-to-use MATLAB tools for the build-ing of communications-system models. With over 100 algorithms for modulation, channel modeling, error correction, MIMO (Multiple Input Multiple Output) techniques, equaliz-ers, and more, the Toolbox allows us to focus on communications system design rather than software engineering. It also includes many standards-based examples in order to allow a quick start
- MATLAB and Simulink are ideal environments for dynamic and large-scale simulations.
- MATLAB enables simulation to be accelerated.
- MATLAB allows implementation workflow gaps to be addressed, using:
  - Automatic MATLAB to C/C++ and HDL (Hardware Description Language) code generation
  - Hardware-in-the-loop verification.

We can divide these capabilities into four categories: algorithm development, modeling and simulation, simulation acceleration, and path to implementation. In this chapter, following a quick introduction to MATLAB and Simulink as core products, we will introduce three cate-gories of capability:

- Tools for modeling and simulation
- Tools for acceleration of the simulation speed
- Tools that enable a path to implementation.

The modeling and simulation capabilities, including various toolboxes, enable users to create simulation models of communications standards, including wireless and mobile standards. Running these simulation models enables the designer to gauge the performance of the entire system and of individual algorithms and to determine the effects of channel degradations and other real-time conditions.

3.3 Focus

The focus of this book is on LTE PHY (Physical Layer) modeling as MATLAB programs. For example, we discuss modeling and simulation of the LTE standard only in the FDD (Frequency Division Duplex) mode. Without any loss of generality and with some modification of MATLAB code, the reader can then adopt our MATLAB program for TDD (Time Division Duplex) mode. We will not cover topics related to control-plane processing, roaming or random access, or multimedia broadcast frames, nor will we cover detailed MATLAB programs related to multicast mode or multi-user MIMO. We will focus instead on a general scenario in which a mobile unit is assigned to a cell and we fully detail user-plane data processing.

3.4 Approach

Starting from the simplest component of the LTE (i.e., the modulators), we will create a series of MATLAB programs that progressively add other components such as scrambling and channel coding to the signal processing chains. At each stage, we will compute performance measures such as Bit Error Rate (BER) to ensure that the combinations of components are properly modeled in MATLAB. We will continue this process and develop MATLAB programs to model OFDM and MIMO operations specified in the standard. In so doing, we will also generate multiple subfunctions that help match the model to the LTE standard. At the end of this process, we will have MATLAB programs and Simulink models that represent important signal processing operations of various downlink modes of the LTE standard.

3.5 PHY Models in MATLAB

In this book we will iteratively and systematically build up the necessary components of the LTE PHY in MATLAB for downlink transmission. However, in order to make the discussion fit a book of this size, we have to be selective in the details we highlight. The pedagogic value of iterative and gradual design can be more beneficial than adherence to all parameters and details specified in the standard. As the title of the book attests, we are aiming for a creating understanding of LTE by augmentation of technical discussions with software programs that can be executed in MATLAB. The ability to run and execute the software and simulate the system adds another dimension to the level of understanding.

Next, we will highlight various products that helps users model, simulate, prototype, and implement wireless systems in MATLAB.

3.6 MATLAB

MATLAB is a widely-used programming language for algorithm development, data analysis, visualization, and numerical computation. If the volume of technical papers and publications mentioning it is any indication, MATLAB has a long history in communications system design and is used by both academics and practitioners. It lets designers focus on algorithms rather than low-level programming. Many of its features and capabilities are perfect for modeling wireless systems: (i) it has an interactive program and environment that matches the exploratory nature of science; (ii) it provides seamless access to data and algorithms; and (iii) it has tools for visualization, algorithm development, and data analysis.

Matrix as fundamental data type: The fundamental data type in MATLAB is the matrix. Since most algorithms used in communications systems are based on block-based or frame-based processing of data, expressing these algorithms is natural in MATLAB. This means the mathematical formula expressed in the matrix format is immediately expressed in MATLAB. For example, in MIMO systems the relationship between the received and the transmitted data is expressed by a system of linear equations of the form $y = Ax + n$. These kinds of relationship can be expressed easily by a single line of MATLAB code. Compare that to a typical C code representing the same algorithms, which will look like a double for loop.

Linear algebra and Fourier analysis: MATLAB contains mathematical, statistical, and engineering functions that support all common engineering and science operations. These functions, developed by experts in mathematics, are the foundation of the MATLAB language. The core math functions use the LAPACK and BLAS linear algebra subroutine libraries and the FFTW Discrete Fourier Transform library. Mathematical functions for linear algebra, statistics, Fourier analysis, filtering, optimization, and numerical integration are implemented as fast and accurate functions in MATLAB.

Visualization for design validation: Most graphical features required to visualize engineering and scientific data are available in MATLAB. These include 2D and 3D plotting functions, 3D volume-visualization functions, tools for the interactive creation of plots, and the exporting of results to all popular graphics formats. Plots can be customized by a variety of methods.

Complex numbers and a range of data types: Simulation of communications systems relies on the extensive use of complex data and random number generators. MATLAB enables you to perform arithmetic operations on a wide range of data types, including doubles, singles, and integers. MATLAB also has optimized functions for random number generators. Functions such as *randn* (which models random numbers with normal distributions), *rand* (for uniform distribution), and *randi* (for discrete integer random distributions) have favorable properties in terms of periodicity and efficiency [1].

3.7   MATLAB Toolboxes

MATLAB's add-on software tools are called toolboxes. These provide specialized mathematical functionalities in areas including signal processing and communications. They complement the core MATLAB library and provide application-specific functions and objects that accelerate the process of modeling and building algorithms and systems. These algorithmic building blocks enable the user to focus on their area of expertise instead of having to reinvent and implement the basics.

Four system toolboxes – DSP System Toolbox [2], Communications System Toolbox [3], Phased Array System Toolbox [4], and Computer Vision System Toolbox [5] – are particularly suitable for system modeling in different application areas. Not only do they provide algorithms for the design, simulation, and verification of various application areas, but they provide components that facilitate the creation of simulation test benches for the modeling of dynamic systems. In later sections we will review some of these system toolboxes in further detail.

3.8 Simulink

Simulink requires MATLAB and provides an environment for multidomain simulation and model-based design for dynamic and embedded systems [6]. It provides an interactive graphical environment and a customizable set of block libraries. With an easy-to-use graphical design environment, Simulink allows us to design, simulate, implement, and test a variety of time-varying systems, including communications, control, signal processing, and video processing.

With Simulink, we can create, model, and maintain a detailed block diagram of our system using a comprehensive set of predefined blocks. Simulink provides tools for hierarchical modeling, data management, and subsystem customization. Additional blocksets or system toolboxes extend Simulink with specific functionality for aerospace, communications, radio frequency, signal processing, video, image processing, and other applications; these features are particularly useful for the modeling and simulation of communications systems.

Integration with MATLAB: MATLAB functions can be called within Simulink models in order to implement algorithms that can analyze data and verify a design. Use of the MATLAB function block in Simulink allows MATLAB code to be integrated into Simulink. Simulink will first use its code-generation capabilities to translate the MATLAB code to C code, then compile the C code as a MEX (MATLAB Executables) function and call the resulting MEX function when executing the Simulink model.

Signal attributes and data-type support: Like MATLAB, Simulink defines the following signal and parameter attributes: data types – single, double, signed, or unsigned 8, 16, or 32 bit integers; Boolean and fixed-point; dimension – scalar, vector, matrix, or N-D arrays; values – real or complex. This enables us, for example, to monitor the effects of finite word lengths on the accuracy of the computation in an algorithm.

Simulation capabilities: After building a model in Simulink, we can simulate its dynamic behavior and view the results. Simulink provides several features and tools that ensure the speed and accuracy of a simulation, including fixed-step and variable-step solvers, a graphical debugger, and a model profiler.

Using solvers: Solvers are numerical-integration algorithms that compute the system dynamics over time using information contained in the model. Simulink provides solvers to support the simulation of continuous-time (analog), discrete-time (digital), hybrid (mixed-signal), and multirate systems.

Executing a simulation: Once we have set the simulation options for a model, we can run the simulation interactively, using the Simulink GUI (Graphical User Interface), or systematically, by running it in batch mode from the MATLAB command line. The following simulation modes can be used:

- Normal (default), which interpretively simulates the model
- Accelerator, which speeds model execution by creating compiled target code, while still allowing the model parameters to be changed
- Rapid accelerator, which can simulate models faster than the accelerator mode but with less interactivity, by creating an executable separate from Simulink that can run on a second processing core.

3.9 Modeling and Simulation

Most algorithm development for various systems and components starts in MATLAB. With a library of digital signal-processing, linear algebra, and mathematical operators, designs can be expressed easily in MATLAB as algorithms composed of a pertinent sequence of operations. As individual algorithms are developed and connected to each other, this forms the basis of a system model. System modeling can best be done in either MATLAB or Simulink. As we saw earlier, Simulink allows MATLAB algorithms and functions to be integrated seamlessly as system components. By using various add-on toolboxes, we can expand the scope of the system and simulate it to verify that it behaves according to specifications. In this section we will introduce some of the MATLAB and Simulink add-on toolboxes that help with this process.

3.9.1 DSP System Toolbox

The DSP System Toolbox provides algorithms and tools for foundational signal processing operations. It comes with a slew of specialized filter design capabilities, FFTs (Fast Fourier Transforms), and multirate processing abilities and features algorithms captured as System objects that make the task of processing streaming data and creating real-time prototypes easier. The DSP System Toolbox has specialized tools for connecting to audio files and devices, performing spectral analysis, and using other interactive visualization techniques that enable the analysis of system behavior and performance. All of these components support automatic C/C++ code generation, most support fixed-point data, and a few generate HDL code.

3.9.2 Communications System Toolbox

The Communications System Toolbox provides algorithms and tools for the design, simulation, and analysis of communications systems. This toolbox is specially designed to model the PHY of communications systems. It contains a library of components including ones for source coding, channel coding, interleaving, modulation, equalization, synchronization, MIMO, and channel modeling. These components are provided as MATLAB functions, MATLAB System objects, and Simulink blocks, so they can be used as part of MATLAB or Simulink system models. All support C/C++ code generation, most support fixed-point data arithmetic, and a few generate HDL code for FPGA or ASIC hardware implementation.

3.9.3 Parallel Computing Toolbox

The Parallel Computing Toolbox [7] can help accelerate computationally and data-intensive problems using multicore processors, GPUs (Graphics Progressing Units), and computer clusters. Features such as parallelized for loops, special array types, and parallelized numerical algorithms allow the parallelization of MATLAB applications. The toolbox can be used with Simulink to run multiple simulations of a model in parallel. Two main approaches to simulation acceleration can be identified:

Multicore or cluster processing: Some applications can be sped up by organizing them into independent tasks and executing several at the same time on different processing units. This class of task-parallel application includes simulations for design optimization, BER testing, and Monte Carlo simulations. As one of its easy-to-use and intuitive capabilities, the toolbox offers *parfor*, a parallel for-loop construct that can automatically distribute independent tasks to multiple MATLAB workers. A MATLAB worker is a MATLAB computational engine that runs independently of the desktop MATLAB session. MATLAB can automatically detect the presence of workers and will revert to serial behavior if only the desktop session is present. Task execution can also be set up using other methods, such as manipulation of task objects in the toolbox.

GPU processing: The Parallel Computing Toolbox provides a special array type that allows computations to be performed on CUDA-enabled NVIDIA GPUs direct from MATLAB. Supported functions include FFT, element-wise operations, and several linear algebra operations. The toolbox also provides a mechanism that allows existing CUDA-based GPU kernels to be used directly from MATLAB. The Communication System Toolbox has many specialized algorithms that support GPU processing. The Parallel Computing Toolbox can be used to execute many communications algorithms directly on the GPU.

3.9.4   Fixed-Point Designer

Fixed-Point Designer [8], previously Fixed-Point Toolbox, provides fixed-point data types, operations, and algorithms in MATLAB. Using Fixed-Point Designer, the effects of finite word lengths can be modeled for variables in various algorithms. The toolbox allows fixed-point algorithms to be designed using MATLAB syntax and the results to be compared with the floating-point implementation of the same algorithm. These algorithms can be reused in Simulink and can pass fixed-point data to and from Simulink models. The toolbox provides a suite of tools that make it easier to convert an algorithm from a floating-point to a fixed-point implementation.

3.10   Prototyping and Implementation

Various MathWorks products can help elaborate a design from concept to embeddable code while staying within the MATLAB environment. The MATLAB algorithm must first be refined based on design constraints such as finite word lengths, limitations on memory and complexity, and so on. It can then be integrated and simulated as part of a larger system model, and bit-true test sequences can be generated to verify that software and hardware implementations match the golden reference results in MATLAB. Finally, C and HDL code can be generated for hardware implementation. With this step, the errors introduced by manual coding can be avoided by maintaining a single design source in MATLAB. Some of these products are presented in this section.

3.10.1   MATLAB Coder

MATLAB Coder [9] generates standalone C and C++ code from MATLAB code. The generated source code is portable and readable. MATLAB Coder supports code generation for a large subset of MATLAB language, including program control constructs, functions, and matrix operations. It also supports code generation for the functions and System objects of various toolboxes and System toolboxes. With MATLAB Coder we can generate:

- MEX functions that let us accelerate computationally intensive portions of MATLAB code and verify the behavior of the generated code

- Readable and portable C/C++ code for integration with existing C/C++ source codes and environments
- Dynamic and static libraries for integration with C-based tools and environments
- C/C++ executable for prototyping of algorithms and provision of proofs-of-concept.

3.10.2   Hardware Implementation

A design for a communication system can be realized as either embedded software or embedded hardware. An embedded software implementation targets DSP and general-purpose processors. The path from a MATLAB model to an embedded software implementation involves two steps: (i) C/C++ code generation from MATLAB and (ii) compiling or hand-coding of the C code as assembly code on the target. MATLAB Coder can be used for the first step and the compilers of various software simulators for hardware targets can be used for the second.

An embedded hardware implementation targets the design on FPGAs and ASICs. The process of realizing a design from a MATLAB model to the final FPGA or ASIC prototype involves two steps: (i) VHDL or Verilog code generation of MATLAB functions or Simulink models with the HDL Coder and (ii) post-processing by the integrated simulation environments to convert the RTL (Register Transfer Level) Verilog and VHDL code into a fully synthesized FPGA or ASIC design. HDL Coder [10] generates portable, synthesizable VHDL and Verilog code from MATLAB functions and Simulink models. It can be used to perform the first step of the implementation. Another MathWorks HDL tool, HDL Verifier, automates Verilog and VHDL design verification using HDL simulators and FPGA hardware-in-the-loop. HDL Verifier [11] can be used to bring an RTL design into MATLAB and to verify it by comparing the outputs of the VHDL and Verilog code with detailed implementations of the same algorithm in MATLAB and Simulink. Since in this book we focus on modeling, simulation, and software prototyping of the LTE standard, discussions regarding hardware implementations and realization of a design as HDL code are outside our scope.

3.11   Introduction to System Objects

In this book we highlight many features of the Communications System Toolbox, particularly we will introduce the new System objects used in the product. With a very intuitive user interface, System objects make the task of expressing communications systems easier and make the resulting MATLAB code more readable and sharable. System objects can be used as part of both MATLAB programs and Simulink models. They are MATLAB objects that represent time-based and executable algorithms and they are organized as objects to make them easy to use and virtually self-documenting. Since in the rest of this book we rely on System objects to express LTE-system models in MATLAB, a short tutorial on how to use these algorithmic components is presented in this section.

3.11.1   System Objects of the Communications System Toolbox

System objects of the Communications System Toolbox belong to the communications (*comm*) package and their names start with the common prefix "*comm*." In order to access all of the System objects of the Communications System Toolbox, type "comm." followed by a Tab key at the MATLAB command prompt:

>> comm.<Tab>

This will produce an alphabetical list of all the System objects available in the toolbox. As of the latest release of MATLAB, the Communications System Toolbox contains a total of 123 algorithms provided as System objects.

Let us choose one of these System objects, for example comm.QPSKModulator, and create one instance of this type of modulator. Let us call this instance "Modulator":

>> Modulator=comm.QPSKModulator

A QPSK (Quadratue Phase Shift Keying) modulator will be created and a description of this object will appear in the MATLAB workspace (Figure 3.1).

Every System object contains properties and methods. Its default properties appear when they are created; this self-documentation is a useful feature of System objects. From looking at the property list of a given System object, we know what parameters it can take and what values are typically assigned to them. For example, the phase-offset property of the QPSK modulator is by default set to $\pi/4$. Let us change this parameter to $\pi/2$. There are two ways to modify properties:

- Create an object with default values and then change a property using dot notation. For example:

  >> Modulator = comm.QPSKModulator;
  >> Modulator.PhaseOffset = pi/2

- Set different properties as they are created using property–values pairs. For example:

  >>Modulator = comm.QPSKModulator ('PhaseOffset',pi/2);

If properties are expressed as a string of characters, a convenient list of possible values appears when we want to set a particular property. For example, when we type "Modulator.SymbolMapping=" followed by a Tab, a list of mapping choices appears to facilitate setting of the property to any of the several options, in this case "Binary" and "Gray."

```
Command Window
>> Modulator = comm.QPSKModulator

Modulator =

  System: comm.QPSKModulator

  Properties:
        PhaseOffset: 0.785398163397448
           BitInput: false
      SymbolMapping: 'Gray'
     OutputDataType: 'double'
```

Figure 3.1 Creating a System object from the Communications System Toolbox

The step method is the main method of execution of a System object. After an object is created and configured, it can be passed an input (or multiple inputs) and its step method can be called to produce its output (or multiple outputs). There are two syntaxes available by which to execute the step method of a System object. We can:

- Use dot notation to call the System object: y = Modulator.step (u)
- Use the step method as a function and make the System object the first function argument: y = step (Modulator, u).

In Figure 3.2, a 10 × 1 column vector of bits (variable u) is created using the MATLAB *randi* function and then passed as input to the Modulator System object. By calling its step method, a 5 × 1 output vector (y) of modulated symbols representing the modulated bits using the QPSK algorithms based on specified properties is created.

```
>> u=randi([0 1], 10, 1)

u =

      1
      1
      0
      1
      1
      0
      0
      1
      1
      1

>> y=step(Modulator,u);
>> y=Modulator.step(u)

y =

      0.7071 - 0.7071i
     -0.7071 + 0.7071i
     -0.7071 - 0.7071i
     -0.7071 + 0.7071i
      0.7071 - 0.7071i
```

Figure 3.2   Executing a System object by calling its step method

Now that we have seen how to access, create, set the properties of, configure, and call a System object to perform computations, let us create a simple script that uses a few System objects to express a simple communications system.

3.11.2   Test Benches with System Objects

Following is a MATLAB script, otherwise known as a testbench, that uses System objects to perform BER analysis of a simple transceiver system. The transceiver is composed of a QPSK modulator, an Additive White Gaussian Noise (AWGN) channel, and a QPSK demodulator. Note that this code employs four System objects from the Communications System Toolbox: *comm.QPSKModulator*, *comm.AWGNChannel*, *comm.QPSKDemodulator*, and *comm.ErrorRate*.

Algorithm

MATLAB script

```
%% Constants
FRM=2048;
MaxNumErrs=200;MaxNumBits=1e7;
EbNo_vector=0:10;BER_vector=zeros(size(EbNo_vector));
%% Initializations
Modulator    = comm.QPSKModulator('BitInput',true);
AWGN         = comm.AWGNChannel;
DeModulator  = comm.QPSKDemodulator('BitOutput',true);
BitError     = comm.ErrorRate;
%% Outer Loop computing Bit-error rate as a function of EbNo
for EbNo = EbNo_vector
   snr = EbNo + 10*log10(2);
   AWGN.EbNo=snr;
   numErrs = 0; numBits = 0;results=zeros(3,1);
   %% Inner loop modeling transmitter, channel model and receiver for each EbNo
   while ((numErrs < MaxNumErrs) && (numBits < MaxNumBits))
      % Transmitter
      u       = randi([0 1], FRM,1);        % Generate random bits
      mod_sig = step(Modulator,  u);        % QPSK Modulator
      % Channel
      rx_sig  = step(AWGN,      mod_sig);   % AWGN channel
      % Receiver
      y =    step(DeModulator, rx_sig);     % QPSK Demodulator
      results = step(BitError,  u, y);      % Update BER
      numErrs = results(2);
      numBits = results(3);
   end
   % Compute BER
   ber = results(1); bits= results(3);
   %% Clean up & collect results
```

```
    reset(BitError);
    BER_vector(EbNo+1)=ber;
end
%% Visualize results
EbNoLin = 10.^(EbNo_vector/10);
theoretical_results = 0.5*erfc(sqrt(EbNoLin));
semilogy(EbNo_vector, BER_vector)
grid;title('BER vs. EbNo - QPSK modulation');
xlabel('Eb/No (dB)');ylabel('BER');hold;
semilogy(EbNo_vector,theoretical_results,'dr');hold;
legend('Simulation','Theoretical');
```

The script is organized in four sections. In the initialization section, the System objects are created and some parameters are set. The second section contains the processing testbench that iterates through Eb/N0 values and computes the corresponding BER measures. The third section is the transceiver processing loop, where the step methods of the System objects are called in order to modulate the input signal, add channel noise to the modulated signal, and demodulate to produce the received signal and compute BER. Finally, in the fourth section, the simulation is cleaned up and terminated and the BER performance results are visualized.

Figure 3.3 verifies that the simulated results match the analytical results.

By running this script we obtain the BER curves of a QPSK modulation system using an AWGN channel. Theoretical results for a QPSK modulation scheme processed by an AWGN channel are expressed in the following equation:

$$BER = \frac{1}{2}\text{erfc}\left(\sqrt{\frac{E_b}{N_0}}\right) \tag{3.1}$$

The use of System objects results in a modular and easy-to-understand MATLAB code and fosters a structure that can be expanded for more complex systems. We will follow this four-step process of initialization, processing loop, termination, and visualization throughout this book. One way of improving MATLAB programs and making them more readable is to separate testbench and visualization operations from the algorithms and system description. Next we will show how to achieve this by capturing our transceiver as a MATLAB function and separating our algorithmic component from the testbench script.

3.11.3   Functions with System Objects

The MATLAB function *chap3_ex02_qpsk* performs the algorithmic portion of our simple QPSK transceiver system. This function has three input variables:

- The first input is a scalar number that corresponds to Eb/N0
- The second input is one of the stopping criteria, based on the maximum number of errors that can be observed before the simulation is stopped
- The third input is the other stop criterion, based on the maximum number of bits that can be processed before the simulation is stopped.

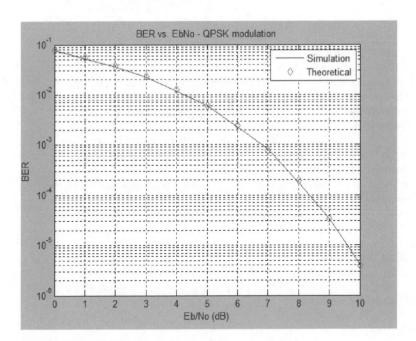

Figure 3.3 BER curves for QPSK modulation under an AWGN channel; simulated and theoretical results match

For each $Eb/N0$ value, the code runs in a while loop until either the specified maximum number of errors is observed or the maximum number of bits is processed. The code executes each System object by calling the step method. It computes two outputs:

• BER, defined as the ratio of the observed number of errors to the number bits processed
• The number of bits processed, based on the stopping criterion defined by the second and third input variables.

Algorithm

MATLAB function

```
function [ber, bits]=chap3_ex02_qpsk(EbNo, maxNumErrs, maxNumBits)
%% Initializations
persistent Modulator AWGN DeModulator BitError
if isempty(Modulator)
   Modulator    = comm.QPSKModulator('BitInput',true);
   AWGN         = comm.AWGNChannel;
   DeModulator = comm.QPSKDemodulator('BitOutput',true);
   BitError     = comm.ErrorRate;
end
%% Constants
FRM=2048;
```

```
M=4; k=log2(M);
snr = EbNo + 10*log10(k);
AWGN.EbNo=snr;
%% Processsing loop modeling transmitter, channel model and receiver
numErrs = 0; numBits = 0;results=zeros(3,1);
while ((numErrs < maxNumErrs) && (numBits < maxNumBits))
    % Transmitter
    u           = randi([0 1], FRM,1);              % Random bits generator
    mod_sig     = Modulator.step(u);                % QPSK Modulator
    % Channel
    rx_sig      = AWGN.step(mod_sig);               % AWGN channel
    % Receiver
    demod       = DeModulator.step(rx_sig);     % QPSK Demodulator
    y           = demod(1:FRM);                     % Compute output bits
    results     = BitError.step(u, y);              % Update BER
    numErrs     = results(2);
    numBits     = results(3);
end
%% Clean up & collect results
ber = results(1); bits= results(3);
reset(BitError);
```

To avoid the overhead involved in creating and releasing System objects every time we call a function, the System objects in a function are denoted by persistent variables. Using a persistent variable allows us to perform such operations as creating System objects only the first time a function is called. This adds to the efficiency of function calls and improves the simulation speed as we call the function in a loop.

3.11.4   Bit Error Rate Simulation

The Communication System Toolbox provides BERTool as an integrated testbench for the performance of BER simulations. BERTool is a graphical application that computes a series of simulated bit error rates and compares the results with known analytical results.

For example, to visualize the simulated BERs for the function *chap3_ex02_qpsk.m*, as depicted in Figure 3.4, go to the Monte Carlo tab, specify the file as the simulation MATLAB file, and specify the $Eb/N0$ values and stopping criteria. BERTool will compute the BER for the range of $Eb/N0$ values provided and will automatically display the results. These simulated results can be compared with theoretical results by going to the Theoretical tab and specifying the modulation and coding schemes used. Figure 3.5 shows an example of the comparison plots generated by the BERTool testbench.

3.12   MATLAB Channel Coding Examples

In this section, using a pedagogic approach and a series of MATLAB programs, we will examine what the Toolbox offers in terms of channel coding. First we will model a system that

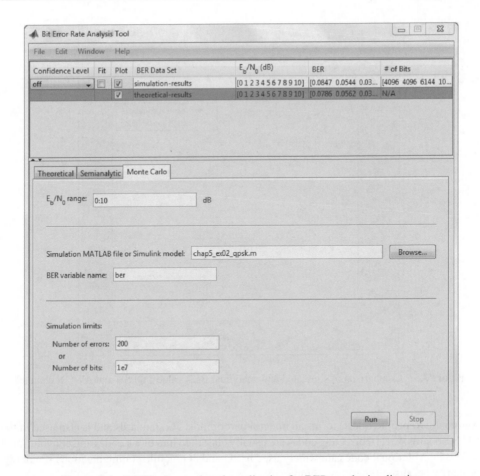

Figure 3.4   BERTool: a testbench application for BER-result visualization

uses convolutional encoding and Viterbi decoding based on hard-decision decoding. Then we will update the algorithm to use soft-decision decoding. Finally, we will replace convolutional coding with a turbo-coding algorithm and compare the performance at each stage. With these simple exercises, not only will we learn how easy it is to use MATLAB and the Communications System Toolbox to add more complexity to our mobile communication model but we will clearly see that the substantial improvement in BER performance explains the predominant role turbo coding plays in the channel coding of the LTE standard.

3.12.1   Error Correction and Detection

Channel coding comprises error detection and error correction. With error detection using the CRC (Cyclic Redundancy Check) detector, the receiver can request the repeat of a transmission, in what is known as an automatic repeat request. Forward-error-correction coding allows errors to be corrected based on the redundancy bits that are included with the transmitted signal. A hybrid of error detection and forward error correction known as HARQ (Hybrid

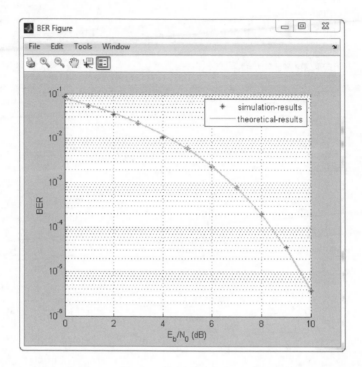

Figure 3.5 Comparison of simulated and analytical BER values: QPSK and AWGN channel

Automatic Repeat Request) forms an integral part of most 3G standards and is also used in the LTE standards. Error-correcting codes are usually classified into block codes and convolutional codes. Convolutional codes are widely used in 2G and 3G mobile communications standards, primarily because of their low complexity.

In this section, we will elaborate our growing MATLAB model, which already includes modulation, to include channel coding. As a perfect vehicle for explaining the value and motivation of the use of turbo coding in the LTE standard, we will compare the performances of convolutional and turbo coding. Furthermore, to explain the tradeoff involved in the use of receiver designs, we will compare the performance of modulation-coding combinations with and without soft-decision decoding.

3.12.2 Convolutional Coding

Convolutional codes are generated by the convolution of the input sequence with the impulse response of the encoder. The encoder accepts blocks of k-bit input samples and, by operating on the current block of data and the m previous input blocks, produces an n-bit block of output samples. The coding rate of the encoder is given by the ratio $Rc = k/n$ and the convolutional encoder is specified by these three parameters (n,k,m). Figure 3.6 illustrates a convolutional encoder.

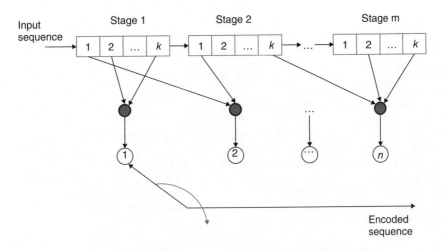

Figure 3.6   Structure of an (n,k,m) convolutional encoder

3.12.3   Hard-Decision Viterbi Decoding

In the first iteration of this exercise, we modify the MATLAB function in the last section to add a channel-coding scheme to the modulation. When a channel-coding scheme is used, the transmitter sends redundancy bits along with message bits through the wireless channel. The receiver accepts the transmitted signal and uses the redundancy bits to detect and correct some of the errors introduced by the channel. Let us start by adding a convolutional encoder and a Viterbi decoder to the communications system. This communications system uses hard-decision Viterbi decoding, where the demodulator maps the received signal to bits and then passes the bits to the Viterbi decoder for error correction. The following MATLAB function (*chap3_ex03_qpsk_viterbi*) uses QPSK modulation and hard-decision Viterbi decoding with an AWGN channel.

Algorithm

MATLAB function

```
function [ber, bits]=chap3_ex03_qpsk_viterbi(EbNo, maxNumErrs, maxNumBits)
%% Initializations
persistent Modulator AWGN DeModulator BitError ConvEncoder Viterbi
if isempty(Modulator)
    Modulator   = comm.QPSKModulator('BitInput',true);
    AWGN        = comm.AWGNChannel;
    DeModulator = comm.QPSKDemodulator('BitOutput',true);
    BitError    = comm.ErrorRate;
    ConvEncoder=comm.ConvolutionalEncoder(...
        'TerminationMethod','Terminated');
    Viterbi=comm.ViterbiDecoder('InputFormat','Hard',...
```

```
        'TerminationMethod','Terminated');
end
%% Constants
FRM=2048;
M=4; k=log2(M); codeRate=1/2;
snr = EbNo + 10*log10(k) + 10*log10(codeRate);
AWGN.EbNo=snr;
%% Processsing loop modeling transmitter, channel model and receiver
numErrs = 0; numBits = 0;results=zeros(3,1);
while ((numErrs < maxNumErrs) && (numBits < maxNumBits))
    % Transmitter
    u           = randi([0 1], FRM,1);              % Random bits generator
    encoded     = ConvEncoder.step(u);              % Convolutional encoder
    mod_sig     = Modulator.step(encoded);      % QPSK Modulator
    % Channel
    rx_sig      = AWGN.step(mod_sig);               % AWGN channel
    % Receiver
    demod       = DeModulator.step(rx_sig);      % QPSK Demodulator
    decoded     = Viterbi.step(demod);             % Viterbi decoder
    y           = decoded(1:FRM);                    % Compute output bits
    results     = BitError.step(u, y);               % Update BER
    numErrs     = results(2);
    numBits     = results(3);
end
%% Clean up & collect results
ber = results(1); bits= results(3);
reset(BitError);
```

By running this function within BERTool, we can gauge the performance of hard-decision Viterbi decoding and compare it with the upper-bound theoretical results. Examining the results in Figure 3.7, we can see that the simulated BER curve falls below the theoretical upper-bound values, which is consistent with our expectations. These results indicate that in order to arrive at a better performance we need to improve our decoding algorithm.

3.12.4   Soft-Decision Viterbi Decoding

In this iteration, we improve BER performance results by using a soft-decision decoding algorithm. In soft-decision decoding, the demodulator maps the received signal to log-likelihood ratios. These probability measures are based on the logarithm of the likelihood that the correct data are received instead of corrupted data. When log-likelihood ratios are provided as the input to the Viterbi decoder, the BER performance of the decoder is improved. An algorithm can be made to perform soft-decision Viterbi decoding by changing a few demodulator and Viterbi-decoder System-object parameters. The following MATLAB function (*chap3_ex04_qpsk_viterbi_soft*) has been updated to use soft-decision Viterbi decoding.

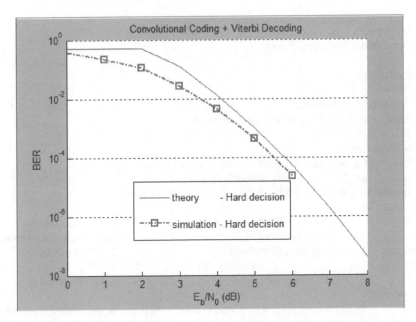

Figure 3.7 Performance with hard-decision Viterbi decoding: QPSK modulation with AWGN channel

Algorithm

MATLAB function

```
function [ber, bits]=chap3_ex04_qpsk_viterbi_soft(EbNo, maxNumErrs, maxNumBits)
%% Initializations
persistent Modulator AWGN DeModulator BitError ConvEncoder Viterbi Quantizer
if isempty(Modulator)
    Modulator    = comm.QPSKModulator('BitInput',true);
    AWGN         = comm.AWGNChannel;
    DeModulator = comm.QPSKDemodulator('BitOutput',true,...
       'DecisionMethod','Log-likelihood ratio',...
       'VarianceSource', 'Input port');
    BitError     = comm.ErrorRate;
    ConvEncoder=comm.ConvolutionalEncoder(...
       'TerminationMethod','Terminated');
    Viterbi=comm.ViterbiDecoder(...
       'InputFormat','Soft',...
       'SoftInputWordLength', 4,...
       'OutputDataType', 'double',...
       'TerminationMethod','Terminated');
    Quantizer=dsp.ScalarQuantizerEncoder(...
       'Partitioning', 'Unbounded',...
```

```
          'BoundaryPoints', -7:7,...
          'OutputIndexDataType','uint8');
end
%% Constants
FRM=2048;
M=4; k=log2(M); codeRate=1/2;
snr = EbNo + 10*log10(k) + 10*log10(codeRate);
noise_var = 10.^(-snr/10);
AWGN.EbNo=snr;
%% Processsing loop modeling transmitter, channel model and receiver
numErrs = 0; numBits = 0;results=zeros(3,1);
while ((numErrs < maxNumErrs) && (numBits < maxNumBits))
    % Transmitter
    u            = randi([0 1], FRM,1);                  % Random bits generator
    encoded   = ConvEncoder.step(u);                     % Convolutional encoder
    mod_sig   = Modulator.step(encoded);                 % QPSK Modulator
    % Channel
    rx_sig     = AWGN.step(mod_sig);                     % AWGN channel
    % Receiver
    demod     = DeModulator.step(rx_sig, noise_var);   % Soft-decision QPSK Demodulator
    llr         = Quantizer.step(-demod);               % Quantize Log-Likelihood Ratios
    decoded   = Viterbi.step(llr);                       % Viterbi decoder with LLRs
    y            = decoded(1:FRM);                        % Compute output bits
    results     = BitError.step(u, y);                   % Update BER
    numErrs   = results(2);
    numBits   = results(3);
end
%% Clean up & collect results
ber = results(1); bits= results(3);
reset(BitError);
```

Theoretically, we expect a 2 dB improvement in the results, and that is exactly what is shown by the simulated curves in Figure 3.8. Next we examine turbo coding to see whether it can offer any improvements to BER results.

3.12.5   Turbo Coding

Turbo codes substantially improve BER performance over soft-decision Viterbi decoding. Turbo coding uses two convolutional encoders in parallel at the transmitter and two A Posteriori Probability (APP) decoders in series at the receiver. This example uses a rate 1/3 turbo coder. For each input bit, the output has one systematic bit and two parity bits, for a total of three bits.

The following MATLAB function has been updated to use turbo encoders and decoders. Note that turbo decoding is an iterative iteration, where performance improves as more iterations are gone through. In this example, we have chosen six as the number of iterations the decoder performs.

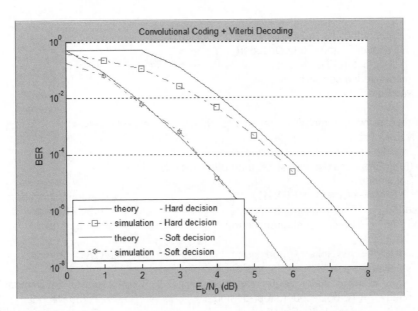

Figure 3.8 Performance with soft-decision Viterbi decoding: QPSK modulation with AWGN channel

Algorithm

MATLAB function

```
function [ber, bits]=chap5_ex05_qpsk_turbo(EbNo, maxNumErrs, maxNumBits)
%% Constants
FRM=2048;
Trellis=poly2trellis(4, [13 15], 13);
Indices=randperm(FRM);
M=4;k=log2(M);
R= FRM/(3* FRM + 4*3);
snr = EbNo + 10*log10(k) + 10*log10(R);
noise_var = 10.^(-snr/10);
%% Initializations
persistent Modulator AWGN DeModulator BitError TurboEncoder TurboDecoder
if isempty(Modulator)
    Modulator    = comm.QPSKModulator('BitInput',true);
    AWGN         = comm.AWGNChannel;
    DeModulator  = comm.QPSKDemodulator('BitOutput',true,...
        'DecisionMethod','Log-likelihood ratio',...
        'VarianceSource', 'Input port');
    BitError     = comm.ErrorRate;
    TurboEncoder=comm.TurboEncoder(...
        'TrellisStructure',Trellis,...
```

```
        'InterleaverIndices',Indices);
    TurboDecoder=comm.TurboDecoder(...
        'TrellisStructure',Trellis,...
        'InterleaverIndices',Indices,...
        'NumIterations',6);
end
%% Processsing loop modeling transmitter, channel model and receiver
AWGN.EbNo=snr;
numErrs = 0; numBits = 0;results=zeros(3,1);
while ((numErrs < maxNumErrs) && (numBits < maxNumBits))
    % Transmitter
    u            = randi([0 1], FRM,1);                      % Random bits generator
    encoded    = TurboEncoder.step(u);                       % Turbo Encoder
    mod_sig    = Modulator.step(encoded);                    % QPSK Modulator
    % Channel
    rx_sig     = AWGN.step(mod_sig);                         % AWGN channel
    % Receiver
    demod    = DeModulator.step(rx_sig, noise_var);   % Soft-decision QPSK Demodulator
    decoded    = TurboDecoder.step(-demod);                  % Turbo Decoder
    y            = decoded(1:FRM);                            % Compute output bits
    results    = BitError.step(u, y);                        % Update BER
    numErrs    = results(2);
    numBits    = results(3);
end
%% Clean up & collect results
ber = results(1); bits= results(3);
reset(BitError);
```

Figure 3.9 illustrates the results for turbo coding in QPSK modulation under an AWGN channel. Note that at 1 dB we have a BER value that occurs at 5 dB for hard-decision and at 3 dB for soft-decision decoding. This clearly indicates the superiority of the turbo-coding algorithm. Bear in mind that this performance gain comes at the cost of an increase in computational complexity. Our turbo decoder went through iterative decoding six times to arrive at this performance. We will study this compromise between performance and complexity for turbo decoders in later chapters.

3.13   Chapter Summary

MATLAB, Simulink, and their toolboxes provide capabilities that can be used to model, simulate, assess the performance of, and eventually generate and implement code for communication systems. For modeling and simulation, we can use algorithmic building blocks from the Communications System Toolbox, either as System objects or as Simulink blocks. Many aspects of PHY processing of mobile standards can be modeled and simulated more efficiently in MATLAB because instead of focusing on creating building blocks, such as modulators and coders, we can focus on introducing more advanced functionality to a system model. Of special

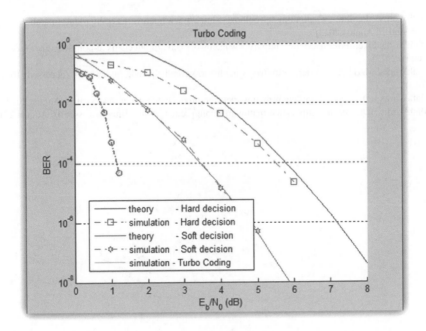

Figure 3.9 Performance with turbo coding: QPSK modulation with AWGN channel

interest are the System objects of the Communications System Toolbox. System objects are modeling and simulation components that are designed to be self-documented, easy-to-use, and custom-designed for the modeling of block-based streaming systems.

When simulating complex systems, we need to access various acceleration techniques. These techniques help us process larger sets of test data and obtain statistically correct assessments in reasonable simulation times. MATLAB toolboxes such as the Parallel Processing Toolbox and MATLAB Coder can speed up simulations. Finally, in order to implement a design in software or hardware we can use code-generation products to gain access to the exact implementation detail via automatic C or HDL code generation.

References

[1] MathWorks Documentation Center, http://www.mathworks.com/help/MATLAB/random-number-genera-tion.html (accessed 16 August 2013).

[2] MathWorks DSP System Toolbox, http://www.mathworks.com/products/dsp-system (accessed 16 August 2013).

[3] MathWorks Communications System Toolbox, http://www.mathworks.com/products/communications (accessed 16 August 2013).

[4] MathWorks Phased Array System Toolbox, http://www.mathworks.com/products/phased-array (accessed 16 August 2013).

[5] MathWorks Computer Vision System Toolbox, http://www.mathworks.com/products/computer-vision (accessed 16 August 2013).

[6] MathWorks Simulink, http://www.mathworks.com/products/simulink (accessed 16 August 2013).

[7] MathWorks Parallel Computing Toolbox, http://www.mathworks.com/products/parallel-computing (accessed 16 August 2013).

[8] MathWorks Fixed-Point Designer, http://www.mathworks.com/products/fixed-point-designer (accessed 16 August 2013).

[9] MathWorks MATLAB Coder, http://www.mathworks.com/help/coder/index.html (accessed 16 August 2013).

[10] MathWorks HDL Coder, http://www.mathworks.com/products/hdl-coder (accessed 16 August 2013).

[11] MathWorks HDL Verifier, http://www.mathworks.com/products/hdl-verifier (accessed 16 August 2013).

4

Modulation and Coding

The LTE (Long Term Evolution) downlink PHY (Physical Layer) chain can be viewed as the combination of processing applied to the Downlink Shared Channel (DLSCH) and Physical Downlink Shared Channel (PDSCH). DLSCH processing is also known as Downlink Transport Channel (TrCH) processing. It includes steps involving Cyclic Redundancy Check (CRC) code attachment, data subblock processing, channel coding based on turbo coders, rate matching, Hybrid Automatic Repeat Request (HARQ) processing, and the reconstruction of codewords. The codewords are inputs for the PDSCH processing, which involves scrambling, modulation, multi-antenna Multiple Input Multiple Output (MIMO) processing, time–frequency resource mapping, and Orthogonal Frequency Division Multiplexing (OFDM) transmission. We have subdivided the components of this two-step DLSCH and PDSCH processing chain into three segments, which are discussed in the next three chapters.

In this chapter, we examine the modulation and coding schemes used in the LTE standard. These include all the combined DLSCH and PDSCH processing steps, excluding the MIMO and OFDM operations. Discussion regarding OFDM and MIMO is presented in the next two chapters. First, we will examine the first couple of operations performed in PDSCH processing, including scrambling and modulation. Then we will examine TrCH processing, comprising a series of operations that map logical channels and user bit payload to codewords, which are passed to the shared physical channel.

We will create MATLAB® programs that completely specify the TrCH processing in the transmitter and the receiver. We will use the MATLAB functions to study the effects of different modulation schemes and different coding rates on the Bit Error Rate (BER) performance with an Additive White Gaussian Noise (AWGN) channel model. These operations completely specify how user data bits are processed to produce the input symbols for the subsequent MIMO and OFDM functional blocks for transmission. Details of MIMO and OFDM are then examined in the next two chapters.

Understanding LTE with MATLAB®: From Mathematical Modeling to Simulation and Prototyping, First Edition.
Houman Zarrinkoub.
© 2014 John Wiley & Sons, Ltd. Published 2014 by John Wiley & Sons, Ltd.

4.1   Modulation Schemes of LTE

The modulation schemes used in the LTE standard include QPSK (Quadrature Phase Shift Keying), 16QAM (Quadrature Amplitude Modulation), and 64QAM. Figure 4.1 shows the constellation diagrams of these three modulation schemes.

In the case of QPSK modulation, each modulation symbol can have one of four different values, which are mapped to four different positions in the constellation diagram. QPSK needs 2 bits to encode each of its four different modulation symbols. The 16QAM modulation involves using 16 different signaling choices and thus utilizes 4 bits of information to encode each modulation symbol. The 64QAM modulation involves 64 different possible signaling values and thus requires 6 bits to represent a single modulation symbol.

The availability of multiple modulation schemes is instrumental in implementing adaptive modulation based on channel conditions. When the radio link is relatively clean – that is, the Signal-to-Noise Ratio (SNR) is relatively high – we can use modulation schemes of denser constellations, such as 64QAM. In such a case, sending a single symbol results in the transmission of 6 bits and therefore can increase our throughput. However, as the channel becomes noisier, we should resort to using modulation schemes with more intersymbol separation, such as QPSK. This in turn will reduce the number of bits per sample and reduce the throughput.

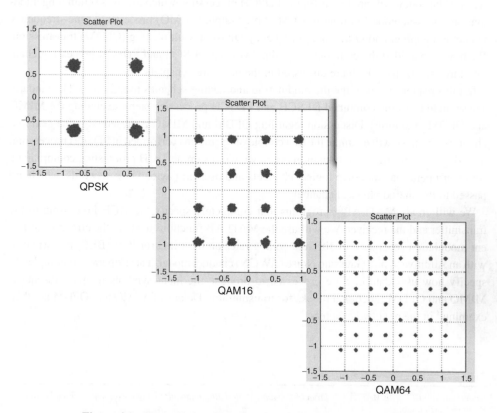

Figure 4.1   Constellation diagrams of QPSK, 16QAM, and 64QAM

Table 4.1   Mapping for an LTE QPSK modulator

Payload bit pattern (2 bits)	Modulated symbol	
	In-phase (I)	Quadrature (Q)
00	$1/\sqrt{2}$	$1/\sqrt{2}$
01	$1/\sqrt{2}$	$-1/\sqrt{2}$
10	$-1/\sqrt{2}$	$1/\sqrt{2}$
11	$-1/\sqrt{2}$	$-1/\sqrt{2}$

Table 4.2   Mapping for an LTE 16QAM modulator

Payload bit pattern (4 bits)	Modulated symbol	
	In-phase (I)	Quadrature (Q)
0000	$1/\sqrt{10}$	$1/\sqrt{10}$
0001	$1/\sqrt{10}$	$3/\sqrt{10}$
0010	$3/\sqrt{10}$	$1/\sqrt{10}$
0011	$3/\sqrt{10}$	$3/\sqrt{10}$
0100	$1/\sqrt{10}$	$-1/\sqrt{10}$
0101	$1/\sqrt{10}$	$-3/\sqrt{10}$
0110	$3/\sqrt{10}$	$-1/\sqrt{10}$
0111	$3/\sqrt{10}$	$-3/\sqrt{10}$
1000	$-1/\sqrt{10}$	$1/\sqrt{10}$
1001	$-1/\sqrt{10}$	$3/\sqrt{10}$
1010	$-3/\sqrt{10}$	$1/\sqrt{10}$
1011	$-3/\sqrt{10}$	$3/\sqrt{10}$
1100	$-1/\sqrt{10}$	$-1/\sqrt{10}$
1101	$-1/\sqrt{10}$	$-3/\sqrt{10}$
1110	$-3/\sqrt{10}$	$-1/\sqrt{10}$
1111	$-3/\sqrt{10}$	$-3/\sqrt{10}$

The LTE modulation mappers, which specify how the modulation symbols are assigned to each bit sequence, are shown in Table 4.1 for QPSK and in Table 4.2 for 16QAM. Due to its large size, we refer the reader to Reference [1] for the standard document on 64QAM modulation mapping.

We note that the mapping of bits to symbols is based on neither a typical binary nor a gray-coded method. Rather, the LTE specification defines a custom constellation mapping. LTE also defines modulation symbols in such a way that the average signal power is normalized to unity.

4.1.1   MATLAB Examples

As the first step in modeling the LTE downlink processing chain, we start with the LTE modulation schemes. The following two MATLAB functions show how you can easily implement the LTE modulation and demodulation algorithms, with all their specifications, using System objects of the Communications System Toolbox.

Algorithm

MATLAB function

```
function y=Modulator(u, Mode)
%% Initialization
persistent QPSK QAM16 QAM64
if isempty(QPSK)
    QPSK        = comm.PSKModulator(4, 'BitInput', true, ...
                    'PhaseOffset', pi/4, 'SymbolMapping', 'Custom', ...
                    'CustomSymbolMapping', [0 2 3 1]);
    QAM16     = comm.RectangularQAMModulator(16, 'BitInput',true,...
                    'NormalizationMethod','Average power',
                    'SymbolMapping', 'Custom', ...
                    'CustomSymbolMapping',
                    [11 10 14 15 9 8 12 13 1 0 4 5 3 2 6 7]);
    QAM64     = comm.RectangularQAMModulator(64, 'BitInput',true,...
                    'NormalizationMethod','Average power',
                    'SymbolMapping', 'Custom',
                    'CustomSymbolMapping',
                    [47 46 42 43 59 58 62 63 45 44 40 41 ...
                    57 56 60 61 37 36 32 33 49 48 52 53 39
                    38  34 35 51 50 54 55 7 6 2 3 19 18 22 23 5
                    4 0 1 17 16 20 21 13 12 8 9 25 24 28 29 15
                    14 10 11 27 26 30 31]);
end
%% Processing
switch Mode
    case 1
      y=step(QPSK, u);
    case 2
      y=step(QAM16, u);
    case 3
      y=step(QAM64, u);
end
```

The Modulator function has two input arguments: the input bit stream (u) and a parameter representing the modulation mode (*Mode*). As its output, the function computes the modulated symbols. The function implements the three different types of modulator used in the LTE standard. For example, in the case of QPSK, we use *a comm.PSKModulator* System object and set its modulation order to 4. Similarly, in the case of 16QAM and 64QAM we use the *comm.RectangulatQAMModulator* System objects and set their modulation orders to 16 and 64, respectively. Depending on the value of the modulation mode, we process the input bits to generate the modulated symbols as the output.

To ensure that the System object exactly matches what the LTE standard specifies, we can set other properties:

1. We can set BitInput = true. This means that the modulator inputs are interpreted as a vector of bit values. For example, in case of a QPSK modulator since every 2 bits are mapped to one modulation symbol, the size of the output vector is half that of the input vector.
2. We can set PhaseOffset = pi/4. This means that the modulated symbols correspond to four points in the complex plane with unity length, whose angles are chosen from the following set: $[3\pi/4, \pi/4, -\pi/4, -3\pi/4]$.
3. Using the CustomSymbolMapping property, we can ensure that the bit patterns specified in LTE result in corresponding output symbols.

Note that the LTE, like other mobile standards, does not make any recommendations concerning the operations performed in the receiver. Hence, all the receiver specifications presented in this book can be considered typical "inverse" operations of operations specified in the transmitter. These proposed inverse operations represent best efforts to recover the estimate of the transmitted bits. Although not specified in the standard, it is necessary to include these receiver-side inverse operations in order to evaluate the accuracy and performance of the system.

As demodulation is the inverse operation for modulation, we now present some typical approaches to demodulation. In the Demodulator function, we use the same three modulation types used in LTE, and depending on the modulation mode, we process the input symbols to generate the demodulated output. As discussed in the previous section, demodulation can be based on either hard-decision decoding or soft-decision decoding. In hard-decision decoding, the input symbols of a demodulator are mapped to estimated bits, whereas in soft-decision decoding the output is a vector of log-likelihood ratios (LLRs).

The function *DemodulatorHard.m* shows a demodulation implementation that employs hard-decision decoding. The function takes as inputs the received modulated symbols (u) and the modulation mode (*Mode*). The function output comprises the demodulated bits.

Algorithm

MATLAB function

```
function y=DemodulatorHard(u, Mode)
%% Initialization
persistent QPSK QAM16 QAM64
if isempty(QPSK)
   QPSK = comm.PSKDemodulator(...
      'ModulationOrder', 4, ...
      'BitOutput', true, ...
      'PhaseOffset', pi/4, 'SymbolMapping', 'Custom', ...
      'CustomSymbolMapping', [0 2 3 1]);

   QAM16 = comm.RectangularQAMDemodulator(...
      'ModulationOrder', 16, ...
      'BitOutput', true, ...
      'NormalizationMethod', 'Average power', 'SymbolMapping', 'Custom', ...
      'CustomSymbolMapping', [11 10 14 15 9 8 12 13 1 0 4 5 3 2 6 7]);
```

```
QAM64 = comm.RectangularQAMDemodulator(...
'ModulationOrder', 64, ...
'BitOutput', true, ...
'NormalizationMethod', 'Average power', 'SymbolMapping', 'Custom', ...
'CustomSymbolMapping', ...
[47 46 42 43 59 58 62 63 45 44 40 41 57 56 60 61 37 ...
36 32 33 49 48 52 53 39 38 34 35 51 50 54 55 7 6 2 3 ...
19 18 22 23 5 4 0 1 17 16 20 21 13 12 8 9 25 24 28 29 ...
15 14 10 11 27 26 30 31]);
end
%% Processing
switch Mode
    case 1
        y=QPSK.step(u);
    case 2
        y=QAM16.step(u);
    case 3
        y=QAM64.step(u);
    otherwise
        error('Invalid Modulation Mode. Use {1,2, or 3}');
end
```

The function *DemodulatorSoft.m* employs soft-decision decoding to perform demodulation. The function has three input arguments: the received modulated symbol stream (u), the estimate of the noise variance in the current subframe (*NoiseVar*), and a parameter representing the modulation mode (*Mode*). As its output, the function computes the LLRs. Examining the differences between the functions, we can see that by setting a couple of properties in the demodulator System objects, including the property called the *DecisionMethod*, we can implement soft-decision demodulation.

Algorithm

MATLAB function

```
function y=DemodulatorSoft(u, Mode, NoiseVar)
%% Initialization
persistent QPSK QAM16 QAM64
if isempty(QPSK)
    QPSK = comm.PSKDemodulator(...
        'ModulationOrder', 4, ...
        'BitOutput', true, ...
        'PhaseOffset', pi/4, 'SymbolMapping', 'Custom', ...
        'CustomSymbolMapping', [0 2 3 1],...
        'DecisionMethod', 'Approximate log-likelihood ratio', ...
        'VarianceSource', 'Input port');
```

```
QAM16 = comm.RectangularQAMDemodulator(...
    'ModulationOrder', 16,  ...
    'BitOutput', true, ...
    'NormalizationMethod', 'Average power', 'SymbolMapping', 'Custom', ...
    'CustomSymbolMapping', [11 10 14 15 9 8 12 13 1 0 4 5 3 2 6 7],...
    'DecisionMethod',  'Approximate log-likelihood ratio', ...
    'VarianceSource', 'Input port');

QAM64 = comm.RectangularQAMDemodulator(...
'ModulationOrder', 64, ...
'BitOutput', true, ...
'NormalizationMethod', 'Average power', 'SymbolMapping', 'Custom', ...
'CustomSymbolMapping', ...
[47 46 42 43 59 58 62 63 45 44 40 41 57 56 60 61 37 36 32 33 ...
49 48 52 53 39 38 34 35 51 50 54 55 7 6 2 3 19 18 22 23 5 4 0 1 ...
17 16 20 21 13 12 8 9 25 24 28 29 15 14 10 11 27 26 30 31],...
'DecisionMethod',  'Approximate log-likelihood ratio', ...
 'VarianceSource', 'Input port');

end
%% Processing
switch Mode
    case 1
        y=step(QPSK, u, NoiseVar);
    case 2
        y=step(QAM16,u, NoiseVar);
    case 3
        y=step(QAM64, u, NoiseVar);
    otherwise
        error('Invalid Modulation Mode. Use {1,2, or 3}');
end
```

4.1.2   BER Measurements

The motivation for using multiple modulation methods in LTE is to provide higher data rates within a given transmission bandwidth. The bandwidth utilization is expressed in bits/s/Hz. Compared to the QPSK, the bandwidth utilization of 16QAM and 64QAM is two and three times higher, respectively. However, higher-order modulation schemes are subject to reduced robustness to channel noise. Compared to the QPSK, modulation schemes such as 16QAM or 64QAM require a higher value for $Eb/N0$ at the receiver for a given bit error probability.

The following MATLAB functions illustrate the first in a series of functions that will eventually implement a realistic transceiver for the LTE PHY modeling in MATLAB. We start with this simple system, which is composed of a modulator, a demodulator, and an AWGN channel, and which computes the BER as a function of the $Eb/N0$ ratio. By running this function with a series of $Eb/N0$ values and changing the *ModulationMode* parameter, we can visualize the relationship between modulation order and robustness to channel noise.

The first function, *chap4_ex01.m*, uses a demodulator based on hard-decision demodulation, while the second function, *chap4_ex02.m*, is based on soft-decision decoding. The elements that are common in these two functions, which represent patterns of the series of functions to come, are as follows:

- Signature and input–output arguments
- Size of user payload (input to the PHY) (specified with a parameter (FRM) denoting the number of input bits in one frame of data)
- Stopping criteria
- Transmitter, channel model, receiver, and measurement sections
- Computation of BER at the end of the simulation.

Algorithm

MATLAB function

```
function [ber, numBits]=chap4_ex01(EbNo, maxNumErrs, maxNumBits)
%% Constants
FRM=2400;      % Size of bit frame
%% Modulation Mode
% ModulationMode=1;                      % QPSK
% ModulationMode=2;                      % QAM16
ModulationMode=3;                        % QAM64
k=2*ModulationMode;                      % Number of bits per modulation symbol
snr = EbNo + 10*log10(k);
%% Processing loop: transmitter, channel model and receiver
numErrs = 0;
numBits = 0;
while ((numErrs < maxNumErrs) && (numBits < maxNumBits))
    % Transmitter
    u  = randi([0 1], FRM,1);                          %  Randomly generated input bits
    t0 = Modulator(u, ModulationMode);       % Modulator
    % Channel
    c0 =  AWGNChannel(t0, snr);                     % AWGN channel
    % Receiver
    r0 =  DemodulatorHard(c0, ModulationMode);  % Demodulator, Hard-decision
    y  = r0(1:FRM);                                       % Recover output bits
    % Measurements
    numErrs  = numErrs + sum(y~=u);                % Update number of bit errors
    numBits  = numBits + FRM;                          % Update number of bits processed
end
%% Clean up & collect results
ber = numErrs/numBits;                               % Compute Bit Error Rate (BER)
```

In order to attain a certain quality of transmission – that is, for a given bit error rate – the E/N value required becomes progressively higher as we move from QPSK to 16QAM and 64QAM modulation. This suggests that lower-order modulation schemes such as QPSK are used in channels with a high degree of degradation in order to lower the probability of error, at

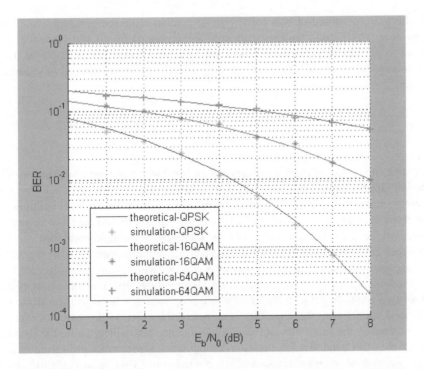

Figure 4.2   Bit error rate as a function of *Eb/N0*: QPSK, 16QAM, and 64QAM

the cost of running at lower data rates. Higher-order modulation schemes such as 64QAM are employed in cleaner channels and can provide a boost in the data rate. The results captured in Figure 4.2 were obtained by running the *chap4_ex01.m* function with different values of E/N and different modulation-mode parameters. The results compare the theoretical and simulation BER curves for modulation schemes used in the LTE standard.

In Chapter 7, we will discuss various methods available to the scheduler for changing the choice of modulation scheme in each scheduling interval based on channel conditions. So far, we have only discussed modulation schemes and have not added coding and scrambling to the mix. Next we will introduce bit-level scrambling, its motivation, and its implementation in MATLAB. Then we will introduce error-correction coding based on turbo coding and an error-detection mechanism represented by CRC-detection processing.

4.2   Bit-Level Scrambling

In LTE downlink processing, the codeword bits generated as the outputs of the channel coding operation are scrambled by a bit-level scrambling sequence. Different scrambling sequences are used in neighboring cells to ensure that the interference is randomized and that transmissions from different cells are separated prior to decoding. In order to achieve this, data bits are scrambled with a sequence that is unique to each cell by initializing the sequence generators in the cell based on the PHY cell identity. Bit-level scrambling is applied to all LTE TrCHs and to the downlink control channels.

Scrambling is composed of two parts: pseudo-random sequence generation and bit-level multiplication. The pseudo-random sequences are generated by a Gold sequence with the length set to 31. The output sequence is defined as the output of an exclusive-or operation applied to a specified pair of sequences. The polynomials specifying this pair of sequences are as follows:

$$p_1(x) = x^{31} + x^3 + 1$$
$$p_2(x) = x^{31} + x^3 + x^2 + x + 1 \tag{4.1}$$

The initialization value of the first sequence is specified with a unit impulse function of length 31. The initialization value for the second random sequence depends on such parameters as the cell identity, number of codewords, and subframe index. Finally, the bit-level multiplication is implemented as an exclusive-or operation between the input bits and the Gold sequence bits. The output of the scrambler is a vector with the same size as the input codeword.

In the receiver, the descrambling operation inverts the operations performed by the scrambler. The same pseudo-random sequence generator is used. However, there is a difference between bit-level scrambling and bit-level descrambling. Descrambling operations can be implemented in one of two ways. If, prior to the descrambling operation, a hard-decision demodulation is performed, the input to the scrambler is represented by bits. In this case, an exclusive-or operation between the input bits and the Gold-sequence bits will generate the descrambler output. On the other hand, if a soft-decision demodulation is performed prior to descrambling, the input signal is no longer composed of bits but rather of LLRs. In that case, descrambling is performed as a multiplication operation between the input log-likelihood values and Gold-sequence bits transformed to coefficient values. A zero-valued Gold-sequence bit is mapped to 1 and a 1-valued bit is mapped to −1.

4.2.1   MATLAB Examples

The following two MATLAB functions show how you can implement the LTE scrambling and descrambling operations with components of the Communications System Toolbox. The Scrambler function has two input arguments: the input bit stream (u) and a parameter representing the subframe index in the current frame (nS). As its output, the function computes the scrambled version of the input bit stream.

In the *Scrambler* function, we first create a Gold Sequence Generator System object. We then assign various properties of the Gold sequence object based on exact specifications of the LTE standard. Both the first and second polynomials are set to MATLAB expressions representing the coefficients specified by the standard. The initialization of the first polynomial is carried out with a specified constant vector. The initialization of the second polynomial happens at the start of each subframe and is specified by a variable called c_*init*. The value of this variable depends on such parameters as the cell identity, the number of codewords, and the subframe index. Finally, the scrambling output is obtained by multiplying each input sample with samples of the Gold sequence generator. Since the input signal to the scrambler is made up of the output bits of the channel coder, the multiplication is implemented as an exclusive-or operation.

Algorithm

MATLAB function

```
function y = Scrambler(u, nS)
%   Downlink scrambling
persistent hSeqGen hInt2Bit
if isempty(hSeqGen)
   maxG=43200;
   hSeqGen = comm.GoldSequence('FirstPolynomial',[1 zeros(1, 27) 1 0 0 1],...
                   'FirstInitialConditions', [zeros(1, 30) 1], ...
                   'SecondPolynomial', [1 zeros(1, 27) 1 1 1 1],...
                   'SecondInitialConditionsSource', 'Input port',...
                   'Shift', 1600,...
                   'VariableSizeOutput', true,...
                   'MaximumOutputSize', [maxG 1]);
   hInt2Bit = comm.IntegerToBit('BitsPerInteger', 31);
end
% Parameters to compute initial condition
RNTI = 1;
NcellID = 0;
q =0;
% Initial conditions
c_init = RNTI*(2^14) + q*(2^13) + floor(nS/2)*(2^9) + NcellID;

% Convert initial condition to binary vector
iniStates = step(hInt2Bit, c_init);

% Generate the scrambling sequence
nSamp = size(u, 1);
seq = step(hSeqGen, iniStates, nSamp);
seq2=zeros(size(u));
seq2(:)=seq(1:numel(u),1);

% Scramble input with the scrambling sequence
y = xor(u, seq2);
```

In the *Descrambler* function, we use the same Gold sequence generator to invert the scrambling operation. The descrambler initialization is synchronized with that of the scrambler. Since operations in the receiver, including descrambling, are not specified in the standard, we can develop two different descramblers. The difference lies in whether or not the preceding demodulator operation, which generates the input to the descrambler, is based on soft-decision or hard-decision decoding. The *DescramlerSoft* function operates on the LLR outputs of the demodulator, converting the Gold-sequence bits into bipolar values of either 1 or −1 and implementing descrambling as a multiplication of demodulator outputs and bipolar sequence values.

Algorithm

MATLAB function

```
function y = DescramblerSoft(u, nS)
%   Downlink descrambling
persistent hSeqGen hInt2Bit;
if isempty(hSeqGen)
    maxG=43200;
    hSeqGen = comm.GoldSequence('FirstPolynomial',[1 zeros(1, 27) 1 0 0 1],...
                    'FirstInitialConditions', [zeros(1, 30) 1], ...
                    'SecondPolynomial', [1 zeros(1, 27) 1 1 1 1],...
                    'SecondInitialConditionsSource', 'Input port',...
                    'Shift', 1600,...
                    'VariableSizeOutput', true,...
                    'MaximumOutputSize', [maxG 1]);
    hInt2Bit = comm.IntegerToBit('BitsPerInteger', 31);
end
% Parameters to compute initial condition
RNTI = 1;
NcellID = 0;
q=0;
% Initial conditions
c_init = RNTI*(2^14) + q*(2^13) + floor(nS/2)*(2^9) + NcellID;
% Convert to binary vector
iniStates = step(hInt2Bit, c_init);
% Generate scrambling sequence
nSamp = size(u, 1);
seq = step(hSeqGen, iniStates, nSamp);
seq2=zeros(size(u));
seq2(:)=seq(1:numel(u),1);
% If descrambler inputs are log-likelihood ratios (LLRs) then
% Convert sequence to a bipolar format
seq2 = 1-2.*seq2;
% Descramble
y = u.*seq2;
```

The *DescramblerHard* function assumes that the demodulator output generates input values to the descrambler based on hard-decision decoding. The function operates on the bit-stream outputs of the demodulator and implements descrambling with the MATLAB exclusive-or operation (*xor*) applied to the demodulator outputs and Gold-sequence values.

Algorithm

MATLAB function

```matlab
function y = DescramblerHard(u, nS)
%   Downlink descrambling
persistent hSeqGen hInt2Bit;
if isempty(hSeqGen)
   maxG=43200;
   hSeqGen = comm.GoldSequence('FirstPolynomial',[1 zeros(1, 27) 1 0 0 1],...
                   'FirstInitialConditions', [zeros(1, 30) 1], ...
                   'SecondPolynomial', [1 zeros(1, 27) 1 1 1 1],...
                   'SecondInitialConditionsSource', 'Input port',...
                   'Shift', 1600,...
                   'VariableSizeOutput', true,...
                   'MaximumOutputSize', [maxG 1]);
   hInt2Bit = comm.IntegerToBit('BitsPerInteger', 31);
end
% Parameters to compute initial condition
RNTI = 1;
NcellID = 0;
q=0;
% Initial conditions
c_init = RNTI*(2^14) + q*(2^13) + floor(nS/2)*(2^9) + NcellID;
% Convert to binary vector
iniStates = step(hInt2Bit, c_init);
% Generate scrambling sequence
nSamp = size(u, 1);
seq = step(hSeqGen, iniStates, nSamp);
% Descramble
y = xor(u(:,1), seq(:,1));
```

4.2.2   BER Measurements

The following illustrate the second in our series of functions that eventually implement a realistic transceiver for the LTE PHY modeling in MATLAB. In this example, *chap4_ex02.m*, we add the scrambling operation before modulation and follow the demodulation with descrambling. We use soft-decision demodulation and the corresponding descrambling function, which operates on soft-decision outputs. To compare the output with the input bits, we convert soft-decision values to bit values.

Algorithm

MATLAB function

```
function [ber, numBits]=chap4_ex02(EbNo, maxNumErrs, maxNumBits)
%% Constants
FRM=2400;      % Size of bit frame
%% Modulation Mode
% ModulationMode=1;              % QPSK
% ModulationMode=2;              % QAM16
ModulationMode=2;                % QAM64
k=2*ModulationMode;              % Number of bits per modulation symbol
snr = EbNo + 10*log10(k);
noiseVar = 10.^(0.1.*(-snr));            % Compute noise variance
%% Processing loop: transmitter, channel model and receiver
numErrs = 0;
numBits = 0;
nS=0;
while ((numErrs < maxNumErrs) && (numBits < maxNumBits))
    % Transmitter
    u  = randi([0 1], FRM,1);                          % Randomly generated input bits
    t0 = Scrambler(u, nS);                             % Scrambler
    t1 = Modulator(t0, ModulationMode);                % Modulator
    % Channel
    c0 = AWGNChannel(t1, snr);                         % AWGN channel
    % Receiver
    r0 = DemodulatorSoft(c0, ModulationMode, noiseVar); % Demodulator
    r1 = DescramblerSoft(r0, nS);                      % Descrambler
    y  = 0.5*(1-sign(r1));                             % Recover output bits
    % Measurements
    numErrs  = numErrs + sum(y~=u);                    % Update number of bit errors
    numBits  = numBits + FRM;                          % Update number of bits processed
    % Manage slot number with each subframe processed
    nS = nS + 2;
    nS = mod(nS, 20);
end
%% Clean up & collect results
ber = numErrs/numBits;                                 % Compute Bit Error Rate (BER)
```

Since a scrambling operation does not affect the sensitivity to the channel noise, the results obtained earlier by running the *chap4_ex01.m* function in Figure 4.2 are obtained again by running the *chap4_ex02.m* function with different values of E/N and modulation-mode parameters.

Table 4.3 Channel-coding schemes for various components of the transport channel (TrCH)

Transport channel (TrCH)	Coding scheme	Coding rate
DL-SCH	Turbo coding	1/3
UL-SCH		
PCH		
MCH		
BCH	Tail biting Convolutional coding	1/3

4.3 Channel Coding

So far we have discussed modulation and scrambling operations performed in physical channel processing. Now we will combine TrCH processing – that is, channel coding – with modulation and scrambling. We will introduce error-correction coding based on turbo coding and an error-detection mechanism represented by CRC detection. Table 4.3 summarizes the channel-coding schemes of various TrCHs. Most physical channels are subject to turbo coding, with the exception of the Broadcast Channel (BCH), which is based on convolutional coding.

Turbo coding is the basis of channel coding as specified in the LTE standard. Although turbo coding has been used in many previous standards, it has always been regarded as an optional component alongside other convolutional coding schemes. However, in LTE turbo coding is the driving component of the channel-coding mechanism. Based on our pedagogic approach, we will gradually build the TrCH processing of the LTE standard, in five steps. First, we implement a turbo-coding algorithm with a 1/3 coding rate. Then we add the early-termination mechanism to the turbo decoder. This makes the computational complexity of the turbo decoder scalable. We then introduce the rate-matching operation, which provides encoding of any given rate by operating on the 1/3-rate turbo-coder output. We introduce functions related to the subblock segmentation and codeword reconstruction. Finally, we put all the components together to implement the processing chain of the TrCH processing. In this book, we omit the introduction of MATLAB functions related to HARQ processing. HARQ processing is quite important, as it essentially reduces the number of retransmissions and enhances performance following transport-block error detections. This omission is in line with our stated scope, which focuses on steady-state user-plane processing.

4.4 Turbo Coding

Turbo coders belong to a category of channel-coding algorithms known as parallel concatenated convolutional coding [2]. As this name would suggest, a turbo code is formed by concatenating two conventional encoders in parallel and separating them by an interleaver. Many

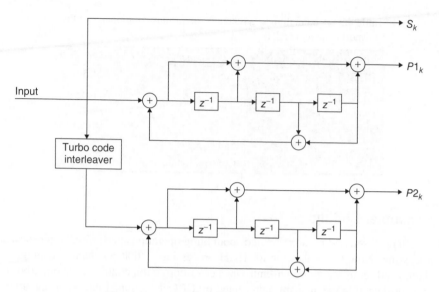

Figure 4.3 Block diagram of a turbo encoder

factors led to the choice of turbo coding in LTE. The first is the near-Shannon-bound performance of turbo coders. Given a sufficient number of iterations in turbo decoding, turbo codes can have a BER performance far in exceeds of those of conventional convolutional coders. Furthermore, they lend themselves to adaptation, due to the use of an innovative rate-matching mechanism, which will be discussed shortly.

4.4.1   Turbo Encoders

As depicted in Figure 4.3, LTE uses turbo coding with a base rate of 1/3 as the cornerstone of its channel-coding scheme. The LTE turbo encoder is based on a parallel concatenation of two 8-state constituent encoders separated by an internal interleaver. The output of the turbo encoder is composed of three streams. The bits of the first stream are usually referred to as Systematic bits. The bits of the second and third streams – that is, the outputs of the two constituent encoders – are usually referred to as Parity 1 and Parity 2 bit streams, respectively. Each constituent encoder is independently terminated by tail bits. This means that for an input block size of K bits, the output of a turbo encoder consists of three streams of length K + 4 bits, due to trellis termination. This makes the coding rate of the turbo coder slightly less than 1/3. Since the tail bits are multiplexed at the end of each stream, the Systematic and Parity 1 and Parity 2 bit streams all are of size K + 4.

To completely specify the turbo encoder, we need to specify the trellis structure of the constituent encoders and the turbo code internal interleaver. The LTE interleaver is based on a simple Quadratic Polynomial Permutation (QPP) scheme. The interleaver permutes the indices of the input bits. The relationship between the output index $p(i)$ and the input index i is described by the following quadratic polynomial expression:

$$p(i) = (f_1 \cdot i + f_2 \cdot i^2) \mathrm{mod}(k) \tag{4.2}$$

where K is the size of the input block and f_1 and f_2 are constants that depend on the value of K. The LTE allows 188 different values for the input block size K. The smallest block size is 40 and largest is 6144. These block sizes and the corresponding f_1 and f_2 constants are summarized in Reference [3].

The LTE turbo coder is a contention-free coder that uses a QPP interleaver, which substantially improves the turbo code performance by streamlining the memory access in interleaving operation. The trellis structure of the constituent encoder is described by the two following polynomials:

$$G_0(z) = 1 + z^{-2} + z^{-3}$$
$$G_1(z) = 1 + z^{-1} + z^{-3} \tag{4.3}$$

This describes a 1/3 turbo encoder with four states and with a trellis structure at each constituent encoder represented by both feed-forward and feedback connection polynomials, with octave values of 13 and 15, respectively.

4.4.2   Turbo Decoders

In the receiver, the turbo decoder inverts the operations performed by the turbo encoder. A turbo decoder is based on the use of two A Posteriori Probability (APP) decoders and two interleavers in a feedback loop. The same trellis structure found in the turbo encoder is used in the APP decoder, as is the same interleaver. The difference is that turbo decoding is an iterative operation. The performance and the computational complexity of a turbo decoder directly relate to the number of iterations performed.

At the receiver, the turbo decoder performs the inverse operation of a turbo encoder. By processing its input signal, which is the output of a demodulator and descrambler, the turbo decoder will recover the best estimate of the TrCH transmitted bits. Note that the turbo decoder input needs to be expressed in LLRs. As we saw earlier, LLRs are generated by the demodulator if soft-decision demodulation is performed.

4.4.3   MATLAB Examples

The following two MATLAB functions show implementations of the LTE turbo encoders and decoders with all their specifications, using System objects of the Communications System Toolbox. In the *TurboEncoder* function, we use *a comm.TurboEncoder* System objects and set the trellis structure and the interleaver properties to implement the functionality as specified in the LTE standard. By calling the step method of the System object, we process the input bits to generate the turbo-encoded bits as the output.

Algorithm

MATLAB function

```
function y=TurboEncoder(u, intrlvrIndices)
%#codegen
persistent Turbo
if isempty(Turbo)
    Turbo = comm.TurboEncoder('TrellisStructure', poly2trellis(4, [13 15], 13), ...
        'InterleaverIndicesSource','Input port');
end
y=step(Turbo, u, intrlvrIndices);
```

Similarly, the *TurboDecoder* function operates on its first input signal (*u*), which is the LLR output of the demodulator and descrambler. The turbo decoder will recover the best estimate of the transmitted bits. The function also takes as inputs the interleaving indices (*intrlvrIndices*) and the maximum number of iterations used in the decoder (*maxIter*).

Algorithm

MATLAB function

```
function y=TurboDecoder(u, intrlvrIndices, maxIter)
%#codegen
persistent Turbo
if isempty(Turbo)
    Turbo = comm.TurboDecoder('TrellisStructure', poly2trellis(4, [13 15], 13),...
        'InterleaverIndicesSource','Input port', ...
        'NumIterations', maxIter);
end
y=step(Turbo, u,  intrlvrIndices);
```

To set the trellis structure, we use the *ploy2trellis* function of the Communications System Toolbox. This function transforms the encoder connection polynomials to a trellis structure. As the LTE trellis structure has both feed-forward and feedback connection polynomials, we first build a binary-number representation of the polynomials and then convert the binary representation into an octal representation. From examining the block diagram of the turbo encoder in Figure 4.3, we can see that this encoder has a constraint length of 4, a generator polynomial matrix of [13 15], and a feedback connection polynomial of 13. Therefore, in order to set the trellis structure, we need to use the *poly2trellis(4, [13 15],13)* function.

To construct the LTE interleaver based on the QPP scheme, we use the *lteIntrlvrIndices* function. This function looks up the LTE interleaver table based on the only allowable 188 input sizes, finds the corresponding f_1 and f_2 constants, and computes the permutation vector as described in the standard.

Algorithm

MATLAB function

```
function indices = IntrlvrIndices(blkLen)
%#codegen
[f1, f2] = getf1f2(blkLen);
Idx     = (0:blkLen-1).';
indices = mod(f1*Idx + f2*Idx.^2, blkLen) + 1;
```

The *comm.TurboEncoder* and *comm.TurboDecoder* System objects are among those that express the algorithms based on direct MATLAB implementations. Therefore, using the MAT-LAB *edit* command, we can inspect the MATLAB code that is executed every time these System objects are used. The creation and authoring of MATLAB-based System objects is beyond the scope of this book; for more information on this topic, the reader is referred to the MATLAB documentation [4]. To illustrate how the MATLAB implementation matches what we expect, we can inspect the *stepimpl* function of this System object.

The *comm.TurboEncoder stepimpl* function performs two convolutional coding operations, first on the input signal and then on the interleaved version of the signal. It then captures the extra samples related to trellis termination and appends them to the end of the Systematic and Parity streams. The *comm.TurboDecoder stepimpl* repeats a sequence of operations, including two APP decoders and interleavers, N times. The value of N corresponds to the maximum number of iterations in a turbo decoder. At the end of each processing iteration, the turbo decoder uses the results to update its best estimate.

4.4.4 BER Measurements

The performance of any turbo coder depends on the number of iterations performed in the decoding operation. This means that for a given turbo encoder (e.g., the one specified in the LTE standard), the BER performance becomes successively better with a greater number of iterations. The function *chap4_ex03_nIter* illustrates this point by computing the BER performance as a function of the number of iterations.

Algorithm

MATLAB function

```
function [ber, numBits]=chap4_ex03_nIter(EbNo, maxNumErrs, maxNumBits, nIter)
%% Constants
FRM=2432;                              % Size of bit frame
Indices = lteIntrlvrIndices(FRM);
M=4;k=log2(M);
R= FRM/(3* FRM + 4*3);
snr = EbNo + 10*log10(k) + 10*log10(R);
noiseVar = 10.^(-snr/10);
```

```
ModulationMode=1;                          % QPSK
%% Processing loop modeling transmitter, channel model and receiver
numErrs = 0; numBits = 0; nS=0;
while ((numErrs < maxNumErrs) && (numBits < maxNumBits))
   % Transmitter
   u = randi([0 1], FRM,1);                            % Randomly generated input bits
   t0 = TurboEncoder(u, Indices);                        % Turbo Encoder
   t1 = Scrambler(t0, nS);                               % Scrambler
   t2 = Modulator(t1, ModulationMode);                   % Modulator
   % Channel
   c0 = AWGNChannel(t2, snr);                           % AWGN channel
   % Receiver
   r0 = DemodulatorSoft(c0, ModulationMode, noiseVar);    % Demodulator
   r1 = DescramblerSoft(r0, nS);                          % Descrambler
   y = TurboDecoder(-r1, Indices,  nIter);               % Turbo Decoder
   % Measurements
   numErrs   = numErrs + sum(y~=u);                    % Update number of bit errors
   numBits   = numBits + FRM;                          % Update number of bits processed
   % Manage slot number with each subframe processed
   nS = nS + 2; nS = mod(nS, 20);
end
%% Clean up & collect results
ber = numErrs/numBits;                        % Compute Bit Error Rate (BER)
```

To compare the performance of a turbo coder with that of a traditional convolutional coder, we also run a function called *chap4_ex03_viterbi.m*, which uses a 1/3-rate convolutional coder, a Viterbi decoder, and soft-decision demodulation.

Algorithm

MATLAB function

```
function [ber, numBits]=chap4_ex03_viterbi(EbNo, maxNumErrs, maxNumBits)
%% Constants
FRM=2432;                          % Size of bit frame
M=4;k=log2(M);
R= FRM/(3* (FRM+6));
snr = EbNo + 10*log10(k) + 10*log10(R);
noiseVar = 10.^(-snr/10);
ModulationMode=1;                  % QPSK
%% Processing loop modeling transmitter, channel model and receiver
numErrs = 0; numBits = 0; nS=0;
while ((numErrs < maxNumErrs) && (numBits < maxNumBits))
   % Transmitter
   u = randi([0 1], FRM,1);                            % Randomly generated input bits
   t0 = ConvolutionalEncoder(u);                       % Convolutional Encoder
```

```
t1 = Scrambler(t0, nS);                              % Scrambler
t2 = Modulator(t1, ModulationMode);                  % Modulator
% Channel
c0 = AWGNChannel(t2, snr);                            % AWGN channel
% Receiver
r0 = DemodulatorSoft(c0, ModulationMode, noiseVar);  % Demodulator
r1 = DescramblerSoft(r0, nS);                         % Descrambler
r2 = ViterbiDecoder(r1);                              % Viterbi Deocder
y=r2(1:FRM);
% Measurements
numErrs   = numErrs + sum(y~=u);                     % Update number of bit errors
numBits   = numBits + FRM;                           % Update number of bits processed
% Manage slot number with each subframe processed
nS = nS + 2; nS = mod(nS, 20);
end
%% Clean up & collect results
ber = numErrs/numBits;                               % Compute Bit Error Rate (BER)
```

Figure 4.4 compares the BER performance of a turbo decoder when one, three, or five iterations of turbo decoding are used with that of a typical Viterbi decoder with the same coding rate. As we increase the number of iterations from one to three and then to five, we

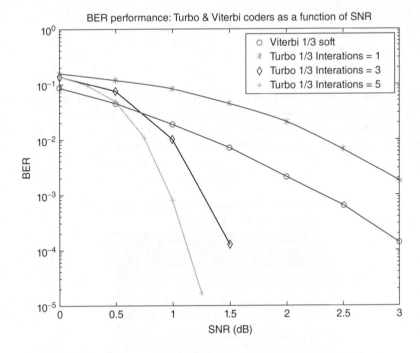

Figure 4.4   Performance of turbo coders as a function of number of iterations

see that the shape of the BER curve reflects the near-optimum quality of a turbo decoder. The curve shows a steep slope after a certain value of E/N. For example, with five as the maximum number of iterations, the LTE turbo decoder combined with QPSK and a soft-decision demodulator becomes capable of reaching a BER value of $2e^{-4}$ with an SNR value of 1.25 dB.

This profile of performance for turbo coding can explain why turbo coding has been selected as the mandatory channel-coding mechanism for user data in the LTE standard.

By executing the following testbench (*chap4_ex03_nIter*), we can measure the transceiver computation time as a function of number of iterations. The computation time is an estimate of the computational complexity of turbo encoding and decoding operations.

Algorithm

MATLAB script

```
%% Computation time of turbo coder
%% as a function of number of iterations
EbNo=1;
maxNumErrs=1e6;
maxNumBits=1e6;
for nIter=1:6
    clear functions
    tic;
    ber=chap4_ex03_nIter(EbNo, maxNumErrs, maxNumBits , nIter);
    toc;
end
```

Table 4.4 summarizes the results. As expected, the complexity and thus the time it takes to complete the decoding operations is proportional to the number of iterations.

To see what function contributes most to the complexity of the transceiver we have developed so far (*chap4_ex03_nIter*), we execute the following profiling script.

Table 4.4 Transceiver computation time as a function of number of iterations

Maximum number of iterations in turbo coding	Elapsed time (s)
1	5.83
2	8.54
3	11.27
4	13.66
5	16.41
6	18.96

Algorithm

MATLAB script

```
%% Profiling the turbo coder system model
EbNo=1;
maxNumErrs=1e6;
maxNumBits=1e6;
profile on
ber=chap4_ex03_nIter(EbNo, maxNumErrs, maxNumBits , 1);
profile viewer
```

The execution times for each line of the system model are summarized in the profiling report shown in Figure 4.5.

The result shows that performing turbo decoding with a fixed value of iterations takes about 86% of the entire system simulation time. The turbo decoder can thus be regarded as one of the bottlenecks of the system. In order to overcome this problem, the LTE standard provides a mechanism in the LTE encoder that enables early termination of turbo decoding without having a severe effect on the performance of the turbo coding. This early-termination mechanism is discussed in the next section.

4.5 Early-Termination Mechanism

The number of iterations performed in a turbo decoder is one of its main characteristics. In implementing an efficient turbo decoder, we face a clear tradeoff. On one hand, the accuracy and performance of the turbo decoder directly relates to its number of iterations. The more iterations, the more accurate the results. On the other hand, the computational complexity of a turbo decoder is also proportional to its number of iterations.

Profile Summary
Generated 08-Jul-2013 17:07:57 using cpu time.

Function Name	Calls	Total Time	Self Time*	Total Time Plot (dark band = self time)
chap4_ex03_nIter	1	18.580 s	0.030 s	▬▬▬▬▬▬▬
TurboDecoder	412	15.990 s	0.030 s	▬▬▬▬▬▬
TurboDecoder>TurboDecoder.stepImpl	412	15.960 s	15.960 s	▬▬▬▬▬▬
GoldSequence>GoldSequence.stepImpl	824	1.010 s	0.040 s	▮
GoldSequenceXor>GoldSequenceXor.stepImpl	824	0.970 s	0.950 s	▮
AWGNChannel	412	0.840 s	0.040 s	▮
DescramblerSoft	412	0.600 s	0.030 s	▮
Scrambler	412	0.520 s	0.080 s	▮

Figure 4.5 Profiling results for a system model, showing the turbo decoder to be the bottleneck

LTE specification allows for an effective way of resolving this tradeoff by devising an early termination. This mechanism is integrated with the turbo encoder. By appending a CRC-checking syndrome to the input of the turbo encoder, we can detect the presence or absence of any bit errors at the end of the iteration in the turbo decoder. Instead of following through with a fixed number of decoding iterations, we now have the option of stopping the decoding early when the CRC check indicates that no errors are detected. This very simple solution manages to reduce the computational complexity of a turbo decoder substantially without severely penalizing its performance.

4.5.1 MATLAB Examples

The following MATLAB function (*TurboDecoder_crc*) shows an implementation of the LTE turbo decoder that examines the CRC bits at the end of the input frame in order to optionally terminate the decoding operations before the maximum number of iterations is performed. As we can see, in this function we use the *LTETurboDecoder* System object instead of the *comm.TurboDecoder* System object.

Algorithm

MATLAB function

```
function [y, flag, iters]=TurboDecoder_crc(u, intrlvrIndices)
%#codegen
MAXITER=6;
persistent Turbo
if isempty(Turbo)
    Turbo = commLTETurboDecoder('InterleaverIndicesSource', 'Input port', ...
        'MaximumIterations', MAXITER);
end
[y, flag, iters] = step(Turbo, u, intrlvrIndices);
```

In the LTETurboDecoder, similar operations are performed to those in the regular turbo decoder. However, at the end of each decoding iteration the last 24 samples of the output that correspond to the CRC bits are examined for error detection. If no errors are detected, we branch out of the loop and terminate the turbo decoding operation. In this case, although the maximum number of iterations has not been executed, an early termination can occur. If we detect errors in the CRC bits, we continue the operations and enter the next decoding iteration until either no more errors are detected in the iteration or we reach the maximum allowable number of iterations.

The following MATLAB function (*CbCRCGenerator*) adds the 24-bit CRC syndrome to the end of the transport block before turbo encoding is performed.

Algorithm

MATLAB function

```
function y = CbCRCGenerator(u)
%#codegen
persistent hTBCRCGen
if isempty(hTBCRCGen)
    hTBCRCGen = comm.CRCGenerator('Polynomial',[1 1 zeros(1, 16) 1 1 0 0 0 1 1]);
end
% Transport block CRC generation
y = step(hTBCRCGen, u);
```

The following MATLAB function (*CbCRCDetector*) extracts the 24-bit CRC syndrome to the end of the transport block after turbo decoding is performed.

Algorithm

MATLAB function

```
function y = CbCRCDetector(u)
%#codegen
persistent hTBCRC
if isempty(hTBCRC)
    hTBCRC = comm.CRCDetector('Polynomial', [1 1 zeros(1, 16) 1 1 0 0 0 1 1]);
end
% Transport block CRC generation
y = step(hTBCRC, u);
```

4.5.2   BER Measurements

To examine the effectiveness of the early termination algorithm, we now compare two implementations of turbo decoding with and without CRC-based early termination. The following function (*chap4_ex04*) performs a combination of CRC generation, turbo coding, scrambling, and modulation and their inverse operations without implementing the early-termination mechanism.

Algorithm

MATLAB function

```
function [ber, numBits]=chap4_ex04(EbNo, maxNumErrs, maxNumBits)
%% Constants
FRM=2432-24;                                % Size of bit frame
Kplus=FRM+24;
Indices = lteIntrlvrIndices(Kplus);
ModulationMode=1;
k=2*ModulationMode;
maxIter=6;
CodingRate=Kplus/(3*Kplus+12);
snr = EbNo + 10*log10(k) + 10*log10(CodingRate);
noiseVar = 10.^(-snr/10);
%% Processing loop modeling transmitter, channel model and receiver
numErrs = 0; numBits = 0; nS=0;
while ((numErrs < maxNumErrs) && (numBits < maxNumBits))
    % Transmitter
    u  = randi([0 1], FRM,1);                        % Randomly generated input bits
    data= CbCRCGenerator(u);                         % Code block CRC generator
    t0 = TurboEncoder(data, Indices);                % Turbo Encoder
    t1 = Scrambler(t0, nS);                          % Scrambler
    t2 = Modulator(t1, ModulationMode);              % Modulator
    % Channel
    c0 = AWGNChannel(t2, snr);                        % AWGN channel
    % Receiver
    r0 = DemodulatorSoft(c0, ModulationMode, noiseVar);   % Demodulator
    r1 = DescramblerSoft(r0, nS);                         % Descrambler
    r2  = TurboDecoder(-r1, Indices,  maxIter);           % Turbo Decoder
    y   = CbCRCDetector(r2);                              % Code block CRC detector
    % Measurements
    numErrs    = numErrs + sum(y~=u);                % Update number of bit errors
    numBits    = numBits + FRM;                      % Update number of bits processed
    % Manage slot number with each subframe processed
    nS = nS + 2; nS = mod(nS, 20);
end
%% Clean up & collect results
ber = numErrs/numBits;                          % Compute Bit Error Rate (BER)
```

The following function (*chap4_ex04_crc*) performs the same transceiver while implementing the early-termination mechanism. In the case of algorithm-deploying early termination, we record the actual number of iterations in each subframe and then compute a histogram.

Algorithm

MATLAB function

```
function [ber, numBits, itersHist]=chap6_ex04_crc(EbNo, maxNumErrs, maxNumBits)
%% Constants
FRM=2432-24;                        % Size of bit frame
Kplus=FRM+24;
Indices = IteIntrlvrIndices(Kplus);
ModulationMode=1;
k=2*ModulationMode;
maxIter=6;
CodingRate=Kplus/(3*Kplus+12);
snr = EbNo + 10*log10(k) + 10*log10(CodingRate);
noiseVar = 10.^(-snr/10);
Hist=dsp.Histogram('LowerLimit', 1, 'UpperLimit', maxIter, 'NumBins', maxIter,
'RunningHistogram', true);
%% Processing loop modeling transmitter, channel model and receiver
numErrs = 0; numBits = 0; nS=0;
while ((numErrs < maxNumErrs) && (numBits < maxNumBits))
    % Transmitter
    u = randi([0 1], FRM,1);                    % Randomly generated input bits
    data= CbCRCGenerator(u);                    % Transport block CRC code
    t0 = TurboEncoder(data, Indices);           % Turbo Encoder
    t1 = Scrambler(t0, nS);                      % Scrambler
    t2 = Modulator(t1, ModulationMode);         % Modulator
    % Channel
    c0 = AWGNChannel(t2, snr);                   % AWGN channel
    % Receiver
    r0 = DemodulatorSoft(c0, ModulationMode, noiseVar);   % Demodulator
    r1 = DescramblerSoft(r0, nS);               % Descrambler
    [y, ~, iters] = TurboDecoder_crc(-r1, Indices);       % Turbo Decoder
    % Measurements
    numErrs = numErrs + sum(y~=u);              % Update number of bit errors
    numBits = numBits + FRM;                    % Update number of bits processed
    itersHist = step(Hist, iters);              % Update histogram
of iteration numbers
    % Manage slot number with each subframe processed
    nS = nS + 2; nS = mod(nS, 20);
end
%% Clean up & collect results
ber = numErrs/numBits;                          % Compute Bit Error Rate (BER)
```

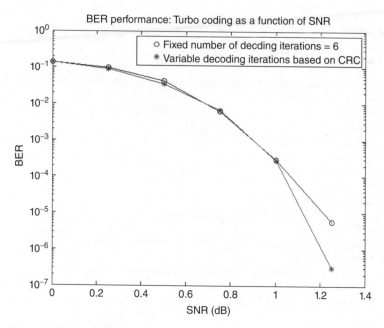

Figure 4.6   Comparison of BER results for cases of turbo coding with and without CRC-based early terminations

The BER results in Figure 4.6 indicate that we get similar BER performance for the range of SNR values with early termination (trace in red) and without early termination (trace in blue).

4.5.3   Timing Measurements

In this experiment, we compare the execution times of a transceiver that employs turbo decoding without a CRC-based early-stopping mechanism (*chap4_ex04.m*) with those of one that does employ a CRC-based early-stopping mechanism (*chap4_ex04_crc.m*). The experiment is performed by calling the following MATLAB testbench.

Algorithm

MATLAB script

```
EbNo=1; maxNumErrs=1e7; maxNumBits=1e7;
tic; [a,b]=chap4_ex04(EbNo,maxNumErrs, maxNumBits); toc;
tic; [a,b]=chap4_ex04_crc(EbNo,maxNumErrs, maxNumBits); toc;
```

```
>> EbNo=1; maxNumErrs=1e7; maxNumBits=1e7;
tic; [a,b]=chap4_ex04(EbNo,maxNumErrs, maxNumBits); toc
tic; [a,b]=chap4_ex04_crc(EbNo,maxNumErrs, maxNumBits); toc
Elapsed time is 178.771266 seconds.
Elapsed time is 131.273779 seconds.
```

Figure 4.7   Typical saving in execution time with early-termination turbo decoding

The first line of the script forces both transceiver functions to process 10 million bits per call for a given $Eb/N0$ value of 1 dB. The second line uses the MATLAB functions *tic* and *toc* to obtain the elapsed time for an algorithm without early termination. The third line records the elapsed time for an algorithm with early termination.

The results printed in the MATLAB command line are shown in Figure 4.7. It takes considerably less time to process the same number of input frames with early termination (131.27 seconds) than it does without early termination (178.77 seconds).

4.6   Rate Matching

So far we have only considered a turbo coding operation with a base coding rate of 1/3. Rate matching is instrumental in the implementation of adaptive coding, an important feature of modern communications standards. It helps augment the throughput based in the channel conditions. In low-distortion channels, we can code the data with coding rates near unity, which reduces the number of bits transmitted for forward error coding. On the other hand, in degraded channels we can use smaller coding rates and increase the number of error-correction bits.

In channel coding with rate matching, we start with a constant 1/3-rate turbo coder and use rate matching to arrive at any desired rate by repeating or puncturing. If a rate lower than 1/3 is requested, we repeat the turbo coder output bits. For rates higher than 1/3, we puncture or remove some of the turbo coder output bits. The puncturing of the code is not the result of simple subsampling but rather is based on an interleaving method. This method is shown to preserve the hamming distances of the resulting higher-rate code [2].

Rate matching is composed of:

- Subblock interleaving
- Parity-bit interlacing
- Bit pruning
- Rate-based bit selection and transmission.

The first operation in rate matching is the subblock interleaving, based on a simple rectangular interleaver. By using a circular buffer concept in rate matching, both the puncturing and the repeating operations that are necessary to increase or decrease (respectively) the rate to the desired level are simply implemented by bit selection operating on a circular buffer. Finally, by concatenating codeblocks, the encoded bits become ready for transfer to the PDSCH for processing.

4.6.1 MATLAB Examples

Staying true to our pedagogic approach of moving from simple to more complex, we will first study rate matching before going on to introduce all the details of transport block channel coding in the LTE standard. This MATLAB function implements the three features of rate matching as specified in the LTE standard: subblock interleaving, Parity bit interlacing, and rate matching with a circular buffer bit selection.

This MATLAB function shows the sequence of interleaving, interlacing, and bit-selection operations that defines the LTE rate-marching algorithm. The input of the rate marcher is the output of a 1/3 turbo encoder. So, for an input block of size K, the input to the rate marcher has a size of $3(K + 4)$, comprising three streams of Systematic and two Parity streams. First, we subdivide each of the three streams into 32 bit sections and interleave each of these sections. Since each stream may not be divisible by 32, we add dummy bits to the beginnings of the streams, such that the resulting vector can be subdivided into some integer number of 32 bits. The subblock interleavings used for Systematic and Parity 1 bits are the same, but the subblock interleaved for Parity 2 bits is different.

We then create an output vector composed of dummy padded Systematic bits and the interlacing of dummy padded Parity 1 and Parity 2 bits. Finally, by removing the dummy bits, we generate the circular buffer used for the rate-necking operation. The last step in rate matching is a bit selection, where the dummy bits in the circular buffer are removed and the first few bits are selected. The ratio of selected bits to the input length of the turbo encoder is the new rate after rate matching.

Algorithm

MATLAB function

```
function y= RateMatcher(in, Kplus, Rate)
% Rate matching per coded block, with and without the bit selection.
D = Kplus+4;
if numel(in)~=3*D, error('Kplus (2nd argument) times 3 must be size of input 1.');end
% Parameters
colTcSb = 32;
rowTcSb = ceil(D/colTcSb);
Kpi = colTcSb * rowTcSb;
Nd = Kpi - D;
% Bit streams
d0 = in(1:3:end); % systematic
d1 = in(2:3:end); % parity 1st
d2 = in(3:3:end); % parity 2nd
i0=(1:D)';
Index=indexGenerator(i0,colTcSb, rowTcSb, Nd);
Index2=indexGenerator2(i0,colTcSb, rowTcSb, Nd);
```

```matlab
% Sub-block interleaving - per stream
v0 = subBlkInterl(d0,Index);
v1 = subBlkInterl(d1,Index);
v2 = subBlkInterl(d2,Index2);
vpre=[v1,v2].';
v12=vpre(:);
%   Concat 0, interleave 1, 2 sub-blk streams
% Bit collection
wk = zeros(numel(in), 1);
wk(1:D) = remove_dummy_bits( v0 );
wk(D+1:end) = remove_dummy_bits( v12);
% Apply rate matching
N=ceil(D/Rate);
y=wk(1:N);
end
function v = indexGenerator(d, colTcSb, rowTcSb, Nd)
% Sub-block interleaving - for d0 and d1 streams only
colPermPat = [0, 16, 8, 24, 4, 20, 12, 28, 2, 18, 10, 26, 6, 22, 14, 30,...
        1, 17, 9, 25, 5, 21, 13, 29, 3, 19, 11, 27, 7, 23, 15, 31];
% For 1 and 2nd streams only
y = [NaN*ones(Nd, 1); d];        % null (NaN) filling
inpMat = reshape(y, colTcSb, rowTcSb).';
permInpMat = inpMat(:, colPermPat+1);
v = permInpMat(:);
end
function v = indexGenerator2(d, colTcSb, rowTcSb, Nd)
% Sub-block interleaving - for d2 stream only
colPermPat = [0, 16, 8, 24, 4, 20, 12, 28, 2, 18, 10, 26, 6, 22, 14, 30,...
        1, 17, 9, 25, 5, 21, 13, 29, 3, 19, 11, 27, 7, 23, 15, 31];
pi = zeros(colTcSb*rowTcSb, 1);
for i = 1 : length(pi)
    pi(i) = colPermPat(floor((i-1)/rowTcSb)+1) + colTcSb*(mod(i-1, rowTcSb)) + 1;
end
% For 3rd stream only
y = [NaN*ones(Nd, 1); d];        % null (NaN) filling
inpMat = reshape(y, colTcSb, rowTcSb).';
ytemp = inpMat.';
y = ytemp(:);
v = y(pi);
end
function out = remove_dummy_bits( wk )
%UNTITLED5 Summary of this function goes here
%out = wk(find(~isnan(wk)));
out=wk(isfinite(wk));
end
function out=subBlkInterl(d0,Index)
out=zeros(size(Index));
IndexG=find(~isnan(Index)==1);
IndexB=find(isnan(Index)==1);
```

```
out(IndexG)=d0(Index(IndexG));
Nd=numel(IndexB);
out(IndexB)=nan*ones(Nd,1);
end
```

In the *RateDematcher* function we perform the inverse operations to those in the rate matching. We create a vector composed of dummy padded Systematic and Parity bits, place the available samples of the input vectors in the vector, and by de-interlacing and de-interleaving create the right number of LLR samples to become inputs to the 1/3 turbo decoder.

Algorithm

MATLAB function

```
function out = RateDematcher(in, Kplus)
% Undoes the Rate matching per coded block.
%#codegen

% Parameters
colTcSb = 32;
D = Kplus+4;
rowTcSb = ceil(D/colTcSb);
Kpi = colTcSb * rowTcSb;
Nd = Kpi - D;

tmp=zeros(3*D,1);
tmp(1:numel(in))=in;

% no bit selection - assume full buffer passed in
i0=(1:D)';
Index= indexGenerator(i0,colTcSb, rowTcSb, Nd);
Index2= indexGenerator2(i0,colTcSb, rowTcSb, Nd);
Indexpre=[Index,Index2+D].';
Index12=Indexpre(:);

% Bit streams
tmp0=tmp(1:D);
tmp12=tmp(D+1:end);
v0 = subBlkDeInterl(tmp0, Index);
d12=subBlkDeInterl(tmp12, Index12);
v1=d12(1:D);
v2=d12(D+(1:D));
```

```
% Interleave 1, 2, 3 sub-blk streams - for turbo decoding
temp = [v0 v1 v2].';
out = temp(:);
end

function v = indexGenerator(d, colTcSb, rowTcSb, Nd)
% Sub-block interleaving - for d0 and d1 streams only

colPermPat = [0, 16, 8, 24, 4, 20, 12, 28, 2, 18, 10, 26, 6, 22, 14, 30,...
        1, 17, 9, 25, 5, 21, 13, 29, 3, 19, 11, 27, 7, 23, 15, 31];

% For 1 and 2nd streams only
y = [NaN*ones(Nd, 1); d];      % null (NaN) filling
inpMat = reshape(y, colTcSb, rowTcSb).';
permInpMat = inpMat(:, colPermPat+1);
v = permInpMat(:);

end

function v = indexGenerator2(d, colTcSb, rowTcSb, Nd)
% Sub-block interleaving - for d2 stream only

colPermPat = [0, 16, 8, 24, 4, 20, 12, 28, 2, 18, 10, 26, 6, 22, 14, 30,...
        1, 17, 9, 25, 5, 21, 13, 29, 3, 19, 11, 27, 7, 23, 15, 31];
pi = zeros(colTcSb*rowTcSb, 1);
for i = 1 : length(pi)
    pi(i) = colPermPat(floor((i-1)/rowTcSb)+1) + colTcSb*(mod(i-1, rowTcSb)) + 1;
end

% For 3rd stream only
y = [NaN*ones(Nd, 1); d];       % null (NaN) filling
inpMat = reshape(y, colTcSb, rowTcSb).';
ytemp = inpMat.';
y = ytemp(:);
v = y(pi);

end

function out=subBlkDeInterl(in,Index)
out=zeros(size(in));
IndexG=find(~isnan(Index)==1);
IndexOut=Index(IndexG);
out(IndexOut)=in;
end
```

4.6.2   BER Measurements

We will now examine the effects of using a coding rate other than 1/3 for the turbo coding algorithm. The function *chap6_ex05_crc* implements the transceiver algorithm that performs a combination of CRC generation, turbo coding, scrambling, and modulation and their inverse operations, while implementing the early-termination mechanism and applying rate-matching operations.

Algorithm

MATLAB function

```
function [ber, numBits, itersHist]=chap6_ex05_crc(EbNo, maxNumErrs, maxNumBits)
%% Constants
FRM=2432-24;                            % Size of bit frame
Kplus=FRM+24;
Indices = IteIntrlvrIndices(Kplus);
ModulationMode=1;
k=2*ModulationMode;
CodingRate=1/2;
snr = EbNo + 10*log10(k) + 10*log10(CodingRate);
noiseVar = 10.^(-snr/10);
Hist=dsp.Histogram('LowerLimit', 1, 'UpperLimit', maxIter, 'NumBins', maxIter,
'RunningHistogram', true);
%% Processing loop modeling transmitter, channel model and receiver
numErrs = 0; numBits = 0; nS=0;
while ((numErrs < maxNumErrs) && (numBits < maxNumBits))
    % Transmitter
    u  = randi([0 1], FRM,1);                    % Randomly generated input bits
    data= CbCRCGenerator(u);                     % Transport block CRC code
    t0 = TurboEncoder(data, Indices);               % Turbo Encoder
    t1= RateMatcher(t0, Kplus, CodingRate);         % Rate Matcher
    t2 = Scrambler(t1, nS);                         % Scrambler
    t3 = Modulator(t2, ModulationMode);             % Modulator
    % Channel
    c0 = AWGNChannel(t3, snr);                       % AWGN channel
    % Receiver
    r0 = DemodulatorSoft(c0, ModulationMode, noiseVar);   % Demodulator
    r1 = DescramblerSoft(r0, nS);                     % Descrambler
    r2 = RateDematcher(r1, Kplus);                   % Rate Matcher
    [y, ~, iters]  = TurboDecoder_crc(-r2, Indices);    % Turbo Decoder
    % Measurements
    numErrs   = numErrs + sum(y~=u);                  % Update number of bit errors
    numBits   = numBits + FRM;                 % Update number of bits processed
    itersHist   = step(Hist, iters);                  % Update histogram
of iteration numbers
```

```
% Manage slot number with each subframe processed
  nS = nS + 2; nS = mod(nS, 20);
end
%% Clean up & collect results
ber = numErrs/numBits;                              % Compute Bit Error Rate (BER)
```

By adding the rate-matching operation after the turbo encoder and the rate-dematching operation after the decoder we can simulate the effects of using any coding rate higher than 1/3. Of
course, lower coding rates are used in transmission scenarios dealing with cleaner channels,
where less error correction is desirable.

By modifying the variable *CodingRate* in the function, we activate the rate-matching operations and can examine the BER performance as a function of SNR for multiple values of target
coding rates. The results in Figure 4.8 show that, as expected, the performance of a 1/3-rate
transceiver is superior to that of the ½ transceiver.

4.7   Codeblock Segmentation

In LTE, a transport block connects the MAC layer and the PHY. The transport block usually contains a large amount of data bits, which are transmitted at the same time. The first
set of operations performed on a transport block is channel coding, which is applied to each

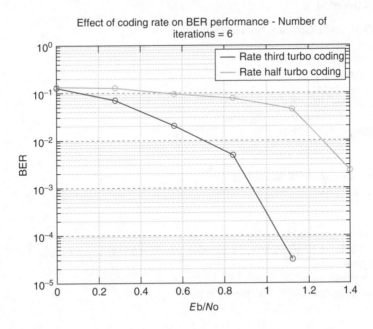

Figure 4.8   Effect of rate matching on turbo coding BER performance

codeblock independently. If the input frame to the turbo encoder exceeds the maximum size, the transport block is usually divided into multiple smaller blocks known as codeblocks. Since the internal interleaver of the turbo encoder is only defined for 188 input block sizes, the sizes of these codeblocks need to match the set of codeblock sizes supported by the turbo coder. A combination of codeblock CRC attachment, turbo coding, and rate matching is then applied to each codeblock independently.

4.7.1 MATLAB Examples

In the following segmentation function we search for the best subblock size to satisfy two properties: (i) it is one among 188 valid block sizes; and (ii) it is an exact integer multiple of the subblock size. The number of subblocks contained in a codeblock is known as parameter C and the size of each subblock is known as Kplus. We also need to compute a parameter E for the codeword. The output of channel coding is known as the codeword; the size of the codeword is the product of C subblocks and the output size per subblock E. The total codeword size is determined by the scheduler, depending on the number of available resources. The effective coding rate is then the ratio of codeword size to subblock size.

Algorithm

MATLAB function

```
function [C, Kplus] = CblkSegParams(tbLen)
%#codegen
%% Code block segmentation
blkSize = tbLen + 24;
maxCBlkLen = 6144;
if (blkSize <= maxCBlkLen)
    C = 1;        % number of code blocks
    b = blkSize;   % total bits
else
    L = 24;
    C = ceil(blkSize/(maxCBlkLen-L));
    b = blkSize + C*L;
end

% Values of K from table 5.1.3-3
validK = [40:8:512 528:16:1024 1056:32:2048 2112:64:6144].';
% First segment size
temp = find(validK >= b/C);
Kplus = validK(temp(1), 1);    % minimum K
```

The following MATLAB function calculates the sizes of subblocks and determines how many are processed in parallel to reconstitute the channel coding outputs. First we divide the total number of codeword bits by the number of subblocks. For each subblock, we ensure that the number of output bits is divisible by the number of modulation bits and the resulting number of multi-antenna layers.

Algorithm

MATLAB function

```
function E = CbBitSelection(C, G, Nl, Qm)
%#codegen
% Bit selection parameters
% G = total number of output bits
% Nl   Number of layers a TB is mapped to (Rel10)
% Qm    modulation bits
Gprime = G/(Nl*Qm);
gamma = mod(Gprime, C);
E=zeros(C,1);
% Rate matching with bit selection
for cbIdx=1:C
    if ((cbIdx-1) <= (C-1-gamma))
        E(cbIdx) = Nl*Qm*floor(Gprime/C);
    else
        E(cbIdx)  = Nl*Qm*ceil(Gprime/C);
    end
end
```

In the receiver, in order to correctly perform the inverse of matching operations, we need the parameters C and *Kplus* (the number of subblocks and the size of each subblock, respectively).

4.8   LTE Transport-Channel Processing

Figure 4.9 shows a block diagram of TrCH processing. Five functional components characterize transport block processing:

- Transport-block CRC attachment
- Codeblock segmentation and codeblock CRC attachment
- Turbo coding based on a 1/3 rate
- Rate matching to handle any requested coding rates
- Codeblock concatenation.

4.8.1   MATLAB Examples

In the following MATLAB function, we need to distinguish between the case where the transport contains only a single codeblock and the cases where it contains more than one codeblock, since in the first case we do not need to apply the CRC attachment to the codeblock as the transport block already contains a CRC attachment.

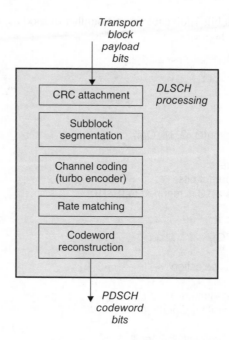

Figure 4.9   Structure of transport-channel processing

Algorithm

MATLAB function

```
function [out, Kplus, C] = TbChannelCoding(in, prmLTE)
% Transport block channel coding
%#codegen
inLen = size(in, 1);
[C, ~, Kplus] = CblkSegParams(inLen-24);
intrlvrIndices = IteIntrlvrIndices(Kplus);
G=prmLTE.maxG;
E_CB=CbBitSelection(C, G, prmLTE.NumLayers, prmLTE.Qm);
% Initialize output
out = false(G, 1);
% Channel coding the TB
if (C==1) % single CB, no CB CRC used
    % Turbo encode
    tEncCbData = TurboEncoder( in, intrlvrIndices);
    % Rate matching, with bit selection
    rmCbData = RateMatcher(tEncCbData, Kplus,  G);
    % unify code paths
    out = logical(rmCbData);
else % multiple CBs in TB
```

```
   startIdx = 0;
   for cbIdx = 1:C
      % Code-block segmentation
      cbData = in((1:(Kplus-24)) + (cbIdx-1)*(Kplus-24));
      % Append checksum to each CB
      crcCbData = CbCRCGenerator( cbData);
      % Turbo encode each CB
      tEncCbData = TurboEncoder(crcCbData, intrlvrIndices);
      % Rate matching with bit selection
      E=E_CB(cbIdx);
      rmCbData = RateMatcher(tEncCbData, Kplus,  E);
      % Code-block concatenation
      out((1:E) + startIdx) = logical(rmCbData);
      startIdx = startIdx + E;
   end
end
```

The sequence of operations performed in channel decoding can be regarded as the inverse of those performed in channel coding, as follows:

- Iteration over each codeblock
- Rate dematching (from target rate to 1/3 rate) composed of:
  - Bit selection and insertion
  - Parity-bit deinterlacing
  - Subblock deinterleaving
  - Recovery of Systematic and Parity bits for turbo decoding
- Codeblock 1/3-rate turbo decoding with early termination based on CRC.

Here we are using CRC of the entire transport block as another early stopping criterion and as a mechanism for updating the state of HARQ. The following MATLAB function summarizes the operations in the TrCH decoder.

Algorithm

MATLAB function

```
function [decTbData, crcCbFlags, iters] = TbChannelDecoding( in, Kplus, C, prmLTE)
% Transport block channel decoding.
%#codegen
intrlvrIndices = IteIntrlvrIndices(Kplus);
% Make fixed size
G=prmLTE.maxG;
E_CB=CbBitSelection(C, G, prmLTE.NumLayers, prmLTE.Qm);
% Channel decoding the TB
```

```
if (C==1) % single CB, no CB CRC used
   % Rate dematching, with bit insertion
   deRMCbData = RateDematcher(-in, Kplus)
   % Turbo decode the single CB
   tDecCbData =TurboDecoder(deRMCbData, intrlvrIndices,  prmLTE.maxIter)
   % Unify code paths
   decTbData  = logical(tDecCbData);
else % multiple CBs in TB
   decTbData  = false((Kplus-24)*C,1); % Account for CB CRC bits
   startIdx = 0;
   for cbIdx = 1:C
      % Code-block segmentation
      E=E_CB(cbIdx);
      rxCbData = in(dtIdx(1:E) + startIdx);
      startIdx = startIdx + E;
      % Rate dematching, with bit insertion
      %   Flip input polarity to match decoder output bit mapping
      deRMCbData = lteCbRateDematching(-rxCbData, Kplus, C, E);
      % Turbo decode each CB with CRC detection
      %   - uses early decoder termination at the CB level
      [crcDetCbData, crcCbFlags(cbIdx), iters(cbIdx)] = ...
              TurboDecoder_crc(deRMCbData, intrlvrIndices);
      % Check the crcCBFlag per CB. If still in error, abort further TB
      % processing for remaining CBs in the TB, as the HARQ process will
      % request retransmission for the whole TB.
      if (~prmLTE.fullDecode)
         if (crcCbFlags(cbIdx)==1) % error
            break;
         end
      end
      % Code-block concatention
      decTbData((1:(Kplus-24)) + (cbIdx-1)*(Kplus-24)) = logical(crcDetCbData);
   end
end
```

4.8.2   BER Measurements

We now measure the bit-error rates of the LTE downlink TrCH in the presence of the AWGN channel noise. The function *chap4_ex06* combines all the TrCH processing operations with scrambling and modulation.

Algorithm

MATLAB function

```
function [ber, numBits]=chap4_ex06(EbNo, maxNumErrs, maxNumBits)
%% Constants
FRM=2432-24;
Kplus=FRM+24;
Indices = IteIntrlvrIndices(Kplus);
ModulationMode=1;
k=2*ModulationMode;
maxIter=6;
CodingRate=1/2;
snr = EbNo + 10*log10(k) + 10*log10(CodingRate);
noiseVar = 10.^(-snr/10);
%% Processing loop modeling transmitter, channel model and receiver
numErrs = 0; numBits = 0; nS=0;
while ((numErrs < maxNumErrs) && (numBits < maxNumBits))
   % Transmitter
   u  = randi([0 1], FRM,1);                           % Randomly generated input bits
   data= CbCRCGenerator(u);                            % Transport block CRC code
   [t1, Kplus, C] = TbChannelCoding(data,Indices,maxIter);   % Transport
Channel encoding
   t2 = Scrambler(t1, nS);                             % Scrambler
   t3 = Modulator(t2, ModulationMode);                % Modulator
   % Channel
   c0 = AWGNChannel(t3, snr);                          % AWGN channel
   % Receiver
   r0 = DemodulatorSoft(c0, ModulationMode, noiseVar);  % Demodulator
   r1 = DescramblerSoft(r0, nS);                       % Descrambler
   [r2,~,~] = TbChannelDecoding(r1, Kplus, C, Indices,maxIter);  % Transport
Channel decoding
   y  = CbCRCDetector(r2);                             % Code block CRC detector
   % Measurements
   numErrs   = numErrs + sum(y~=u);                    % Update number of bit errors
   numBits   = numBits + FRM;                          % Update number of bits processed
   % Manage slot number with each subframe processed
   nS = nS + 2; nS = mod(nS, 20);
end
%% Clean up & collect results
ber = numErrs/numBits;                                 % Compute Bit Error Rate (BER)
```

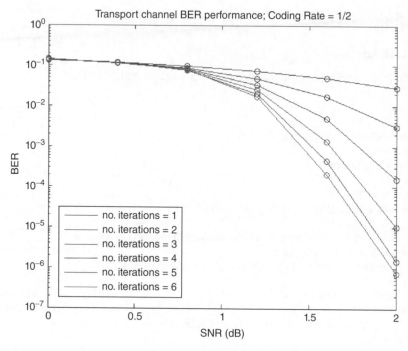

Figure 4.10 BER performance of DLSCH and QPSK as a function of $Eb/N0$ and the number of decoding operations

By executing this function with a range of SNR values, we can verify that the combination of processing applied to the DLSCH and PDSCH without OFDM and MIMO operations is implemented properly. Figure 4.10 illustrates the BER performance of the transceiver. In this experiment, we use rate matching with a coding rate of 1/2 and a QPSK modulator and repeat the operations for a range of values for a maximum number of iterations from one to six. As expected, by providing more decoding iterations we obtain progressively better performance results. This again shows the critical role that early termination can play in making the DLSCH processing specified in the LTE standard more realizable.

4.9   Chapter Summary

So far we have studied the forward error-correction scheme employed in the LTE standard based on a simple channel model (AWGN). The LTE standard uses AWGN environment propagation for static performance measurement. No fading or multipaths exist for this propagation model and it does not take into account the frequency response of real channels. Most real channels add to the transmitted signal various forms of fading and other correlated distortions. These fading profiles introduce intersymbol interference, which must be compensated by using equalization.

We observe that the performance of iterative turbo channel coding depends on the number of iterations used. This motivates the discussion regarding simulation acceleration in Chapter 9.

We also note that studying the performance in an AWGN channel ignores real channel models and the effects of multipath fading. This motivates the discussions in Chapters 5 and 6 on how to combat multipath fading using OFDM and Single-Carrier Frequency Division Multiplexing (SC-FDM) with frequency-domain equalizers.

References

[1] 3GPP (2009) Evolved Universal Terrestrial Radio Access (E-UTRA); Multiplexing and Channel Coding. TS 36.212.
[2] Proakis, J. and Salehi, M. (2007) *Digital Communications*, 5th edn, McGraw-Hill Education, New York.
[3] 3GPP (2011) Evolved Universal Terrestrial Radio Access (E-UTRA); Physical Channels and Modulation Version 10.0.0. TS 36.211, January 2011.
[4] Online: http://www.mathworks.com/help/comm/ug/define-basic-system-objects-1.html (last accessed September 30, 2013).

5

OFDM

So far we have considered the modulation, scrambling, and coding specifications of the LTE standard and have used a simplistic channel model (Additive White Gaussian Noise, AWGN) to undertake performance evaluations. Understanding of Orthogonal Frequency Division Multiplexing (OFDM), which is the fundamental air interface in the standard, necessitates understanding and modeling of more sophisticated channel models.

In this chapter we consider realistic channel models that take into account the dynamic channel responses and fading conditions. Short-term fading effects such as multipath fading and Doppler effects resulting from mobility will lead to frequency-selective channel models. OFDM and Single-Carrier Frequency Division Multiplexing (SC-FDM) multiple-access technologies in LTE, for downlink and uplink respectively, use efficient frequency-domain equalizers to combat frequency-selective fading and contribute to its superior spectral efficiency. In this chapter we will focus on a single-antenna configuration, while in the next chapter we will combine Multiple Input Multiple Output (MIMO) and OFDM.

We will detail the basis of the OFDM technology and discuss OFDM frame structure and implementation in the LTE standard. We will then discuss the time–frequency mapping of the OFDM signal and the various resource element granularities used to adaptively exploit the channel bandwidth, followed by the frequency-domain equalization of the OFDM signal at the receiver. We examine Zero Forcing (ZF), Minimum Mean Squared Error (MMSE), equalizers and details of the interpolation of reference signals or pilots. Finally, we examine the performance of a transceiver composed of components developed so far under various multipath fading and mobility conditions specified by the standard.

5.1 Channel Modeling

Wireless channels are characterized by the availability of different paths of propagation between transmitters and receivers. Besides the direct path between the transmitter and the receiver, which may even not exist, other paths can be formed through reflection, diffraction, scattering, or other propagation scenarios. By going through different paths, different versions of the transmitted signals can be received simultaneously at the receiver. These different

Understanding LTE with MATLAB®: From Mathematical Modeling to Simulation and Prototyping, First Edition.
Houman Zarrinkoub.
© 2014 John Wiley & Sons, Ltd. Published 2014 by John Wiley & Sons, Ltd.

versions exhibit varying profiles of signal power and time delay or phase. Since these received signals are correlated in time, an AWGN model is not the most representative channel model for most wireless connections. Therefore, proper modeling of the characteristics of a wireless channel is an important requirement for the design of mobile communications systems. Channel propagation usually results in a reduced power in received signals relative to the transmitted signal. In general, the power reductions are treated in two categories: (i) signal attenuations or large-scale fading and (ii) fading or small-scale fading.

5.1.1 Large-Scale and Small-Scale Fading

Path loss and shadowing are among the most prominent large-scale fading effects. These large-scale features are taken into account in design and cellular topography [1]. Small-scale fading includes multipath fading and time dispersion due to mobility. These features are of short duration and must be adaptively dealt with. The design of the PHY (Physical Layer) should include techniques that effectively deal with these types of channel impairment [1].

5.1.2 Multipath fading Effects

Multipath fading is characterized by a power-delay profile comprising two components: a vector of relative delays and a vector of average power parameters. Another useful set of scalar measures is either mean excess delay or the Root Mean Square (RMS) delay spread as the first and second moments of relative delays. Multipath fading can be flat or frequency-selective. If the bandwidth is larger than the inverse of the delay spread, channel frequency response will lead to multipath fading.

   In the cellular communication context, signals are received at the mobile terminal following a direct path from the base station. Some signals will also be reflected off buildings or other reflectors and will reach the mobile terminal with a time delay and an attenuated power. Since the mobile receiver gets the linear combination of these signals, the net signal obtained is essentially a convolution of input signal and the impulse response of the channel. In the frequency domain, the frequency response of the channel includes different response patterns at different frequencies; we thus have frequency-selective fading (Figure 5.1).

   In the case of a time-dispersive channel with multipath-propagation characteristics, there will be not only intersymbol interference within a subcarrier but also interference between

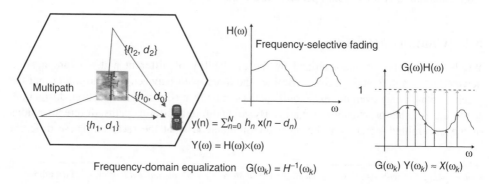

Figure 5.1 Multipath propagation, frequency-selective fading, and frequency-domain equalization

subcarriers. This is because the orthogonality between the subcarriers will be partially lost due to the overlap of the demodulator correlation interval for one path with the symbol boundary of another. The integration interval, used to compute the Fast Fourier Transform (FFT), will not necessarily correspond to an integer number of periods of complex exponentials of that path, as the modulation symbols may differ between consecutive symbol intervals.

5.1.3   Doppler Effects

For mobile systems transmitting over a broad bandwidth, such as the LTE, the predominant channel degradation is a result of short-term fading caused by multipath propagation. We need to account for the effects of a fading channel in order to provide an accurate evaluation of the LTE system performance. As a result of mobile terminal movement, the profile of channel impulse response can vary. Fast- and slow-fading channels reflect the speed of the mobile terminal and are expressed in terms of Doppler frequency shifts [1].

5.1.4   MATLAB® Examples

We can study the effects of channel responses to a transmitted signal by using the various channel models of the Communications System Toolbox. *Rayleigh* and *Rician* channel objects can be used to model a single propagation path and the *comm.MIMOChannel* System object can be used to study the effects of multiple antennas and multiple propagation paths. All of these components use the delay profile and Doppler shift as parameters to model the dynamics of a fading channel.

In order to become familiar with these objects, let us examine four types of channel separately. The differences between these channel models relate to (i) whether or not a frequency-flat or a frequency-selective channel is present and (ii) whether or not a frequency-dispersive channel is present due to the Doppler shift caused by the mobility of the receiver.

5.1.4.1   Low-Mobility Flat-Fading Channels

The first type of channel is a low-mobility flat-fading channel. In this case, the delay profile does not contain multiple shifts in time. It is characterized by a single dominant delay value that denotes the time difference between the transmitter and the receiver. Furthermore, low mobility leads to a Doppler shift of near zero. The following MATLAB function implements such a channel model.

Algorithm

MATLAB function

```
function y = ChanModelFading(in, Chan)
% Static (No mobility) Flat Fading Channel
%#codegen
% Get simulation params
numTx=1;
```

```
numRx=1;
chanSRate = Chan.chanSRate;
PathDelays = Chan.PathDelays;
PathGains  = Chan.PathGains;
Doppler    = Chan.DopplerShift;
% Initialize objects
persistent chanObj
if isempty(chanObj)
   chanObj = comm.MIMOChannel(...
      'SampleRate', chanSRate, ...
      'MaximumDopplerShift', Doppler, ...
      'PathDelays', PathDelays,...
      'AveragePathGains', PathGains,...
      'NumTransmitAntennas', numTx,...
      'TransmitCorrelationMatrix', eye(numTx),...
      'NumReceiveAntennas', numRx,...
      'ReceiveCorrelationMatrix', eye(numRx),...
      'PathGainsOutputPort', false,...
      'NormalizePathGains', true,...
      'NormalizeChannelOutputs', true);
end
y = step(chanObj, in);
```

In order to visualize the effect of this type of channel on a system, we add the fading channel to a system containing coding scrambling and modulation and observe the input signal to the demodulator. Note that by running the experiment in the following MATLAB function, we can observe how even a static flat-fading channel that represents a mild form of channel response significantly degrades the performance.

Algorithm

MATLAB function

```
function [ber, bits] = chap5_ex01(EbNo, maxNumErrs, maxNumBits, prmLTE)
%#codegen
%% Constants
FRM=2432-24;
Kplus=FRM+24;
Indices = lteIntrlvrIndices(Kplus);
ModulationMode=1;
k=2*ModulationMode;
maxIter=6;
CodingRate=1/2;
snr = EbNo + 10*log10(k) + 10*log10(CodingRate);
noiseVar = 10.^(-snr/10);
%% Processing loop modeling transmitter, channel model and receiver
```

```
numErrs = 0; numBits = 0; nS=0;
while ((numErrs < maxNumErrs) && (numBits < maxNumBits))
  % Transmitter
  u  = randi([0 1], FRM,1);                    % Randomly generated input bits
  data= CbCRCGenerator(u);                     % Transport block CRC code
  [t1, Kplus, C] = TbChannelCoding(data, prmLTE);
  t2 = Scrambler(t1, nS);                      % Scrambler
  t3 = Modulator(t2, ModulationMode);          % Modulator
  % Channel & Add AWG noise
  [rxFade, ~] = MIMOFadingChan(t3, prmLTE);
  nVar = 10.^(0.1.*(-EbNo));                    % assume unit sigPower
  c0  = AWGNChannel2(rxFade, nVar );           % AWGN channel
  % Receiver
  r0 = DemodulatorSoft(c0, ModulationMode, noiseVar);   % Demodulator
  r1 = DescramblerSoft(r0, nS);                % Descrambler
  r2 = RateDematcher(r1, Kplus);               % Rate Matcher
  r3 = TurboDecoder(-r2, Indices,  maxIter);   % Turbo Decoder
  y  = CbCRCDetector(r3);                      % Code block CRC detector
  % Measurements
  numErrs  = numErrs + sum(y~=u);              % Update number of bit errors
  numBits  = numBits + FRM;               % Update number of bits processed
  % Manage slot number with each subframe processed
  nS = nS + 2; nS = mod(nS, 20);
end
%% Clean up & collect results
ber = numErrs/numBits;                         % Compute Bit Error Rate (BER)
```

Figure 5.2 shows the frequency response of the transmitted and received signals within the transmission bandwidth. It explains why this is called a flat-fading channel, as throughout the bandwidth at each frequency the response is faded by the same value.

5.1.4.2 High-Mobility Flat-Fading Channels

We now set a non-zero value for the Doppler shift in order to model a high-mobility flat-fading channel. Note that the profile of the channel response is still that of flat fading, but the gain for

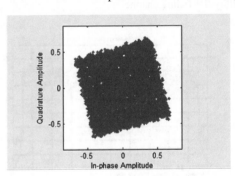
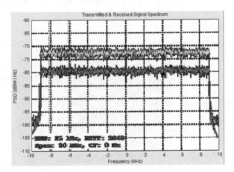

Figure 5.2 Low-mobility flat-fading channel

the entire spectrum varies as a function of time. Also, the constellation of the received signal still resembles a 64QAM (Quadrature Amplitude Modulation) modulation. However, at each time step the constellation rotates based on the phase offset as a result of the Doppler shift. These effects are illustrated in Figure 5.3.

5.1.4.3 Low-Mobility Frequency-Selective Channels

In this section, we examine a frequency-selective channel model that will still have a zero Doppler shift but it will have a vector for the associated delay profile. With the vector of time delays, we observe a gain vector of the same size. This results in a frequency-selective channel response, as observed in Figure 5.4.

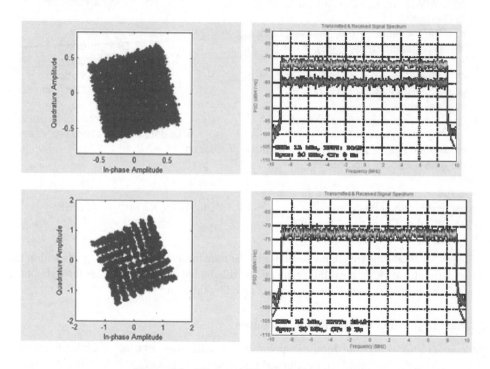

Figure 5.3   High-mobility flat-fading channel

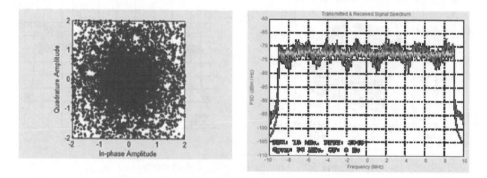

Figure 5.4   Low-mobility frequency-selective fading

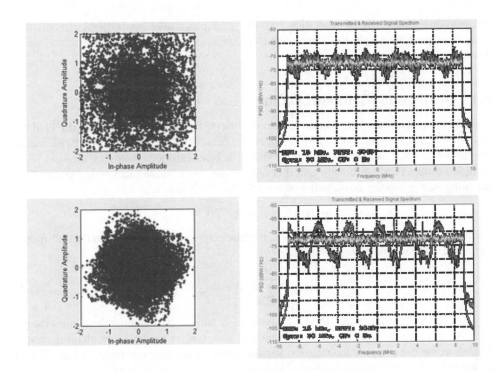

Figure 5.5   High-mobility frequency-selective fading

5.1.4.4   High-Mobility Frequency-Selective Channels

Finally, by setting a non-zero value for the Doppler shift we can model a high-mobility frequency-selective channel. As in the previous high-mobility case, we observe a varying profile for channel gain at different frequency values. We also note that the channel responses vary in time. Figure 5.5 illustrates the magnitude spectra of the channel response computed over two subframes 10 ms apart.

5.2   Scope

In this book, without any loss for generality, we focus on the normal Cyclic Prefix (CP) that leads to a particular time framing (seven OFDM symbols per slot) and subcarrier spacing of 15 kHz. The MATLAB functionality is flexibly parameterized such that by changing a few parameters, extended CP mode can be easily simulated.

In this volume we do not deal with system access procedures, startup, random access, or handoff scenarios. We discuss the steady-state signal processing downlink transmission that takes place once the call within a cell is already established. As such, the synchronization signals and the Broadcast Channels (BCHs) (downlink) and random access channels (uplink) will not be elaborately discussed and the accompanying MATLAB functions will not be developed.

5.3   Workflow

Starting with coding, modulation, and scrambling, we add channel modeling, which includes flat or frequency-selective fading. In this chapter we discuss a single-antenna transmission scenario (either Single Input Single Output, SISO, or Single Input Multiple Output, SIMO).

We focus on reference-signal generation, resource-grid specification, and OFDM transmission. Finally, we put together the testbench for the system model that implements the first mode of downlink transmission.

5.4   OFDM and Multipath Fading

The OFDM modulated signal is computed as the Inverse Fast Fourier Transform (IFFT) of the resource elements associated with different subcarriers. The IFFT output can be considered the sum of complex exponential functions known as basis functions, complex sinusoids, harmonics, or tones of a multitone signal. Let us consider one of these tones or harmonics (i.e., the complex exponential associated with a particular subcarrier) as:

$$x(n)\rfloor_{\omega=k\Delta f} = a_k e^{j2\pi kn/N} \tag{5.1}$$

Equation 5.2 shows how a channel with an impulse response (h_m) operates on the transmitted signal $x(n)$ to provide the received signal $y(n)$.

$$y(n) = \sum_{m=0}^{M} h_m x(n - d_m) \tag{5.2}$$

Now, due to linearity, when the OFDM signal is subject to a multipath fading channel each of its complex exponential components is also subject to the same channel model. Therefore, we can compute the received version of each subcarrier component of the OFDM signal ($y(n)\rfloor_{\omega=k\Delta f}$) as the convolution of that transmitted component and the channel impulse response.

$$y(n)\rfloor_{\omega=k\Delta f} = \sum_{m=0}^{M} h_m x(n)\rfloor_{\omega=k\Delta f} \tag{5.3}$$

The next step explains the necessity of introducing the CP to the OFDM formulation. We can substitute the expression $x(n)\rfloor_{\omega=k\Delta f} = a_k e^{j2\pi kn/N}$ for the complex sinusoidal component in Equation 5.3 if and only if the multipath delay values d_m are less than or equal to the CP length. Otherwise, with even a single delay value outside the CP range, we cross the OFDM symbol boundary and the orthogonality between subcarrier components is lost. Now, assuming that the delay spread is within the CP range, we can obtain the following expression for the received subcarrier component as a function of the transmitted subcarrier component:

$$y(n)\rfloor_{\omega=k\Delta f} = \sum_{m=0}^{M} h_m a_k e^{j2\pi k(n-d_m)/N} \tag{5.4}$$

After some algebraic simplifications, the output can be expressed as:

$$y(n)\rfloor_{\omega=k\Delta f} = a_k e^{\frac{j2\pi kn}{N}} \sum_{m=0}^{M} h_m e^{-\frac{j2\pi kd_m}{N}} \tag{5.5}$$

Note that the last expression, $\sum_{m=0}^{M} h_m e^{-\frac{j2\pi kd_m}{N}}$, is not a function of the time index n but rather can be viewed as a gain that is a function of the subcarrier index k and is applied to complex exponential component. By defining this gain as $H_k = \sum_{m=0}^{M} h_m e^{-\frac{j2\pi kd_m}{N}}$ and substituting it into Equation 5.5, we obtain the following expression for the received OFDM signal component:

$$y(n)\rfloor_{\omega=k\Delta f} = H_k a_k e^{\frac{j2\pi kn}{N}} \tag{5.6}$$

Now we look at OFDM operations at the receiver. Note that after removing the CP, the first operation in the receiver is to compute the Fourier transform, as defined by the following expression:

$$Y(\omega) = \sum_{n=0}^{N} y(n)e^{-j2\pi\omega n/N} \tag{5.7}$$

When we the express the received signal based on its Fourier formulation, $y(n) = \frac{1}{N}\sum_{k=1}^{N} Y(\omega)e^{j2\pi kn/N}$, all inner product terms for the frequencies other than the subcarrier vanish due to the orthogonality of the IFFT basis functions. The only non-zero term that determines the Fourier transform of the received signal at the subcarrier belongs to the received subcarrier component; that is:

$$Y(\omega)\rfloor_{\omega=k\Delta f} = \frac{1}{N}\sum_{n=0}^{N} y(n)\rfloor_{\omega=k\Delta f} e^{-j2\pi\omega_k n/N} \tag{5.8}$$

By substituting the expression for the received OFDM signal component, we obtain the following:

$$Y(\omega)\rfloor_{\omega=k\Delta f} = \frac{1}{N}\sum_{n=0}^{N} H_k a_k e^{j2\pi k\Delta f n/N} e^{-\frac{j2\pi k\Delta f n}{N}} \tag{5.9}$$

Simplification of this expression results in an intuitive formula for the received signal at a given subcarrier component:

$$Y(\omega)\rfloor_{\omega=k\Delta f} = H_k a_k \tag{5.10}$$

This expression indicates that the received signal at any subcarrier is the product of the transmitted symbol a_k and the multipath gain H_k. This simple expression is the basis for defining frequency-domain equalization using the pilot signals.

5.5   OFDM and Channel-Response Estimation

Each transmitted signal component that is subject to the multipath fading channel will arrive at the receiver as a scaled version of the transmitted signal. The gain is characterized by the channel response. Pilot or reference signals can be considered known signals placed at regular subcarriers positions. We can estimate the channel response at those subcarriers by dividing the received version at the subcarrier by the known transmitted value. The channel response at each particular subcarrier is then calculated as:

$$H(\omega)\rfloor_{\omega=k\Delta f} = \frac{Y(\omega)\rfloor_{\omega=k\Delta f}}{X(\omega)\rfloor_{\omega=k\Delta f}}$$

$$H(\omega)\rfloor_{\omega=k\Delta f} = \frac{H_k a_k}{a_k}$$

$$H(\omega)\rfloor_{\omega=k\Delta f} = H_k \tag{5.11}$$

Through various forms of interpolation we can now estimate the channel response not only at known subcarriers but at all subcarriers. This enables us to perform equalization, defined as reversing the effects of the fading channel in the frequency domain. This is more efficient than classical time-domain equalization techniques, which estimate the channel impulse response and use adaptive filtering to equalize the received signals.

5.6   Frequency-Domain Equalization

One of the most important features of the OFDM is its robust and efficient treatment of multipath fading. OFDM compensates for the effect of fading through a frequency-domain approach to equalization. Instead of filtering the received signal in time with the inverse of the channel impulse response, OFDM first constructs a frequency-domain representation of the data and then uses reference signals to invert the frequency response of the channel.

This implies a two-step process. First, the construction of a time–frequency resource grid, where data are aligned with subcarriers in the frequency domain before a series of OFDM symbols is generated in time. This step is also known as the resource element mapping. The types of signal that form the LTE downlink resource grid include the following:

- User data (Physical Downlink Shared Channel, PDSCH)
- Cell-Specific Reference (CSR) signals (otherwise known as pilots)
- Primary Synchronization Signals (PSSs) and Secondary Synchronization Signals (SSSs)
- Physical Broadcast Channels (PBCHs)
- Physical Downlink Control Channels (PDCCHs).

Second, we take the vector of resource elements as input and generate the OFDM symbols. This process involves performing an IFFT operation to generate the OFDM modulated signal and a CP insertion. The use of CP enables the receiver to sample each OFDM symbol for exactly one period in the time domain. The availability of CP helps mitigate the effects of intersymbol interference when the channel delay spread is less than the length of the CP.

Before OFDM signal generation, we need to generate the resource grid based on either a type-1 or a type-2 frame structure. Since we are showcasing Frequency Division Duplex (FDD) duplexing throughout this book, we will use type 1 here. Next we will show how to generate relevant signals to form the resource grid and how to create the OFDM symbols for transmission.

5.7   LTE Resource Grid

Understanding the time–frequency representation of data, organized as the resource grid, is a key step in understanding the LTE transmission scheme. The resource grid is essentially a matrix whose elements are the modulated symbols computed as the outputs of the modulation mapper. In its 2D representation, the y-axis of the grid represents the subcarriers aligned along the frequency dimension and the x-axis represents the OFDM symbols aligned along the time dimension [2].

The placement of data within the resource grid is quite important and reveals some of the design parameters of the LTE physical model. For example, the placement and resolution of pilot signals (CSR) along both axes of the resource grid determines the accuracy of the channel-response estimation in both time and frequency. Similarly, placing the PDCCH control-channel information at the beginning of each subframe helps the receiver decipher important processing parameters (such as the type of modulator and the MIMO mode used) before the system starts decoding the user data in the subframe.

The details involved in placing data within the resource grid can only be understood within the context of time framing and the way in which LTE defines a frame, a subframe, and a slot. Each LTE frame has a duration of 10 ms and is composed of ten 1 ms subframes marked by indices 0 to 9. Each subframe is subdivided into two slots of 0.5 ms duration, with each slot comprising seven OFDM symbols if a normal CP is used and six if an extended CP is used.

The placement of each modulated data type (user data, CSR, DCI, PSS, SSS, and BCH) into the resource grid follows a particular structure in both time and frequency. This structure depends on three parameters: the subcarrier (y-axis) index, the OFDM symbol (x-axis) index, and the index of the 1 ms subframe within a 10 ms frame. All subframes within a frame contain three types of data: user data (PDSCH), pilot CSRs, and downlink control data (PDCCH). The PSS and SSS are only available in subframes 0 and 5 at specific OFDM symbol indices (SSS at fifth symbol and PSS at sixth symbol) and specific subcarrier indices (72 subcarriers around the center of the resource grid). The PBCH is located only within subframe 0 at specific OFDM symbol indices (extending from the seventh to the tenth symbol) and specific subcarrier indices (72 subcarriers around the center of the resource grid). Figure 5.6 illustrates the placement of different modulated data, based on the signal types, within the resource grid.

5.8   Configuring the Resource Grid

Let us discuss the size and composition of the resource grid and how it is updated every subframe. Throughout this book, we process the LTE transceiver (transmitter, channel model, and receiver) one subframe at a time. Since the length of each subframe is 1 ms, processing one second of data involves processing 1000 iterations of the transceiver.

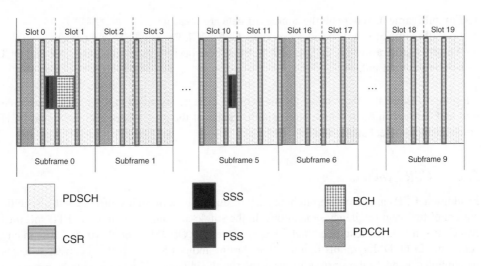

Figure 5.6   LTE resource grid content – entire grid view – featuring six types of signal

In each subframe, the size of the resource grid (N_{total} = the total number of symbols that fill up the grid) is a function of the following four parameters:

$$\begin{cases} N_{rb} & \textit{Number of resource blocks in resource grid} \\ N_{sc} & \textit{Number of subcarriers in resource blocks} \\ N_{sym} & \textit{Number of symbols per slot} \\ N_{slot} & \textit{Number of slots per subframe} \end{cases}$$

The total resource grid size is the product of the number of rows (total number of subcarriers) and number of columns (total number of OFDM symbols per subframe). The total number of subcarriers is the product of the number of resource blocks (N_{rb}) and number of subcarriers per resource block (N_{sc}). The total number of OFDM symbols per subframe is the product of the number of symbols per slot (N_{sym}) and number of slots per subframe (N_{slot}).

$$N_{total} = N_{rb} \cdot N_{sc} \times N_{sym} \cdot N_{slot} \tag{5.12}$$

The number of slots per subframe (N_{slot}) is a constant value of 2. The number of symbols per slot (N_{sym}) depends on whether a normal or an extended CP is used. As throughout this book we will be using a normal CP, the number of symbols per slot will have a constant value of 7. The number of subcarriers per resource block (N_{sc}) also depends on CP type; if we assume a normal CP, it has a constant value of 12. Therefore, the resource grid size completely depends on the number of resource blocks, which is a direct function of the bandwidth.

As discussed in the last section, the resource elements come from six types of data source: user data, CSR, DCI, PSS, SSS, and BCH. Some of these sources are available in all subframes of a frame (user data, CSR, DCI), some are only available in subframes 0 and 5 (PSS and SSS), and some are only available in subframe 0 (BCH). Since the total number of symbols in a resource grid is constant, in each frame we must compute the amount of user data in three different ways:

1. **For subframe 0**: Where all the sources of data are present.
2. **For subframe 5**: Where besides user data, CSR, DCI, PSS, and SSS are present.
3. **All other subframes {1, 2, 3, 4, 6, 7, 8, 9}**: Where besides user data, only CSR and DCI symbols are present.

Figure 5.7 illustrates the relative locations of six different types of data within the resource grid and focuses on six resource blocks in the center of the grid, where PSS, SSS, and BCH are available in select subframes.

5.8.1 CSR Symbols

In addition, CSRs are placed throughout each resource block in each subframe with a specific pattern of time and frequency separations. In the single-antenna configuration, LTE specifies two CSR symbols per resource block in each of the four OFDM symbols {0, 5, 7, 12} in any subframe. In OFDM symbols 0 and 7, the starting indices are the first subcarrier, whereas in symbolks 5 and 12 the starting index is the fourth subcarrier. The separation between two CSR symbols in the frequency domain is six subcarriers. There are a total of $N_{CSR} = 8N_{rb}$ CSR symbols available in the resource grid.

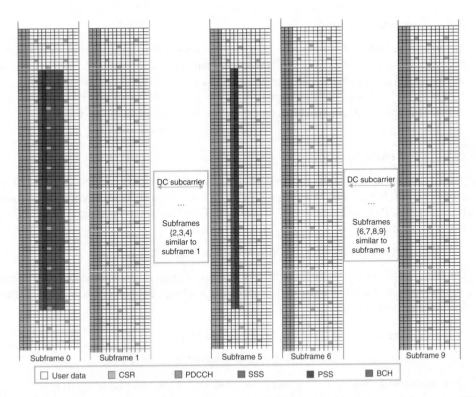

Figure 5.7 LTE resource grid content – focused on eight resource blocks around the center (DC subcarrier) – featuring six different types of data

5.8.2 DCI Symbols

The DCI is placed within the first N OFDM symbols in each subframe, where N is either 1, 2, or 3. The DCI carries the content of the PDCCH, PCFICH (Physical Control Format Indicator Channel), and PHICH (Physical Hybrid ARQ Indicator Channel), and together these occupy all the resource elements of the first and possibly the second and third OFDM symbols in each subframe, with the exception of the CSR data, which are distributed along the first OFDM symbol of each subframe. The size of the DCI per subframe is $N_{DCI} = N_{rb}(10 + 12(N - 1))$. In this chapter we will not generate and fill in the DCI in the resource grid; we will discuss the DCI in some detail in Chapter 7.

5.8.3 BCH Symbols

The PBCH is located within subframe 0 and occupies six central resource blocks from the seventh to the tenth OFDM symbol. Since the seventh OFDM symbol includes CSR symbols, its BCH has a size of only 60 ($72 - 2 \times 6$), whereas in the next three symbols the size is 72. The total BCH size for the whole frame is N_{BCH} $60 + 3 \times 72 = 276$.

5.8.4 Synchronization Symbols

Both the PSS and the SSS are placed within the six resource blocks centered on the DC sub-carrier. The PSS occupies the sixth OFDM symbol and the SSS occupies the fifth symbol in subframes 0 and 5. Since there is no overlap with CSR signals in these symbols, the total number for each of the synchronization signals is $N_{PSS} = N_{SSS} = 72$ per subframe, and since two subframes per frame contain synchronization signals, the total is 144 for the frame.

5.8.5 User-Data Symbols

The total amount of data in the resource grid depends on the number of resource blocks or essentially on the bandwidth. The resource elements come from six types of data source (user data, CSR, DCI, PSS, SSS, and BCH). Therefore, if the bandwidth is constant the resource grid size is constant and is the sum of all these constituents:

$$N_{total} = N_{user\ data} + N_{CSR} + N_{DCI} + N_{PSS} + N_{SSS} + N_{BCH} \tag{5.13}$$

The presence or absence of BCH or synchronization signals in a subframe depends on the subframe index. As a result, the size of the user data in a subframe also depends on the subframe index in the following way:

1. **For subframe 0**: Where all sources of data are present,

$$N_{user\ data} = N_{total} - (N_{CSR} + N_{DCI} + N_{PSS} + N_{SSS} + N_{BCH}) \tag{5.14}$$

2. **For subframe 5**: Where besides user data, CSR, DCI, PSS, and SSS are present,

$$N_{user\ data} = N_{total} - (N_{CSR} + N_{DCI} + N_{PSS} + N_{SSS}) \tag{5.15}$$

3. **For all other subframes {1, 2, 3, 4, 6, 7, 8, 9}**: Where besides user data, only CSR and DCI symbols are present,

$$N_{user\ data} = N_{total} - (N_{CSR} + N_{DCI}) \tag{5.16}$$

The following MATLAB function performs calculations highlighted above and sets some of the parameters of the PDSCH. The function takes three parameters as its input argument: the channel bandwidth (*chanBW*), the number of OFDM symbols dedicated to the control channel in each subframe (*contReg*), and the modulation type used (*modType*). It computes many parameters used in PDSCH processing, including the details of the resource grid.

Algorithm

MATLAB function

```
function p= prmsPDSCH(chanBW, contReg, modType, varargin)
%  Returns parameter structures for LTE PDSCH simulation.
%
%  Assumes a FDD, normal cyclic prefix, full-bandwidth, single-user
%  SISO or SIMO downlink transmission.
%% PDSCH parameters
```

```
switch chanBW
   case 1 % 1.4 MHz
      BW = 1.4e6; N = 128; cpLen0 = 10; cpLenR = 9;
      Nrb = 6; chanSRate = 1.92e6;
   case 2 % 3 MHz
      BW = 3e6; N = 256; cpLen0 = 20; cpLenR = 18;
      Nrb = 15; chanSRate = 3.84e6;
   case 3 % 5 MHz
      BW = 5e6; N = 512; cpLen0 = 40; cpLenR = 36;
      Nrb = 25; chanSRate = 7.68e6;
   case 4 % 10 MHz
      BW = 10e6; N = 1024; cpLen0 = 80; cpLenR = 72;
      Nrb = 50; chanSRate = 15.36e6;
   case 5 % 15 MHz
      BW = 15e6; N = 1536; cpLen0 = 120; cpLenR = 108;
      Nrb = 75; chanSRate = 23.04e6;
   case 6 % 20 MHz
      BW = 20e6; N = 2048; cpLen0 = 160; cpLenR = 144;
      Nrb = 100; chanSRate = 30.72e6;
end
p.BW = BW;              % Channel bandwidth
p.N = N;               % NFFT
p.cpLen0 = cpLen0;        % Cyclic prefix length for 1st symbol
p.cpLenR = cpLenR;         % Cyclic prefix length for remaining
p.Nrb = Nrb;              % Number of resource blocks
p.chanSRate = chanSRate;   % Channel sampling rate
p.contReg = contReg;
if nargin > 3, numTx=varargin{4};else numTx=1;end
if nargin > 4, numRx=varargin{5};else numRx=1;end
p.numTx = numTx;
p.numRx = numRx;
p.numLayers = 1;
p.numCodeWords = 1;
% For Normal cyclic prefix, FDD mode
p.deltaF = 15e3;   % subcarrier spacing
p.Nrb_sc = 12;     % no. of subcarriers per resource block
p.Ndl_symb = 7;    % no. of OFDM symbols in a slot
% Actual PDSCH bits calculation - accounting for PDCCH, PBCH, PSS, SSS
numResources = (p.Nrb*p.Nrb_sc)*(p.Ndl_symb*2);
numCSRRE = 2*2*2 * p.Nrb;          % CSR, RE per OFDMsym/slot/subframe per RB
numContRE = (10 + 12*(p.contReg-1))*p.Nrb;
numBCHRE = 60+72+72+72;            % removing the CSR present in 1st symbol
numSSSRE=72;
numPSSRE=72;
numDataRE=zeros(3,1);
% Account for BCH, PSS, SSS and PDCCH for subframe 0
numDataRE(1)=numResources-numCSRRE-numContRE-numSSSRE - numPSSRE-
numBCHRE;
% Account for PSS, SSS and PDCCH for subframe 5
```

```
numDataRE(2)=numResources-numCSRRE-numContRE-numSSSRE - numPSSRE;
% Account for PDCCH only in all other subframes
numDataRE(3)=numResources-numCSRRE-numContRE;
% Maximum data resources - with no extra overheads (only CSR + data)
p.numResources=numResources;
p.numCSRResources =  numCSRRE;
p.numContRE = numContRE;
p.numBCHRE = numBCHRE;
p.numSSSRE=numSSSRE;
p.numPSSRE=numPSSRE;
p.numDataRE=numDataRE;
p.numDataResources = p.numResources - p.numCSRResources;
% Modulation types , bits per symbol, number of layers per codeword
Qm = 2 * modType;
p.Qm = Qm;
p.numLayPerCW = p.numLayers/p.numCodeWords;
% Maximum data bits - with no extra overheads (only CSR + data)
p.numDataBits = p.numDataResources*Qm*p.numLayPerCW;
numPDSCHBits =numDataRE*Qm*p.numLayPerCW;
p.numPDSCHBits = numPDSCHBits;
p.maxG = max(numPDSCHBits);
```

In this chapter we omit the generation of DCI, BCH, and synchronization signals. Instead we focus on computing the content of CSR and user-data signals to fill up the resource grid and use OFDM transmission to model transmission mode 1 of the LTE standard.

5.9   Generating Reference Signals

To ensure that the transmitter and receiver can generate the same CSR reference sequence, LTE uses a Gold sequence that is initialized based on parameters that are available at both the transmitter and the receiver. These parameters include the cell identity number ($NcellID$), the subframe index (nS), the slot index (i), and the index of OFDM symbols containing the CSR in the slot ($lIdx$).

The function has two input arguments: the subframe index (nS) and the number of transmit antennas ($numTx$). As we are only modeling the single-antenna case in this chapter, the second parameter is set to a value of 1. As a convenience, we will use the same function in the following chapter, where multiple antennas are present. Based on the Gold sequence and for all available antenna ports, the function will generate the number of CSR values needed to provide a channel estimation. The output variable y is a matrix whose size is equal to the product of the number of rows and the number of columns. The number of rows is the maximum number of CSR signals in the resource grid and the number of columns is the number of transmit antennas. The following MATLAB function shows how each element of the CSR signal is generated.

Algorithm

MATLAB function

```
function y = CSRgenerator(nS, numTx)
%  LTE Cell-Specific Reference signal generation.
%  Section 6.10.1 of 3GPP TS 36.211 v10.0.0.
%  Generate the whole set per OFDM symbol, for 2 OFDM symbols per slot,
%  for 2 slots per subframe, per antenna port (numTx).
%  This fcn accounts for the per antenna port sequence generation, while
%  the actual mapping to resource elements is done in the Resource mapper.
%#codegen
persistent hSeqGen;
persistent hInt2Bit;
% Assumed parameters
NcellID = 0;        % One of possible 504 values
Ncp = 1;            % for normal CP, or 0 for Extended CP
NmaxDL_RB = 100;    % largest downlink bandwidth configuration, in resource blocks
y = complex(zeros(NmaxDL_RB*2, 2, 2, numTx));
l = [0; 4];      % OFDM symbol idx in a slot for common first antenna port
% Buffer for sequence per OFDM symbol
seq = zeros(size(y,1)*2, 1); % *2 for complex outputs
if isempty(hSeqGen)
    hSeqGen = comm.GoldSequence('FirstPolynomial',[1 zeros(1, 27) 1 0 0 1],...
                    'FirstInitialConditions', [zeros(1, 30) 1], ...
                    'SecondPolynomial', [1 zeros(1, 27) 1 1 1 1],...
                    'SecondInitialConditionsSource', 'Input port',...
                    'Shift', 1600,...
                    'SamplesPerFrame', length(seq));
    hInt2Bit = comm.IntegerToBit('BitsPerInteger', 31);
end
% Generate the common first antenna port sequences
for i = 1:2 % slot wise
    for lIdx = 1:2 % symbol wise
        c_init = (2^10)*(7*((nS+i-1)+1)+l(lIdx)+1)*(2*NcellID+1) + 2*NcellID + Ncp;
        % Convert to binary vector
        iniStates = step(hInt2Bit, c_init);
        % Scrambling sequence - as per Section 7.2, 36.211
        seq = step(hSeqGen, iniStates);
        % Store the common first antenna port sequences
        y(:, lIdx, i, 1) = (1/sqrt(2))*complex(1-2.*seq(1:2:end), 1-2.*seq(2:2:end));
    end
end
% Copy the duplicate set for second antenna port, if exists
if (numTx>1)
    y(:, :, :, 2) = y(:, :, :, 1);
end
```

```
% Also generate the sequence for l=1 index for numTx = 4
if (numTx>2)
    for i = 1:2 % slot wise
        % l = 1
        c_init = (2^10)*(7*((nS+i-1)+1)+1+1)*(2*NcellID+1) + 2*NcellID + Ncp;
        % Convert to binary vector
        iniStates = step(hInt2Bit, c_init);
        % Scrambling sequence - as per Section 7.2, 36.211
        seq = step(hSeqGen, iniStates);
        % Store the third antenna port sequences
        y(:, 1, i, 3) = (1/sqrt(2))*complex(1-2.*seq(1:2:end), 1-2.*seq(2:2:end));
    end
    % Copy the duplicate set for fourth antenna port
    y(:, 1, :, 4) = y(:, 1, :, 3);
end
```

5.10   Resource Element Mapping

In this section we detail the resource element mapping that places the components of the resource grid in the locations specified in the standard. Mapping is performed essentially by creating indices to the resource grid matrix and placing various information types within the grid. The illustrations of three different types of resource block given in Figures 5.8–5.10 help visualize formulations for these indices. Depending on which subframe is in use, we populate the BCH, PSS, and SSS in either subframe 0 or subframe 5 of the six central resource

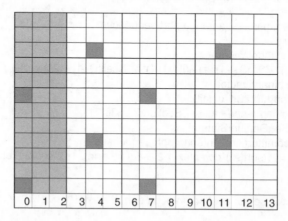

Figure 5.8   Resource element mapping: all resource blocks in subframes 1, 2, 3, 4, 6, 7, 8, and 9 + noncentral resource blocks of subframes 0 and 5. Includes DCI, CSR, and user data

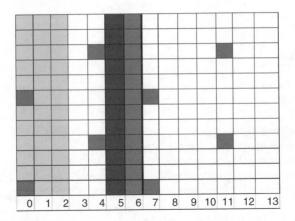

Figure 5.9 Resource element mapping: central resource blocks of subframe 5. Includes PSS, SSS, DCI, CSR, and user data

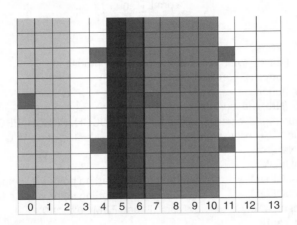

Figure 5.10 Resource element mapping: central resource blocks of subframe 0. Includes BCH, PSS, SSS, DCI, CSR, and user data

blocks around the DC subcarrier. The CSRs are placed in symbols 0 and 5 of each slot, with a frequency-domain separation of six subcarriers.

The following MATLAB function shows resource element mapping. Since MATLAB uses a 1-based indexing notation, we generate indices for various elements in the matrix starting from 1 instead of 0, as specified by the standard. The function takes as input the user data (*in*), CSR signal (*csr*), subframe index (*nS*), and parameters of the PDSCH captured in a structure called *prmLTE*. Depending on the availability of BCH, SSS, PSS, and DCI, the function may take on additional inputs. The output variable *y* is the resource grid matrix. The 2D grid matrix has a number of rows equal to the number of subcarriers and number of columns, totalling 14 (two slots each containing seven OFDM symbols).

Algorithm

MATLAB function

```
function y = REmapper_1Tx(in, csr, nS, prmLTE, varargin)
%#codegen
switch nargin
   case 4, pdcch=[];pss=[];sss=[];bch=[];
   case 5, pdcch=varargin{1};pss=[];sss=[];bch=[];
   case 6, pdcch=varargin{1};pss=varargin{2};sss=[];bch=[];
   case 7, pdcch=varargin{1};pss=varargin{2};sss=varargin{3};bch=[];
   case 8, pdcch=varargin{1};pss=varargin{2};sss=varargin{3};bch=varargin{4};
   otherwise
      error('REMapper has 4 to 8 arguments!');
end
% NcellID = 0;                        % One of possible 504 values
% numTx = 1;                          % prmLTE.numTx;
% Get input params
Nrb = prmLTE.Nrb;                     % either of {6, }
Nrb_sc = prmLTE.Nrb_sc;               % 12 for normal mode
Ndl_symb = prmLTE.Ndl_symb;           % 7   for normal mode
numContSymb   = prmLTE.contReg;  % either {1, 2, 3}
% Initialize output buffer
y = complex(zeros(Nrb*Nrb_sc, Ndl_symb*2));
%% Specify resource grid location indices for CSR, PDCCH, PDSCH, PBCH, PSS, SSS
%% 1st: Indices for CSR pilot symbols
lenOFDM = Nrb*Nrb_sc;
idx      = 1:lenOFDM;
idx_csr0 = 1:6:lenOFDM;               % More general starting point = 1+mod(NcellID, 6);
idx_csr4 = 4:6:lenOFDM;               % More general starting point = 1+mod(3+NcellID, 6);
idx_csr   =[idx_csr0, 4*lenOFDM+idx_csr4, 7*lenOFDM+idx_csr0,
11*lenOFDM+idx_csr4];
%% 2nd: Indices for PDCCH control data symbols
ContREs=numContSymb*lenOFDM;
idx_dci=1:ContREs;
idx_pdcch = ExpungeFrom(idx_dci,idx_csr0);
%% 3rd: Indices for PDSCH and PDSCH data in OFDM symbols where pilots
are present
idx_data0= ExpungeFrom(idx,idx_csr0);
idx_data4 = ExpungeFrom(idx,idx_csr4);
%% Handle 3 types of subframes differently
switch nS
   %% 4th: Indices for BCH, PSS, SSS are only found in specific subframes 0 and 5
   % These symbols share the same 6 center sub-carrier locations (idx_ctr)
   % and differ in OFDM symbol number.
   case 0 % Subframe 0
      % PBCH, PSS, SSS are available + CSR, PDCCH, PDSCH
      idx_6rbs = (1:72);
```

```
    idx_ctr = 0.5* lenOFDM - 36 + idx_6rbs ;
    idx_SSS  = 5* lenOFDM + idx_ctr;
    idx_PSS  = 6* lenOFDM + idx_ctr;
    idx_ctr0 = ExpungeFrom(idx_ctr,idx_csr0);
    idx_bch=[7*lenOFDM + idx_ctr0, 8*lenOFDM + idx_ctr, 9*lenOFDM + idx_ctr,
10*lenOFDM + idx_ctr];
    idx_data5  = ExpungeFrom(idx,idx_ctr);
    idx_data7 = ExpungeFrom(idx_data0,idx_ctr);
    idx_data  = [ContREs+1:4*lenOFDM,  4*lenOFDM+idx_data4, ...
        5*lenOFDM+idx_data5, 6*lenOFDM+idx_data5, 7*lenOFDM+idx_data7,
8*lenOFDM+idx_data5, ...
        9*lenOFDM+idx_data5, 10*lenOFDM+idx_data5, 11*lenOFDM+idx_data4, ...
        12*lenOFDM+1:14*lenOFDM];
    y(idx_csr)=csr(:);              % Insert Cell-Specific Reference signal (CSR) = pilots
    y(idx_data)=in;                 % Insert Physical Downlink Shared Channel
(PDSCH) = user data
    if ˜isempty(pdcch), y(idx_pdcch)=pdcch;end
% Insert Physical Downlink Control Channel (PDCCH)
    if ˜isempty(pss), y(idx_PSS)=pss;end % Insert Primary Synchronization
Signal (PSS)
    if ˜isempty(sss), y(idx_SSS)=sss;end
% Insert Secondary Synchronization Signal (SSS)
    if ˜isempty(bch), y(idx_bch)=bch;end % Insert Broadcast Channel data (BCH)

  case 10 % Subframe 5
    % PSS, SSS are available + CSR, PDCCH, PDSCH
    % Primary and Secondary synchronization signals in OFDM symbols 5 and 6
    idx_6rbs = (1:72);
    idx_ctr = 0.5* lenOFDM - 36 + idx_6rbs ;
    idx_SSS  = 5* lenOFDM + idx_ctr;
    idx_PSS  = 6* lenOFDM + idx_ctr;
    idx_data5 = ExpungeFrom(idx,idx_ctr);
    idx_data  = [ContREs+1:4*lenOFDM, 4*lenOFDM+idx_data4,
5*lenOFDM+idx_data5, 6*lenOFDM+idx_data5, ...
        7*lenOFDM+idx_data0, 8*lenOFDM+1:11*lenOFDM,  11*lenOFDM+idx_data4, ...
        12*lenOFDM+1:14*lenOFDM];
    y(idx_csr)=csr(:);              % Insert Cell-Specific Reference signal (CSR) = pilots
    y(idx_data)=in;                 % Insert Physical Downlink Shared Channel
(PDSCH) = user data
    if ˜isempty(pdcch), y(idx_pdcch)=pdcch;end
% Insert Physical Downlink Control Channel (PDCCH)
    if ˜isempty(pss), y(idx_PSS)=pss;end % Insert Primary Synchronization Signal (PSS)
    if ˜isempty(sss), y(idx_SSS)=sss;end
% Insert Secondary Synchronization Signal (SSS)

  otherwise % other subframes
    % Only CSR, PDCCH, PDSCH
    idx_data = [ContREs+1:4*lenOFDM, 4*lenOFDM+idx_data4, ...
        5*lenOFDM+1:7*lenOFDM, ...
```

```
          7*lenOFDM+idx_data0, ...
          8*lenOFDM+1:11*lenOFDM, ...
          11*lenOFDM+idx_data4, ...
          12*lenOFDM+1:14*lenOFDM];
       y(idx_csr)=csr(:);               % Insert Cell-Specific Reference signal (CSR) = pilots
       y(idx_data)=in;                  % Insert Physical Downlink Shared Channel
(PDSCH) = user data
          if ~isempty(pdcch), y(idx_pdcch)=pdcch;end
% Insert Physical Downlink Control Channel (PDCCH)
end
end
```

5.11   OFDM Signal Generation

OFDM signal generation operates on the resource grid. It takes the OFDM symbols (columns of data in the resource grid matrix) one by one and performs an IFFT operation followed by CP addition to generate the OFDM modulated signal. The following MATLAB function shows how, prior to the IFFT operation, data are packed into the FFT buffer and reordered to exclude the DC subcarrier. Following the IFFT operation we scale the output. The CP addition prepends N last samples of the IFFT output to the beginning of the buffer. The value of the N in the first OFDM symbol is different from that in all other OFDM symbols. The inputs to the function are the resource grid (*in*) and the structure containing the parameters of the PDSCH (*prmLTE*). CPs have different lengths across symbols in a slot. The length of the CP in the first OFDM symbol of each slot (*cpLen0*) is slightly larger than the CP values in the remaining six symbols of the slot (*cpLenR*). This difference is taken into account in the for loop that computes the output signal as it serializes and appends the length of each OFDM modulated signal to the output vector per subframe [3].

The output of the function is a 2D matrix: the size of the first dimension is the output per subframe and the second dimension is the number of antenna ports. Since in this chapter we focus on the single-antenna case, the output will be a column vector with a second dimension equal to one. We do not have to modify this function when we introduce MIMO techniques in the next chapter, as it serves single-channel and multichannel OFDM signal-generation cases equally.

Algorithm

MATLAB function

```
function y = OFDMTx(in, prmLTE)
%#codegen
persistent hIFFT;
if isempty(hIFFT)
    hIFFT = dsp.IFFT;
end
```

```
[len, numSymb, numLayers] = size(in);
% N assumes 15KHz subcarrier spacing
N = prmLTE.N;
cpLen0 = prmLTE.cpLen0;
cpLenR = prmLTE.cpLenR;
slotLen = (N*7 + cpLen0 + cpLenR*6);
subframeLen = slotLen*2;
tmp = complex(zeros(N, numSymb, numLayers));
% Pack data, add DC, and reorder
tmp(N/2-len/2+1:N/2, :, :) = in(1:len/2, :, :);
tmp(N/2+2:N/2+1+len/2, :, :) = in(len/2+1:len, :, :);
tmp = [tmp(N/2+1:N, :, :); tmp(1:N/2, :, :)];
% IFFT processing
x = step(hIFFT, tmp);
x = x.*(N/sqrt(len));
% Add cyclic prefix per OFDM symbol per antenna port
% and serialize over the subframe (equal to 2 slots)
% For a subframe of data
y = complex(zeros(subframeLen, numLayers));
for j = 1:2 % Over the two slots
    % First OFDM symbol
    y((j-1)*slotLen+(1:cpLen0), :) = x((N-cpLen0+1):N, (j-1)*7+1, :);
    y((j-1)*slotLen+cpLen0+(1:N), :) = x(1:N, (j-1)*7+1, :);

    % Next 6 OFDM symbols
    for k = 1:6
        y((j-1)*slotLen+cpLen0+k*N+(k-1)*cpLenR+(1:cpLenR), :) = x(N-cpLenR+1:N,
(j-1)*7+k+1, :);
        y((j-1)*slotLen+cpLen0+k*N+k*cpLenR+(1:N), :) = x(1:N, (j-1)*7+k+1, :);
    end
end
```

5.12   Channel Modeling

The following MATLAB function shows the channel model operating on the OFDM signal. It is derived from the generic SISO channel model we developed earlier in this chapter. The function takes as input the generated OFDM signal (*in*), the structure containing the parameters of the PDSCH (*prmLTE*), and another structure containing parameters of the channel model (*prmMdl*). Based on input parameters, it applies either a frequency-flat or frequency-selective channel to the input signal. The output of the channel model (*y*) is the signal that arrives at the receiver. The second output (*yPg*) is a matrix containing the channel-path gains of the underlying fading process. This signal can be used to estimate the "ideal" channel response. We will present details of ideal channel responses and comparisons with estimated responses in subsequent chapters.

Algorithm

MATLAB function

```
function [y, yPg] = MIMOFadingChan(in, prmLTE, prmMdl)
% MIMOFadingChan
%#codegen
% Get simulation params
numTx      = prmLTE.numTx;
numRx      = prmLTE.numRx;
chanMdl    = prmMdl.chanMdl;
chanSRate  = prmLTE.chanSRate;
corrLvl    = prmMdl.corrLevel;
switch chanMdl
  case 'flat-low-mobility',
      PathDelays = 0*(1/chanSRate);
      PathGains  = 0;
      Doppler=0;
      ChannelType =1;
  case 'flat-high-mobility',
      PathDelays = 0*(1/chanSRate);
      PathGains  = 0;
      Doppler=70;
      ChannelType =1;
  case 'frequency-selective-low-mobility',
      PathDelays = [0 10 20 30 100]*(1/chanSRate);
      PathGains  = [0 -3 -6 -8 -17.2];
      Doppler=0;
      ChannelType =1;
  case 'frequency-selective-high-mobility',
      PathDelays = [0 10 20 30 100]*(1/chanSRate);
      PathGains  = [0 -3 -6 -8 -17.2];
      Doppler=70;
      ChannelType =1;
  case 'EPA 0Hz'
      PathDelays = [0 30 70 90 110 190 410]*1e-9;
      PathGains  = [0 -1 -2 -3 -8 -17.2 -20.8];
      Doppler=0;
      ChannelType =1;
  otherwise
      ChannelType =2;
      AntConfig=char([48+numTx,'x',48+numRx]);
end
% Initialize objects
persistent chanObj;
if isempty(chanObj)
  if ChannelType ==1
```

```
        chanObj = comm.MIMOChannel('SampleRate', chanSRate, ...
           'MaximumDopplerShift', Doppler, ...
           'PathDelays', PathDelays,...
           'AveragePathGains', PathGains,...
           'RandomStream', 'mt19937ar with seed',...
           'Seed', 100,...
           'NumTransmitAntennas', numTx,...
           'TransmitCorrelationMatrix', eye(numTx),...
           'NumReceiveAntennas', numRx,...
           'ReceiveCorrelationMatrix', eye(numRx),...
           'PathGainsOutputPort', true,...
           'NormalizePathGains', true,...
           'NormalizeChannelOutputs', true);
    else
       chanObj = comm.LTEMIMOChannel('SampleRate', chanSRate, ...
           'Profile', chanMdl, ...
           'AntennaConfiguration', AntConfig, ...
           'CorrelationLevel', corrLvl,...
           'RandomStream', 'mt19937ar with seed',...
           'Seed', 100,...
           'PathGainsOutputPort', true);
    end
end
[y, yPg] = step(chanObj, in);
```

Besides the fading channel, the simulation also requires the addition of the AWGN channel. The following function illustrates the AWGN channel used throughout this book. It applies an AWGN to its first input signal (u), based on the value of the noise power (*noiseVar*) given as its second input argument.

Algorithm

MATLAB function

```
function y = AWGNChannel(u, noiseVar )
%% Initialization
persistent AWGN
if isempty(AWGN)
   AWGN        = comm.AWGNChannel('NoiseMethod', 'Variance', ...
      'VarianceSource', 'Input port');
end
y = step(AWGN, u, noiseVar);
end
```

5.13   OFDM Receiver

At the OFDM receiver we perform the inverse operations to those at the transmitter. First the CP is removed and an FFT operation is performed to recover the received data and reference signals at each subcarrier. Different FFT lengths are used based on the channel bandwidth. Through a combination of scaling, reordering, DC subcarrier removal, and unpacking, the received modulated symbols are placed in the same order in which they were placed into the resource grid at the transmitter. The following MATLAB function shows the sequence of operations performed in the OFDM receiver. The inputs are the receiver input signal (*in*) and the structure containing the parameters of the PDSCH (*prmLTE*). The output is the resource grid recovered at the receiver.

Algorithm

MATLAB function

```
function y = OFDMRx(in, prmLTE)
%#codegen
persistent hFFT;
if isempty(hFFT)
    hFFT = dsp.FFT;
end
% For a subframe of data
numDataTones = prmLTE.Nrb*prmLTE.Nrb_sc;
numSymb = prmLTE.Ndl_symb*2;
[~, numLayers] = size(in);
% N assumes 15KHz subcarrier spacing, else N = 4096
N = prmLTE.N;
cpLen0 = prmLTE.cpLen0;
cpLenR = prmLTE.cpLenR;
slotLen = (N*7 + cpLen0 + cpLenR*6);
tmp = complex(zeros(N, numSymb, numLayers));
% Remove CP - unequal lengths over a slot
for j = 1:2 % over two slots
    % First OFDM symbol
    tmp(:, (j-1)*7+1, :) = in((j-1)*slotLen+cpLen0 + (1:N), :);

    % Next 6 OFDM symbols
    for k = 1:6
        tmp(:, (j-1)*7+k+1, :) = in((j-1)*slotLen+cpLen0+k*N+k*cpLenR + (1:N), :);
    end
end
% FFT processing
x = step(hFFT, tmp);
x =  x./(N/sqrt(numDataTones));
% For a subframe of data
y = complex(zeros(numDataTones, numSymb, numLayers));
% Reorder, remove DC, Unpack data
```

```
x = [x(N/2+1:N, :, :); x(1:N/2, :, :)];
y(1:(numDataTones/2), :, :) = x(N/2-numDataTones/2+1:N/2, :, :);
y(numDataTones/2+1:numDataTones, :, :) = x(N/2+2:N/2+1+numDataTones/2, :, :);
end
```

5.14   Resource Element Demapping

Resource element demapping inverts the operations of resource grid mapping. The following MATLAB function illustrates how the reference signal and data are extracted from the recovered resource grid at the receiver. The function has three input arguments: the received resource grid (*in*), the index of the subframe (*nS*), and the PDSCH parameter set. The function outputs extracted user data (*data*), the indices to the user data (*idx_data*), the CSR signals (*csr*), and optionally the DCI (*pdcch*), PSS and SSS (*pss, sss*), and BCH (*bch*) signals. As different subframes contain different content, the second input subframe index parameter (*nS*) enables the function to separate the correct data. The same algorithm used in the resource-mapping function is used here to generate indices in the demapping function.

Algorithm

MATLAB function

```
function [data, csr, idx_data, pdcch, pss, sss, bch] = REdemapper_1Tx(in, nS, prmLTE)
%#codegen
% NcellID = 0;                        % One of possible 504 values
% numTx = 1;                          % prmLTE.numTx;
% Get input params
Nrb = prmLTE.Nrb;                     % either of {6, }
Nrb_sc = prmLTE.Nrb_sc;               % 12 for normal mode
numContSymb   = prmLTE.contReg; % either {1, 2, 3}
%% Specify resource grid location indices for CSR, PDCCH, PDSCH, PBCH, PSS, SSS
%% 1st: Indices for CSR pilot symbols
lenOFDM = Nrb*Nrb_sc;
idx        = 1:lenOFDM;
idx_csr0   = 1:6:lenOFDM;            % More general starting point = 1+mod(NcellID, 6);
idx_csr4   = 4:6:lenOFDM;            % More general starting point = 1+mod(3+NcellID, 6);
idx_csr   =[idx_csr0, 4*lenOFDM+idx_csr4, 7*lenOFDM+idx_csr0, 11*lenOFDM+idx_csr4];
%% 2nd: Indices for PDCCH control data symbols
ContREs=numContSymb*lenOFDM;
idx_dci=1:ContREs;
idx_pdcch = ExpungeFrom(idx_dci,idx_csr0);
%% 3rd: Indices for PDSCH and PDSCH data in OFDM symbols where pilots are present
idx_data0= ExpungeFrom(idx,idx_csr0);
idx_data4 = ExpungeFrom(idx,idx_csr4);
%% Handle 3 types of subframes differently
pss=complex(zeros(72,1));
sss=complex(zeros(72,1));
```

```
bch=complex(zeros(72*4,1));
switch nS
    %% 4th: Indices for BCH, PSS, SSS are only found in specific subframes 0 and 5
    % These symbols share the same 6 center sub-carrier locations (idx_ctr)
    % and differ in OFDM symbol number.
    case 0 % Subframe 0
        % PBCH, PSS, SSS are available + CSR, PDCCH, PDSCH
        idx_6rbs = (1:72);
        idx_ctr = 0.5* lenOFDM - 36 + idx_6rbs ;
        idx_SSS  = 5* lenOFDM + idx_ctr;
        idx_PSS  = 6* lenOFDM + idx_ctr;
        idx_ctr0 = ExpungeFrom(idx_ctr,idx_csr0);
        idx_bch=[7*lenOFDM + idx_ctr0, 8*lenOFDM + idx_ctr, 9*lenOFDM + idx_ctr,
10*lenOFDM + idx_ctr];
        idx_data5  = ExpungeFrom(idx,idx_ctr);
        idx_data7 = ExpungeFrom(idx_data0,idx_ctr);
        idx_data  = [ContREs+1:4*lenOFDM,  4*lenOFDM+idx_data4, ...
            5*lenOFDM+idx_data5, 6*lenOFDM+idx_data5,  7*lenOFDM+idx_data7,
8*lenOFDM+idx_data5, ...
            9*lenOFDM+idx_data5, 10*lenOFDM+idx_data5, 11*lenOFDM+idx_data4, ...
            12*lenOFDM+1:14*lenOFDM];
        pss=in(idx_PSS).';   % Primary Synchronization Signal (PSS)
        sss=in(idx_SSS).';    % Secondary Synchronization Signal (SSS)
        bch=in(idx_bch).';    % Broadcast Channel data (BCH)

    case 10 % Subframe 5
        % PSS, SSS are available + CSR, PDCCH, PDSCH
        % Primary and Secondary synchronization signals in OFDM symbols 5 and 6
        idx_6rbs = (1:72);
        idx_ctr = 0.5* lenOFDM - 36 + idx_6rbs ;
        idx_SSS = 5* lenOFDM + idx_ctr;
        idx_PSS = 6* lenOFDM + idx_ctr;
        idx_data5 = ExpungeFrom(idx,idx_ctr);
        idx_data  = [ContREs+1:4*lenOFDM, 4*lenOFDM+idx_data4,
5*lenOFDM+idx_data5, 6*lenOFDM+idx_data5, ...
            7*lenOFDM+idx_data0, 8*lenOFDM+1:11*lenOFDM,  11*lenOFDM+idx_data4, ...
            12*lenOFDM+1:14*lenOFDM];
        pss=in(idx_PSS).';   % Primary Synchronization Signal (PSS)
        sss=in(idx_SSS).';    % Secondary Synchronization Signal (SSS)

    otherwise % other subframes
        % Only CSR, PDCCH, PDSCH
        idx_data = [ContREs+1:4*lenOFDM, 4*lenOFDM+idx_data4, ...
            5*lenOFDM+1:7*lenOFDM, ...
            7*lenOFDM+idx_data0, ...
            8*lenOFDM+1:11*lenOFDM, ...
            11*lenOFDM+idx_data4, ...
            12*lenOFDM+1:14*lenOFDM];
end
```

```
data=in(idx_data).';      % Physical Downlink Shared Channel (PDSCH) = user data
csr=in(idx_csr).';        % Cell-Specific Reference signal (CSR) = pilots
pdcch = in(idx_pdcch).';  % Physical Downlink Control Channel (PDCCH)
end
```

5.15   Channel Estimation

Channel estimation is performed by examining known reference symbols, also referred to as pilots, inserted at regular intervals within the OFDM time–frequency grid. Using known reference symbols, the receiver can estimate the channel response at the subcarriers where the reference symbols were transmitted. The reference symbols should have a sufficiently high density in both the time and the frequency domains. If so, with appropriate expansion operations we can provide estimates for the entire time–frequency grid.

The following MATLAB function performs channel estimation for a single-antenna transmission. The inputs to the function are the structure containing the parameters of the PDSCH (*prmLTE*), the received resource grid (*Rx*), the CSR (*Ref*), and the bandwidth expansion mode (*Mode*). After reshaping the received version of the resource grid, the received signals are aligned with the corresponding pilot elements stored in the CSR. We then compute an estimate of the channel-response matrix (*hD*) by simply dividing the received pilots by the transmitted reference signals. Following computation of the channel-response matrix over the resource elements that align with CSR signals, we perform a full-bandwidth expansion. Based on a subset of reference signals in the resource grid, we perform expansion by averaging or interpolating to generate the channel-response estimate for the entire resource grid; that is, at each subcarrier and each OFDM symbol in a subframe.

Algorithm

MATLAB function

```
function hD = ChanEstimate_1Tx(prmLTE, Rx, Ref, Mode)
%#codegen
Nrb       = prmLTE.Nrb;      % Number of resource blocks
Nrb_sc    = prmLTE.Nrb_sc;         % 12 for normal mode
Ndl_symb  = prmLTE.Ndl_symb;       % 7   for normal mode
% Assume same number of Tx and Rx antennas = 1
% Initialize output buffer
hD = complex(zeros(Nrb*Nrb_sc, Ndl_symb*2));
% Estimate channel based on CSR - per antenna port
csrRx = reshape(Rx, numel(Rx)/4, 4); % Align received pilots with reference pilots
hp    = csrRx./Ref;          % Just divide received pilot by reference pilot
% to obtain channel response at pilot locations
% Now use some form of averaging/interpolation/repeating to
% compute channel response for the whole grid
% Choose one of 3 estimation methods "average" or "interpolate" or "hybrid"
switch Mode
    case 'average'
```

```
    hD=gridResponse_averageSubframe(hp, Nrb, Nrb_sc, Ndl_symb);
  case 'interpolate'
    hD=gridResponse_interpolate(hp, Nrb, Nrb_sc, Ndl_symb);
  otherwise
    error('Choose the right mode for function ChanEstimate.');
end
end
```

Typical interpolation algorithms involve interpolation between subcarriers in the frequency domain in OFDM symbols that contain CSR signals (subframes 0, 5, 7, and 12). Having computed the channel response over all subcarriers of these particular symbols, we can interpolate in time to find the channel response across the whole grid. The following MATLAB function (*gridResponse_interpolate*) performs this type of expansion algorithm based on interpolation.

Algorithm

MATLAB function

```
function hD=gridResponse_interpolate(hp, Nrb, Nrb_sc, Ndl_symb)
% Interpolate among subcarriers in each OFDM symbol
% containing CSR (Symbols 1,5,8,12)
% The interpolation assumes NCellID = 0.
% Then interpolate between OFDM symbols
hD = complex(zeros(Nrb*Nrb_sc, Ndl_symb*2));
N=size(hp,2);
Separation=6;
Edges=[0,5;3,2;0,5;3,2];
Symbol=[1,5,8,12];
% First: Compute channel response over all resource elements of OFDM symbols 0,4,7,11
for n=1:N
   Edge=Edges(n,:);
   y = InterpolateCsr(hp(:,n),  Separation, Edge);
   hD(:,Symbol(n))=y;
end
% Second: Interpolate between OFDM symbols {0,4} {4,7}, {7, 11}, {11, 13}
for m=[2, 3, 4, 6, 7]
   alpha=0.25*(m-1);
   beta=1-alpha;
   hD(:,m)   = beta*hD(:,1) + alpha*hD(:,  5);
   hD(:,m+7) =beta*hD(:,8) + alpha*hD(:,12);
end
```

Typical averaging algorithms interpolate between subcarriers in the frequency domain in OFDM symbols that contain CSR signals (subframes 0, 5, 7, and 12). First we combine the CSR signals from the first two OFDM symbols (subframes 0 and 5). Instead of a separation of six subcarriers between CSR signals, this produces a separation of three subcarriers. Then we interpolate the values along the frequency axis. Finally, we apply the same channel response to all of the OFDM symbols of the slot or subframe to find the channel response of the whole grid. The following MATLAB function (*gridResponse_averageSubframe*) performs this type of expansion algorithm based on averaging and interpolation.

Algorithm

MATLAB function

```
function hD=gridResponse_averageSubframe(hp, Nrb, Nrb_sc, Ndl_symb)
% Average over the two same Freq subcarriers, and then interpolate between
% them - get all estimates and then repeat over all columns (symbols).
% The interpolation assumes NCellID = 0.
% Time average two pilots over the slots, then interpolate (F)
% between the 4 averaged values, repeat for all symbols in subframe
h1_a1 = mean([hp(:, 1), hp(:, 3)],2);
h1_a2 = mean([hp(:, 2), hp(:, 4)],2);
h1_a_mat = [h1_a1 h1_a2].';
h1_a = h1_a_mat(:);
h1_all = complex(zeros(length(h1_a)*3,1));
for i = 1:length(h1_a)-1
    delta = (h1_a(i+1)-h1_a(i))/3;
    h1_all((i-1)*3+1) = h1_a(i);
    h1_all((i-1)*3+2) = h1_a(i)+delta;
    h1_all((i-1)*3+3) = h1_a(i)+2*delta;
end
% fill the last three - use the last delta
h1_all(end-2) = h1_a(end);
h1_all(end-1) = h1_a(end)+delta;
h1_all(end) = h1_a(end)+2*delta;
% Compute the channel response over the whole grid by repeating
hD = h1_all(1:Nrb*Nrb_sc, ones(1, Ndl_symb*2));
end
```

5.16 Equalizer Gain Computation

A frequency-domain equalizer computes a gain for application to all received resource elements at each subcarrier. Different algorithms can be used for frequency-domain equalization. The simplest is the ZF algorithm, in which the gain is found as a ratio of the transmitted resource element to the estimated channel at each subcarrier. A more sophisticated algorithm is the MMSE estimation, which relies on more detailed knowledge of the channel time/frequency characteristics and computes the gain as a modified ratio that takes into account the effect of the uncorrelated channel noise. After the equalizer gain is found, the best estimate of the resource element is the product of the received resource element and the equalizer gain. The following MATLAB function implements both a ZF and an MMSE equalizer and lets the user choose between them based on an equalization mode.

Algorithm

MATLAB function

```
function [out, Eq] = Equalizer(in, hD, nVar, EqMode)
%#codegen
switch EqMode
    case 1,
        Eq = ( conj(hD))./((conj(hD).*hD));        % Zero forcing
    case 2,
        Eq = ( conj(hD))./((conj(hD).*hD)+nVar);  % MMSE
    otherwise,
        error('Two equalization mode vaible: Zero forcing or MMSE');
end
out=in.*Eq;
```

5.17   Visualizing the Channel

Visualizing various signals can help us to verify whether an OFDM transmission is implemented properly. In OFDM, each modulated symbol is transmitted on one subcarrier (in frequency) of a single OFDM symbol (in time). This enables us to directly observe the effects of fading on the transmitted symbols before and after channel processing. In the following MATLAB function, we showcase a Spectrum Analyzer System object from the DSP System Toolbox that enables us to efficiently look at the spectrum of the data at the transmitter and the receiver. The function input variables *txSig* and *rxSig* represent the OFDM modulated signals before and after channel modeling, respectively. The input variable *yRec* represents the user data after equalization. By visualizing these three variables with the Spectrum Analyzer we observe the effects of the channel model on the transmitted signal and the effect of channel estimation and equalization on recovery of a best estimate of the transmitted signal in the receiver. We also use a Constellation Diagram System object from the Communications System Toolbox to observe the effect of the fading channel on the modulated symbols before and after equalization.

Algorithm

MATLAB function

```
function zVisualize(prmLTE, txSig, rxSig, yRec, dataRx, csr, nS)
% Constellation Scopes & Spectral Analyzers
persistent hScope1 hScope2 hSpecAnalyzer
if isempty(hSpecAnalyzer)
    % Constellation Diagrams
    hScope1 = comm.ConstellationDiagram('SymbolsToDisplay',...
        prmLTE.numDataResources, 'ShowReferenceConstellation', false,...
        'YLimits', [-2 2], 'XLimits', [-2 2], 'Position', ...
        figposition([5 60 20 25]), 'Name', 'Before Equalizer');
    hScope2 = comm.ConstellationDiagram('SymbolsToDisplay',...
```

```
         prmLTE.numDataResources, 'ShowReferenceConstellation', false,...
         'YLimits', [-2 2], 'XLimits', [-2 2], 'Position', ...
         figposition([6 61 20 25]), 'Name', 'After Equalizer');
      % Spectrum Scope
      hSpecAnalyzer = dsp.SpectrumAnalyzer('SampleRate', prmLTE.chanSRate, ...
         'SpectrumType', 'Power density', 'PowerUnits', 'dBW', ...
         'RBWSource', 'Property', 'RBW', 15000,...
         'FrequencySpan', 'Span and center frequency',...
         'Span', prmLTE.BW, 'CenterFrequency', 0,...
         'FFTLengthSource', 'Property', 'FFTLength', prmLTE.N,...
         'Title', 'Transmitted & Received Signal Spectrum', 'YLimits', [-110 -60],...
         'YLabel', 'PSD');
end
% Update Spectrum scope
% Received signal after equalization
yRecGrid = REmapper_1Tx(yRec, csr, nS, prmLTE);
yRecGridSig = lteOFDMTx(yRecGrid, prmLTE);
% Take certain symbols off a subframe only
step(hSpecAnalyzer, ...
   [SymbSpec(txSig, prmLTE), SymbSpec(rxSig, prmLTE),
SymbSpec(yRecGridSig, prmLTE)]);
% Update Constellation Scope
if (nS~=0 && nS~=10)
   step(hScope1, dataRx(:, 1));
   step(hScope2, yRec(:, 1));
end
end
% Helper function
function y = SymbSpec(in, prmLTE)
N = prmLTE.N;
cpLenR = prmLTE.cpLen0;
y = complex(zeros(N+cpLenR, 1));
% Use the first Tx/Rx antenna of the input for the display
y(:,1) = in(end-(N+cpLenR)+1:end, 1);
end
```

5.18   Downlink Transmission Mode 1

In this section we will put together a model of downlink transmission mode 1 of the LTE standard with the functions we have developed in the last two chapters. Mode 1 is based on a single-antenna transmission. We will build two variants of this mode:

1. **The SISO case**: Where only one antenna is available, both at the transmitter and at the receiver.
2. **The SIMO case**: Where we use a single transmitter antenna but multiple receiver antennas, in order to exploit the benefits of receive diversity.

Throughout this book, each of our PHY signal processing models includes a transmitter, a channel model, and a receiver. In this section, transmitter processing includes both Downlink Shared Channel (DLSCH) and PDSCH operations. Channel modeling involves the combination of a fading channel and an AWGN channel. The receiver inverts the operations of the DLSCH and the PDSCH.

The unit of simulation is a subframe. As user data are generated and processed in every subframe, we keep track of subframe indexing in order to perform appropriate operations at different subframe indices. Incrementation of the subframe index proceeds until a full frame is processed. At this point, the subframe index is reset. This process is repeated for multiple frames until the simulation stopping criteria are met. In simulating both variants of LTE mode 1, the operations are subdivided into two sections:

1. **A MATLAB function**: Contains all the operations in the transmitter, channel model, and receiver for a single subframe of data.
2. **A MATLAB script**: Initializes and sets up all the parameters of DLSCH, PDSCH, and channel model, then iterates through multiple subframes and computes the Bit Error Rate (BER) measures, stopping when a maximum number of errors are found or a maximum number of bits are processed.

5.18.1   The SISO Case

The following MATLAB function contains the operations in the transceiver (transmitter, channel model, and receiver) for the SISO case. The signal processing chain in the transmitter is a combination of DLSCH and PDSCH, as follows:

- Generation of payload data for a single subframe (a transport block).
- DLSCH processing, including: transport block Cyclic Redundancy Check (CRC) attachment, codeblock segmentation and CRC attachment, turbo coding based on a 1/3-rate code, rate matching, and codeblock concatenation to generate a codeword input to the PDSCH.
- PDSCH processing, including: scrambling of codeword bits, modulation of scrambled bits, mapping of complex-valued modulation symbols to the resource elements forming the resource grid on a single antenna port, generation of OFDM signal for transmission.

The channel modeling includes a combination of fading channel and AWGN channel. The receiver operation, which inverts the PDSCH operations, includes the following: the OFDM signal receiver generating the resource grid, resource-element demapping to separate the CSR signal from the user data, channel estimation, and frequency-domain equalization based on the CSR signal and soft-decision demodulation and descrambling.

Finally, the inverse operations of the DLSCH are performed, including: codeblock segmentation, rate dematching, and turbo decoding with an early stopping option based on CRC detection. The receiver output variable *data_out* and the transmitter input transport block variable *dataIn* are provided as the first two output arguments of the function. Alongside these variables, a few others are included as outputs to enhance the task of examining the system performance. We will discuss some of the qualitative and quantitative measures of performance shortly.

Algorithm

MATLAB function

```
function [dataIn, dataOut, txSig, rxSig, dataRx, yRec, csr]...
    = commlteSISO_step(nS, snrdB, prmLTEDLSCH, prmLTEPDSCH, prmMdl)
%% TX
% Generate payload
dataIn = genPayload(nS, prmLTEDLSCH.TBLenVec);
% Transport block CRC generation
tbCrcOut1 =CRCgenerator(dataIn);
% Channel coding includes - CB segmentation, turbo coding, rate matching,
% bit selection, CB concatenation - per codeword
[data, Kplus1, C1] = lteTbChannelCoding(tbCrcOut1, nS, prmLTEDLSCH, prmLTEPDSCH);
%Scramble codeword
scramOut = lteScramble(data, nS, 0, prmLTEPDSCH.maxG);
% Modulate
modOut = Modulator(scramOut, prmLTEPDSCH.modType);
% Generate Cell-Specific Reference (CSR) signals
csr = CSRgenerator(nS, prmLTEPDSCH.numTx);
% Resource grid filling
E=8*prmLTEPDSCH.Nrb;
csr_ref=reshape(csr(1:E),2*prmLTEPDSCH.Nrb,4);
txGrid = REmapper_1Tx(modOut, csr_ref, nS, prmLTEPDSCH);
% OFDM transmitter
txSig = OFDMTx(txGrid, prmLTEPDSCH);
%% Channel
% SISO Fading channel
[rxFade, chPathG] = MIMOFadingChan(txSig, prmLTEPDSCH, prmMdl);
idealhD = lteIdChEst(prmLTEPDSCH, prmMdl, chPathG, nS);
% Add AWG noise
nVar = 10.^(0.1.*(-snrdB));
rxSig = AWGNChannel(rxFade, nVar);
%% RX
% OFDM Rx
rxGrid = OFDMRx(rxSig, prmLTEPDSCH);
% updated for numLayers -> numTx
[dataRx, csrRx, idx_data] = REdemapper_1Tx(rxGrid, nS, prmLTEPDSCH);
% MIMO channel estimation
if prmMdl.chEstOn
    chEst = ChanEstimate_1Tx(prmLTEPDSCH, csrRx, csr_ref, 'interpolate');
    hD=chEst(idx_data).';
else
    hD = idealhD;
end
% Frequency-domain equalizer
yRec = Equalizer(dataRx, hD, nVar, prmLTEPDSCH.Eqmode);
% Demodulate
```

```
demodOut = DemodulatorSoft(yRec, prmLTEPDSCH.modType, nVar);
% Descramble both received codewords
rxCW =  lteDescramble(demodOut, nS, 0, prmLTEPDSCH.maxG);
% Channel decoding includes - CB segmentation, turbo decoding, rate dematching
[decTbData1, ~,~] = lteTbChannelDecoding(nS, rxCW, Kplus1, C1,  prmLTEDLSCH,
prmLTEPDSCH);
% Transport block CRC detection
[dataOut, ~] = CRCdetector(decTbData1);
end
```

5.18.1.1   Structure of the Transceiver Model

The following MATLAB script calls the SISO transceiver function just described. First it calls an initialization routine (*commlteSISO_initialize*), which sets all the relevant DLSCH and PDSCH and channel model parameters into three MATLAB structures (*prmLTEDLSCH, prmLTEPDSCH, prmMdl*). Then it sets up a while loop that performs subframe processing iterations. Before the while loop, it initializes the subframe index (*nS*) and ensures that the index resets when a frame of data (10 ms) has been processed. It also contains the criteria for stopping the simulation (maximum number of bits processed or maximum number of errors found). This script also compares the input and output bits in order to compute the BER and calls a visualization function to illustrate the channel response and modulation constellation before and after equalization.

Algorithm

MATLAB script: commlteSISO

```
% Script for SISO LTE (mode 1)
% Single codeword transmission only,
clear all
clear functions
disp('Simulating the LTE Mode 1: Single Tx and Rx antrenna');
%% Set simulation parametrs & initialize parameter structures
commlteSISO_params;
[prmLTEPDSCH, prmLTEDLSCH, prmMdl] = commlteSISO_initialize( chanBW,
contReg,  modType, Eqmode,...
    cRate,maxIter, fullDecode, chanMdl, corrLvl, chEstOn, maxNumErrs, maxNumBits);
clear chanBW contReg numTx numRx modType Eqmode cRate maxIter fullDecode
chanMdl corrLvl chEstOn maxNumErrs maxNumBits;
%%
hPBer = comm.ErrorRate;
iter=numel(prmMdl.snrdBs);
snrdB=prmMdl.snrdBs(iter);
maxNumErrs=prmMdl.maxNumErrs(iter);
maxNumBits=prmMdl.maxNumBits(iter);
%% Simulation loop
```

```
nS = 0; % Slot number, one of [0:2:18]
Measures = zeros(3,1); %initialize BER output
while (( Measures(2)< maxNumErrs) && (Measures(3) < maxNumBits))
  [dataIn, dataOut, txSig, rxSig, dataRx, yRec, csr] = ...
    commlteSISO_step(nS, snrdB, prmLTEDLSCH, prmLTEPDSCH, prmMdl);
  % Calculate  bit errors
  Measures = step(hPBer, dataIn, dataOut);
   % Visualize constellations and spectrum
  if visualsOn, zVisualize( prmLTEPDSCH, txSig, rxSig, yRec, dataRx, csr, nS);end;
  % Update subframe number
  nS = nS + 2; if nS > 19, nS = mod(nS, 20); end;
end
disp(Measures);
```

The following initialization function sets critical simulation parameters. As we are simulating the SISO case, it sets the number of transmit and receive antennas to one. To set PDSCH parameters, the function allows the user to choose a particular bandwidth (*chanBW*) from among six supported, the number of OFDM symbols occupying the control region (*contReg*), one of three modulation types (*modType*), and the type of equalization algorithm used. To set the DLSCH parameter, the function takes as input parameters the coding rate (*cRate*), the maximum number of iterations used in the turbo decoder (*maxIter*), and whether or not a full turbo decoding or early stopping is used within the turbo decoder (*fullDecode*). Finally, the function sets parameters controlling the channel model, including the type of channel mode used (*chanMdl*), the level of correlation between consecutive antenna ports (*corrLvl*), and whether or not estimated or ideal channel estimation is used (*chEstOn*).

Algorithm

MATLAB function

```
function [prmLTEPDSCH, prmLTEDLSCH, prmMdl] = commlteSISO_initialize(chanBW,
contReg, modType, Eqmode,...
                    cRate,maxIter, fullDecode, chanMdl, corrLvl, chEstOn,
maxNumErrs, maxNumBits)
% Create the parameter structures
% PDSCH and DLSCH
prmLTEPDSCH = prmsPDSCH(chanBW, contReg, modType);
prmLTEPDSCH.Eqmode=Eqmode;
prmLTEPDSCH.modType=modType;
prmLTEDLSCH = prmsDLSCH(cRate,maxIter, fullDecode, prmLTEPDSCH);
% Channel parameters
prmMdl.chanMdl = chanMdl;
prmMdl.corrLevel = corrLvl;
prmMdl.chEstOn = chEstOn;
switch modType
```

```
    case 1
       snrdBs=[0:4:8, 9:12];
    case 2
       snrdBs=[0:4:12, 13:16];
    otherwise
       snrdBs=0:4:24;
end
prmMdl.snrdBs=snrdBs;
prmMdl.maxNumBits=maxNumBits*ones(size(snrdBs));
prmMdl.maxNumErrs=maxNumErrs*ones(size(snrdBs));
```

5.18.1.2   Verifying Transceiver Performance

By executing the MATLAB script of the SISO transceiver model (*commlteSISO*) we can look at various signals in order to assess the performance of the system. To run the model script, we need first to set parameters related to various components of the model. The following script (*commlteSISO_params*) sets relevant parameters, including setting the modulation type to a 16QAM modulator.

Algorithm

MATLAB script

```
% PDSCH
numTx      = 1;   % Number of transmit antennas
numRx      = 1;   % Number of receive antennas
chanBW     = 4;   % Index to chanel bandwidth used [1,....6]
contReg    = 1;   % No. of OFDM symbols dedictaed to control information [1,...,3]
modType    = 2;   % Modulation type [1, 2, 3] for ['QPSK','16QAM','64QAM']
% DLSCH
cRate      = 1/3; % Rate matching target coding rate
maxIter    = 6;   % Maximum number of turbo decoding terations
fullDecode = 0;   % Whether "full" or "early stopping" turbo decoding is performed
% Channel model
chanMdl    = 'frequency-selective-high-mobility';
corrLvl    = 'Low';
% Simulation parametrs
Eqmode     = 2;   % Type of equalizer used [1,2] for ['ZF', 'MMSE']
chEstOn    = 1;   % Whether channel estimation is done or ideal channel model used
maxNumErrs = 5e7; % Maximum number of errors found before simulation stops
maxNumBits = 5e7; % Maximum number of bits processed before simulation stops
visualsOn  = 1;   % Whether to visualize channel response and constellations
```

For example, to examine the effects of equalization, we can visualize the constellation diagram of the user data recovered at the receiver before and after equalization. The MATLAB variables *dataRx* and *yRec* are provided as output arguments of the *commlteSISO_step* MAT-LAB function in order to enable visualization. Figure 5.11 illustrates the constellation diagrams, showing that the equalizer can compensate for the effects of fading channel (plot on the left) and results in a constellation that more closely resembles the constellation of the 16QAM modulator used in this experiment (plot on the right).

To examine the effectiveness of the OFDM receiver in combating the effects of multipath fading, we can look at the power spectral density of the transmitted signal and the received signals before and after equalization. Output MATLAB variables (*txSig*, *rxSig*, and *yRec*) enable this visualization. Figure 5.12 illustrates the spectra of the transmitted signal, the received signal before equalization, and the received signal after equalization. The results show that while the

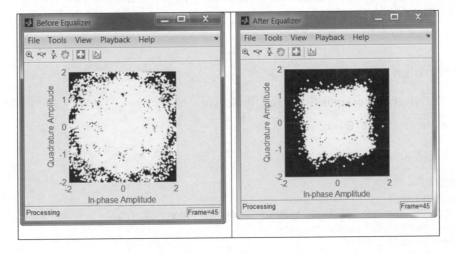

Figure 5.11   LTE SISO model: constellation diagram of the user data before and after equalization

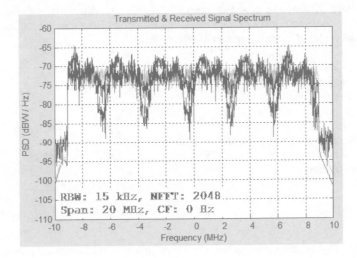

Figure 5.12   LTE SISO model: spectra of transmitted and received signals before and after equalization

transmitted signal has a spectrum with magnitude response normalized to one, the received-signal magnitude spectrum reflects the effects of the response to multipath fading of the channel. After equalization, the effects of the fading are mostly mitigated and the magnitude spectrum shows a more frequency-flat nature, which closely resembles the transmitted spectrum.

5.18.1.3   BER Measurements

In order to verify the BER performance of the transceiver, we create a testbench called *commlteSISO_test_timing_ber.m*. This first initializes the LTE system parameters and then iterates through a range of Signal-to-Noise Ratio (SNR) values and calls the *commlteSISO_fcn* function in the loop in order to compute the corresponding BER values. It also uses a combination of MATLAB *tic* and *toc* functions to measure the time needed to complete the iterations.

Algorithm

MATLAB script: commlteSISO_test_timing_ber

```
% Script for SISO LTE (mode 1)
%
% Single codeword transmission only,
%
clear all
clear functions
disp('Simulating the LTE Mode 1: Single Tx and Rx antrenna');
%% Create the parameter structures
commlteSISO_params;
[prmLTEPDSCH, prmLTEDLSCH, prmMdl] = commlteSISO_initialize( chanBW,
contReg, modType, Eqmode,...
    cRate,maxIter, fullDecode, chanMdl, corrLvl, chEstOn, maxNumErrs, maxNumBits);
clear  chanBW contReg numTx numRx modType Eqmode cRate maxIter fullDecode
chanMdl corrLvl chEstOn maxNumErrs maxNumBits;
%%
zReport_data_rate(prmLTEPDSCH, prmLTEDLSCH);
MaxIter=numel(prmMdl.snrdBs);
ber_vector=zeros(1,MaxIter);
tic;
for n=1:MaxIter
    fprintf(1,'Iteration %2d out of %2d\n', n, MaxIter);
    [ber, ~] = commlteSISO_fcn(n, prmLTEPDSCH, prmLTEDLSCH, prmMdl);
    ber_vector(n)=ber;
end;
toc;
```

When the MATLAB script is executed, messages appear in the command prompt, including transceiver parameters (modulation type, coding rate, channel bandwidth, antenna configuration, maximum data rate), the iteration being executed, and the final tally of elapsed time.

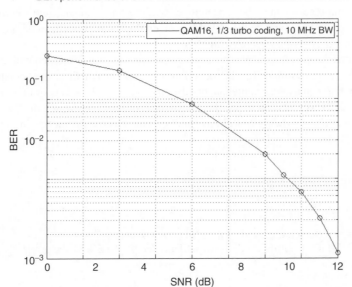

Figure 5.13 BER results: SISO model

Figure 5.13 shows the BER of the transceiver as a function of the SNR value. In this example, we process 50 million bits in each of the eight iterations characterized by a single SNR value. The transceiver uses a 16QAM modulation scheme, with a coding rate of 1/3, a system bandwidth of 10 MHz, and a SISO (1 × 1) antenna configuration. Choosing this parameter set gives a maximum data rate of 9.91 Mbps, as reported by the function *zReport_data_rate.m*.

5.18.2 The SIMO Case

The SIMO mode can be regarded as a general case of the SISO mode. LTE transmission mode 1 is usually regarded as the SIMO mode of transmission. In this mode, the signal processing chain is very similar to the SISO case, with the exception that it employs multiple (in our functions either two or four) receive antennas. Using multiple antennas at the receiver allows us to take advantage of receive diversity. Receive diversity with Maximum Ratio Combining (MRC) results in a system with better BER performance than its SISO counterpart. Modeling receive diversity does not change the transmitter but introduces many changes to channel modeling and receiver operations. All of these changes relate to multichannel processing.

Following transmitter operation, the fading channel processes samples from a single transmit antenna. However, depending on the number of receive antennas, it applies channel modeling to each link (transmitter–receiver pair) separately. The output of the fading channel is now a multichannel matrix with a number of rows equal to the number of transmitted samples and a number of columns equal to the number of receiver antennas. Similarly, the AWGN channel processes the multichannel output of the fading channel and produces an output of the same size with added white noise.

As a multichannel received signal is now the input to the receiver, the first set of operations performed in the receiver must be repeated across different channels (representing different receive antennas). These include the OFDM receiver, the resource-element demapper, and the channel estimator up to the equalizer function.

The estimated data resource elements at each receiver are now combined with a new equalizer to generate a best estimate for the transmitted signal. The equalizer uses either a ZF or an MMSE method to equal at each antenna, but the results are combined according to MRC. This method essentially weights and scales the contribution of each receive antenna according to its power measure. The following MATLAB function contains the operations in the transceiver for the SIMO case.

Algorithm

MATLAB function

```
function [dataIn, dataOut, txSig, rxSig, dataRx, yRec, csr]...
    = commlteSIMO_step(nS, snrdB, prmLTEDLSCH, prmLTEPDSCH, prmMdl)
%% TX
%  Generate payload
dataIn = genPayload(nS,  prmLTEDLSCH.TBLenVec);
% Transport block CRC generation
tbCrcOut1 =CRCgenerator(dataIn);
% Channel coding includes – CB segmentation, turbo coding, rate matching,
% bit selection, CB concatenation – per codeword
[data, Kplus1, C1] = lteTbChannelCoding(tbCrcOut1, nS, prmLTEDLSCH, prmLTEPDSCH);
%Scramble codeword
scramOut = lteScramble(data, nS, 0, prmLTEPDSCH.maxG);
% Modulate
modOut = Modulator(scramOut, prmLTEPDSCH.modType);
% Generate Cell-Specific Reference (CSR) signals
csr = CSRgenerator(nS, prmLTEPDSCH.numTx);
% Resource grid filling
E=8*prmLTEPDSCH.Nrb;
csr_ref=reshape(csr(1:E),2*prmLTEPDSCH.Nrb,4);
txGrid = Remapper_1Tx(modOut, csr_ref, nS, prmLTEPDSCH);
% OFDM transmitter
txSig = OFDMTx(txGrid, prmLTEPDSCH);
%% Channel
% SISO Fading channel
numRx=prmLTEPDSCH.numRx;
[rxFade, chPathG] = MIMOFadingChan(txSig, prmLTEPDSCH, prmMdl);
idealhD = lteIdChEst(prmLTEPDSCH,  prmMdl, chPathG, nS);
% Add AWG noise
nVar = 10.^(0.1.*(-snrdB));
rxSig =  AWGNChannel(rxFade, nVar);
%% RX
% OFDM Rx
```

```
rxGrid = OFDMRx(rxSig, prmLTEPDSCH);
% updated for numLayers -> numTx
[dataRx, csrRx, idx_data] = Redemapper_1Tx(rxGrid, nS, prmLTEPDSCH);
% MIMO channel estimation
if prmMdl.chEstOn
    chEst = ChanEstimate_1Tx(prmLTEPDSCH, csrRx,  csr_ref, 'interpolate');
    hD=complex(zeros(numel(idx_data),numRx));
    for n=1:numRx
        tmp=chEst(:,:,n);
        hD(:,n)=tmp(idx_data).';
    end
else
    hD = idealhD;
end
% Frequency-domain equalizer
% Based on Maximum-Combining Ratio (MCR)
yRec = Equalizer_simo( dataRx, hD, nVar, prmLTEPDSCH.Eqmode);
% Demodulate
demodOut = DemodulatorSoft(yRec, prmLTEPDSCH.modType, nVar);
% Descramble both received codewords
rxCW = lteDescramble(demodOut, nS, 0, prmLTEPDSCH.maxG);
% Channel decoding includes – CB segmentation, turbo decoding, rate dematching
[decTbData1, ~,~] = lteTbChannelDecoding(nS, rxCW, Kplus1, C1,  prmLTEDLSCH,
prmLTEPDSCH);
% Transport block CRC detection
[dataOut, ~] = CRCdetector(decTbData1);
end
```

5.18.2.1   Modified Functions

The modifications needed to enable the SIMO mode affect the following three functions.

Redemapper_1Tx now supports multichannel processing by iterating through receive antennas in a for loop in order to extract data, CSR, and other signals separately in each.

Algorithm

MATLAB function

```
function [data, csr, idx_data, pdcch, pss, sss, bch] = Redemapper_1Tx(in, nS, prmLTE)
%#codegen
% NcellID = 0;                        % One of possible 504 values
% numTx = 1;                          % prmLTE.numTx;
% Get input params
numRx=prmLTE.numRx;                   % number of receive antennas
Nrb = prmLTE.Nrb;                     % either of {6,...,1}
```

```
Nrb_sc = prmLTE.Nrb_sc;            % 12 for normal mode
numContSymb    = prmLTE.contReg;  % either {1, 2, 3}
Npss= prmLTE.numPSSRE;
Nsss=prmLTE.numSSSRE;
Nbch=prmLTE.numBCHRE;
Ncsr=prmLTE.numCSRResources;
Ndci=prmLTE.numContRE;
%% Specify resource grid location indices for CSR, PDCCH, PDSCH, PBCH, PSS, SSS
%% 1st: Indices for CSR pilot symbols
lenOFDM = Nrb*Nrb_sc;
idx       = 1:lenOFDM;
idx_csr0  = 1:6:lenOFDM;            % More general starting point = 1+mod(NcellID, 6);
idx_csr4  = 4:6:lenOFDM;            % More general starting point = 1+mod(3+NcellID, 6);
idx_csr   =[idx_csr0, 4*lenOFDM+idx_csr4, 7*lenOFDM+idx_csr0, 11*lenOFDM+idx_csr4];
%% 2nd: Indices for PDCCH control data symbols
ContREs=numContSymb*lenOFDM;
idx_dci=1:ContREs;
idx_pdcch = ExpungeFrom(idx_dci,idx_csr0);
%% 3rd: Indices for PDSCH and PDSCH data in OFDM symbols where pilots are present
idx_data0= ExpungeFrom(idx,idx_csr0);
idx_data4 = ExpungeFrom(idx,idx_csr4);
switch nS
    %% 4th: Indices for BCH, PSS, SSS are only found in specific subframes 0 and 5
    % These symbols share the same 6 center sub-carrier locations (idx_ctr)
    % and differ in OFDM symbol number.
    Case 0 % Subframe 0
        % PBCH, PSS, SSS are available + CSR, PDCCH, PDSCH
        idx_6rbs = (1:72);
        idx_ctr = 0.5* lenOFDM – 36 + idx_6rbs ;
        idx_SSS  = 5* lenOFDM + idx_ctr;
        idx_PSS  = 6* lenOFDM + idx_ctr;
        idx_ctr0 = ExpungeFrom(idx_ctr,idx_csr0);
        idx_bch=[7*lenOFDM + idx_ctr0, 8*lenOFDM + idx_ctr, 9*lenOFDM + idx_ctr,
10*lenOFDM + idx_ctr];
        idx_data5   = ExpungeFrom(idx,idx_ctr);
        idx_data7 = ExpungeFrom(idx_data0,idx_ctr);
        idx_data  = [ContREs+1:4*lenOFDM,  4*lenOFDM+idx_data4, ...
           5*lenOFDM+idx_data5, 6*lenOFDM+idx_data5, 7*lenOFDM+idx_data7,
8*lenOFDM+idx_data5, ...
           9*lenOFDM+idx_data5, 10*lenOFDM+idx_data5, 11*lenOFDM+idx_data4, ...
           12*lenOFDM+1:14*lenOFDM];
    case 10 % Subframe 5
        % PSS, SSS are available + CSR, PDCCH, PDSCH
        % Primary and Secondary synchronization signals in OFDM symbols 5 and 6
        idx_6rbs = (1:72);
        idx_ctr = 0.5* lenOFDM – 36 + idx_6rbs ;
        idx_SSS  = 5* lenOFDM + idx_ctr;
        idx_PSS  = 6* lenOFDM + idx_ctr;
        idx_data5 = ExpungeFrom(idx,idx_ctr);
```

```
    idx_data = [ContREs+1:4*lenOFDM, 4*lenOFDM+idx_data4, 5*lenOFDM+idx_data5,
6*lenOFDM+idx_data5, ...
        7*lenOFDM+idx_data0, 8*lenOFDM+1:11*lenOFDM, 11*lenOFDM+idx_data4, ...
        12*lenOFDM+1:14*lenOFDM];
  otherwise % other subframes
    % Only CSR, PDCCH, PDSCH
    idx_data = [ContREs+1:4*lenOFDM, 4*lenOFDM+idx_data4, ...
        5*lenOFDM+1:7*lenOFDM, ...
        7*lenOFDM+idx_data0, ...
        8*lenOFDM+1:11*lenOFDM, ...
        11*lenOFDM+idx_data4, ...
        12*lenOFDM+1:14*lenOFDM];
end
%% Handle 3 types of subframes differently
pss=complex(zeros(Npss,numRx));
sss=complex(zeros(Nsss,numRx));
bch=complex(zeros(Nbch,numRx));
data=complex(zeros(numel(idx_data),numRx));
csr=complex(zeros(Ncsr,numRx));
pdcch = complex(zeros(Ndci,numRx));
for n=1:numRx
  tmp=in(:,:,n);
  data(:,n)=tmp(idx_data.');       % Physical Downlink Shared Channel (PDSCH) = user data
  csr(:,n)=tmp(idx_csr.');          % Cell-Specific Reference signal (CSR) = pilots
  pdcch(:,n) = tmp(idx_pdcch.');   % Physical Downlink Control Channel (PDCCH)
  if nS==0
    pss(:,n)=tmp(idx_PSS.');       % Primary Synchronization Signal (PSS)
    sss(:,n)=tmp(idx_SSS.');        % Secondary Synchronization Signal (SSS)
    bch(:,n)=tmp(idx_bch.');       % Broadcast Channel data (BCH)
  elseif nS==10
    pss(:,n)=tmp(idx_PSS.');       % Primary Synchronization Signal (PSS)
    sss(:,n)=tmp(idx_SSS.');        % Secondary Synchronization Signal (SSS)
  end
end
```

The updated function *ChanEstimate_1Tx* now supports multichannel processing by repeating the process of resource-grid generation based on CSR signals across multiple antennas.

Algorithm

MATLAB function

```
function hD = ChanEstimate_1Tx(prmLTE, Rx, Ref, Mode)
%#codegen
Nrb      = prmLTE.Nrb;      % Number of resource blocks
Nrb_sc   = prmLTE.Nrb_sc;          % 12 for normal mode
Ndl_symb = prmLTE.Ndl_symb;       % 7   for normal mode
```

```
numRx = prmLTE.numRx;
% Assume same number of Tx and Rx antennas = 1
% Initialize output buffer
hD = complex(zeros(Nrb*Nrb_sc, Ndl_symb*2,numRx));
% Estimate channel based on CSR – per antenna port
csrRx = reshape(Rx, numel(Rx)/(4*numRx), 4, numRx); % Align received pilots with refer-
ence pilots
for n=1:numRx
    hp= csrRx(:,:,n)./Ref;              % Just divide received pilot by reference pilot
    % to obtain channel response at pilot locations
    % Now use some form of averaging/interpolation/repeating to
    % compute channel response for the whole grid
    % Choose one of 3 estimation methods "average" or "interpolate" or "hybrid"
    switch Mode
        case 'average'
            tmp=gridResponse_averageSubframe(hp, Nrb, Nrb_sc, Ndl_symb);
        case 'interpolate'
            tmp=gridResponse_interpolate(hp, Nrb, Nrb_sc, Ndl_symb);
        otherwise
            error('Choose the right mode for function ChanEstimate.');
    end
    hD(:,:,n)=tmp;
end
```

Unlike the frequency-domain equalizer of the SISO mode, the equalizer in the SIMO mode must combine contributions from multiple channels. The new equalizer (*Equalizer_simo*) employs the MRC method to generate a best estimate of the resource element at the receiver [4].

Algorithm

MATLAB function

```
function [y, num, denum] = Equalizer_simo(in, hD, nVar, prmLTE)
%#codegen
EqMode=prmLTE.Eqmode;
numTx=prmLTE.numTx;
numRx=size(hD,2);
if (numTx>1), error('Equalizer_simo: edicated to single transmit antenna case.');end
if numRx==1
    switch EqMode
        case 1, % Zero forcing
            num = conj(hD);
            denum=conj(hD).*hD;
        case 2, % MMSE
            num = conj(hD);
            denum=conj(hD).*hD+nVar;
    end
```

```
else
   num = conj(hD);
   denum=conj(hD).*hD;
end
y = sum(in .*num,2)./sum(denum,2);
```

5.18.2.2   Verifying Transceiver Performance

In order to observe the effect of receive diversity on performance, we can execute the MAT-LAB script of the SIMO transceiver model (*commlteSIMO*). First, we set parameters related to various component of the model in the script (*commlteSIMO_params*). This is the same script used in the SISO case, except we change the number of receive parameters from one to four.

Algorithm

MATLAB script

```
% PDSCH
numTx      = 1;    % Number of transmit antennas
numRx      = 4;    % Number of receive antennas
chanBW     = 4;    % Index to chanel bandwidth used [1,....6]
contReg    = 1;    % No. of OFDM symbols edicated to control information [1,...,3]
modType    = 2;    % Modulation type [1, 2, 3] for ['QPSK','16QAM','64QAM']
% DLSCH
cRate      = 1/3;  % Rate matching target coding rate
maxIter    = 6;    % Maximum number of turbo decoding terations
fullDecode = 0;    % Whether "full" or "early stopping" turbo decoding is performed
% Channel model
chanMdl    = 'frequency-selective-high-mobility';
corrLvl    = 'Low';
% Simulation parametrs
Eqmode     = 2;    % Type of equalizer used [1,2] for ['ZF', 'MMSE']
chEstOn    = 1;    % Whether channel estimation is done or ideal channel model used
maxNumErrs = 5e7;  % Maximum number of errors found before simulation stops
maxNumBits = 5e7;  % Maximum number of bits processed before simulation stops
visualsOn  = 1;    % Whether to visualize channel response and constellations
```

Figure 5.14 illustrates the constellation diagrams and shows how the SIMO OFDM transceiver compensates for the multipath fading effect and rotates and scales back the corrupted constellation (before equalization) to a constellation that can properly be demodulated (after equalization). Figure 5.15 shows the power spectral density of the transmitted and received signals before and after equalization. The results show that while the transmitted signal has a power spectral magnitude that is normalized to one, the received-signal

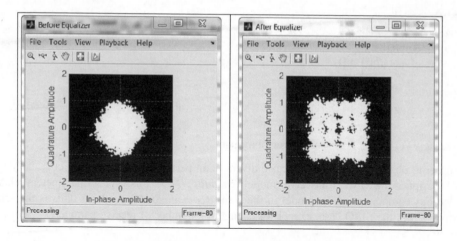

Figure 5.14 LTE SISO model: constellation diagram of the user data before and after equalization

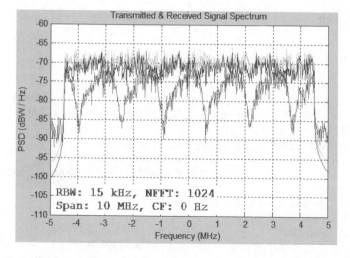

Figure 5.15 LTE SIMO model: spectra of transmitted and received signals before and after equalization

magnitude spectrum reflects the effects of the multipath fading response of the channel. After equalization, the magnitude spectrum shows a more frequency-flat nature, which closely resembles the transmitted spectrum.

5.18.2.3 BER Measurements

In order to verify the BER performance of the transceiver, we create a testbench called *commlteSIMO_test_timing_ber.m*. This first initializes the LTE system parameters and then iterates through a range of SNR values and calls the *commlteSIMO_fcn* function in the loop in order to compute the corresponding BER values.

Algorithm

MATLAB script: *commlteSIMO_test_timing_ber*

```
% Script for SIMO LTE (mode 1)
%
% Single codeword transmission only
%
clear all
clear functions
disp('Simulating the LTE Mode 1: Single Tx and multiple Rx antrennas');
%% Set simulation parametrs & initialize parameter structures
commlteSIMO_params_ber;
[prmLTEPDSCH, prmLTEDLSCH, prmMdl] = commlteSIMO_initialize( chanBW,
contReg,  modType, ...
    Eqmode, numTx, numRx, cRate,maxIter, fullDecode, chanMdl, corrLvl, ...
    chEstOn, maxNumErrs, maxNumBits);
clear chanBW contReg numTx numRx modType Eqmode cRate maxIter fullDecode
chanMdl corrLvl chEstOn maxNumErrs maxNumBits;
%%
zReport_data_rate(prmLTEPDSCH, prmLTEDLSCH);
MaxIter=numel(prmMdl.snrdBs);
ber_vector=zeros(1,MaxIter);
tic;
for n=1:MaxIter
    fprintf(1,'Iteration %2d out of %2d\n', n, MaxIter);
    [ber, ˜] = commlteSIMO_fcn(n, prmLTEPDSCH, prmLTEDLSCH, prmMdl);
    ber_vector(n)=ber;
end;
toc;
semilogy(prmMdl.snrdBs, ber_vector);
title('BER - commlteSISO');xlabel('SNR (dB)');ylabel('ber');grid;
```

When the MATLAB script is executed, messages appear in the command prompt, including transceiver parameters (modulation type, coding rate, channel bandwidth, antenna configuration, and maximum data rate), the iteration being executed, and the final tally of elapsed time.

Figure 5.16 shows the BER of the transceiver as a function of the SNR value. In this example, we process 50 million bits in each of the eight iterations characterized by a single SNR value. The transceiver uses a 16QAM modulation scheme, with a coding rate of 1/3, a system bandwidth of 10 MHz, and SIMO antenna configurations of 1×4. Choosing this parameter set leads to a maximum data rate of 9.91 Mbps, as reported by the function *zReport_data_rate.m*. Running all eight iterations takes about 4025 seconds to complete without any acceleration methods.

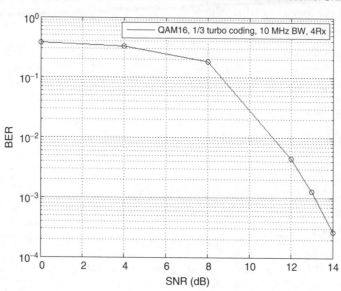

Figure 5.16   BER results: SIMO mode

5.19   Chapter Summary

In this chapter we studied the multicarrier transmission scheme used in the LTE standard. We focused on developing the downlink transceiver based on the OFDM transmission in MATLAB. First we examined a more realistic representation of a mobile communications channel and introduced the multipath fading channel models. Then we presented the functional elements of an OFDM transmission scheme, designed to combat the effects of multipath fading.

We then reviewed the functional elements in the transmitter, including: (i) the time–frequency representation of data leading up to the formation of a resource grid, (ii) the inclusion of OFDM pilot signals (or reference signals) within the resource grid, and (iii) the OFDM signal generation that uses inverse FFT to compute the transmitted data as a time-domain signal that is completely specified in the frequency domain based on a resource-grid representation.

We subsequently reviewed typical functional elements in the receiver, including: (i) the OFDM receiver that computes the received resource grid, (ii) channel estimation based on reference signals, (iii) computation of the channel response for the entire resource grid based on interpolation of channel estimation results, and (iv) frequency-domain equalization based on the estimated channel response, used to recover best estimates for transmitted resource elements.

Finally, we integrated all of the functional elements to create a transceiver model in MAT-LAB for the single-antenna downlink transmission mode of the LTE standard. Otherwise known as LTE transmission mode 1, the transceiver handles both the SISO and SIMO downlink transceiver operations. Through simulations, we performed both qualitative assessments and BER performance measurements. The results show that the transceiver effectively combats the effects of intersymbol interference caused by multipath fading. In the next chapter we will introduce the MIMO multi-antenna schemes, in which more than one antenna is used for transmission.

References

[1] Y.S. Cho, J.K. Kim, W.Y. Yang, C.G. Kang, *MIMO-OFDM Wireless Communications with MATLAB*, John Wiley and Sons (Asia) Pte Ltd, 2010.

[2] 3GPP (2011) Evolved Universal Terrestrial Radio Access (E-UTRA); Physical Channels and Modulation Version 10.0.0. TS 36.211, January 2011.

[3] A. Ghosh, R. Ratasuk, *Essentials of LTE and LTE-A*, Cambridge University Press, 2011.

[4] H. Jafarkhani, *Space-Time Coding; Theory and Practice*, Cambridge University Press, 2005.

6

MIMO

So far we have studied the modulation, scrambling, coding, channel modeling, and multicarrier transmission schemes used in the LTE (Long Term Evolution) standard. In this chapter we focus on its multi-antenna characteristics. The LTE and LTE-Advanced standards achieve high maximum data rates mainly as the result of incorporating many multi-antenna or MIMO (Multiple Input Multiple Output) techniques. LTE can be regarded as a MIMO–OFDM (Orthogonal Frequency Division Multiplexing) system, with MIMO multi-antenna configurations being combined with the OFDM multicarrier transmission scheme.

In general, multi-antenna transmission schemes map modulated data symbols to multiple antennas ports. In the OFDM transmission scheme, each antenna constructs the resource grid, generates the OFDM symbols, and transmits the signal. In a MIMO–OFDM system, the process of resource-grid mapping and OFDM modulation is repeated over multiple transmit antennas. Depending on the MIMO mode used, this multi-antenna extension may result in a boost in data rates or an improvement in the link quality.

In this chapter, we will first review MIMO algorithms of the first four transmission modes of the LTE standard. These transmission modes exploit two main MIMO techniques: (i) transmit diversity (techniques such as Space–Frequency Block Coding, SFBC) and (ii) spatial multiplexing with or without delay-diversity coding. As noted earlier, transmit diversity techniques improve the link quality and reliability but not the data rate or spectral efficiency of a system. On the other hand, spatial multiplexing can bring about in a substantial boost in data rates.

6.1 Definition of MIMO

"MIMO antenna processing" is often used as a general term to refer to all techniques employing multiple transmit and receive antennas. The LTE standard is based on a combination of MIMO multi-antenna techniques and OFDM multicarrier techniques. Essentially, in LTE relationships between multiple transmit and receive antennas are best explained at each individual subcarrier rather than across the entire bandwidth. Figure 6.1 illustrates transmit and receive antenna relationships, together with the channel gains linking each antenna pair.

Understanding LTE with MATLAB®: From Mathematical Modeling to Simulation and Prototyping, First Edition.
Houman Zarrinkoub.
© 2014 John Wiley & Sons, Ltd. Published 2014 by John Wiley & Sons, Ltd.

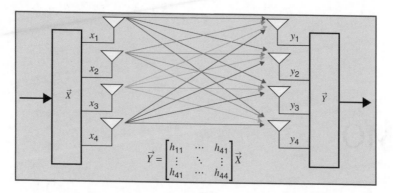

Figure 6.1   Block diagram of a MIMO transmitter, receiver, and channel

At each subcarrier, the relationship between the received and transmitted resource elements on different antennas is expressed by a system of linear equations. In this system, the vector of received resource elements on receive antennas results from the multiplication of the MIMO channel matrix by the vector of transmitted resource elements on transmit antennas. As indicated by the MIMO system of equations, in order to recover the best estimate of the transmitted resource element at a given subcarrier, we need not only the vector of received resource elements but also the channel response (or the CSI, Channel State information) connecting each pair of transmit and receive antennas.

6.2   Motivation for MIMO

Theoretically, the best way to increase data rates over a communications link is to increase the overall received signal power for a given transmit power [1]. An effective way of increasing the received power is to use additional antennas at the transmitter and/or the receiver. This represents a class known as multi-antenna or MIMO techniques. Impressive improvements in capacity and Bit Error Rates (BERs) brought about by the use of MIMO techniques have spurred a lot of interest in multi-antenna radio systems. Along with the gains, however, comes added computational complexity. The complexity of a MIMO technique is usually in proportion to the number of antennas used.

Among various MIMO techniques, spatial multiplexing introduces a multi-antenna methodology that achieves a linear capacity growth with the number of antennas [1]. Given that typical methods of increasing capacity such as increasing power only lead to a logarithmic improvement, the promise of substantial capacity gains from the use of MIMO techniques represents a historical step forward in wireless communications.

6.3   Types of MIMO

LTE takes extensive advantage of MIMO techniques, for example by introducing many forms of multi-antenna technique in each of its nine downlink transmission modes. LTE-Advanced provides multiple transmit-antenna configurations of up to eight antennas at a time.

Let us examine the mathematical foundation of MIMO systems. A successful implementation of a MIMO system hinges on solving systems of linear equations at the receiver in order to correctly recover the transmitted data. In the presence of channel degradations, the full spectrum exhibits a frequency-selective response. At each sub-band, however, the channel response is flatter and may be approximated by a scalar gain value. In a MIMO system, at each subcarrier the relationship between any pair of transmitted and received symbols can be expressed with a single gain value. This means that the relationship between multiple transmitters and receivers can be expressed with a MIMO system of linear equations, which are solved at the receiver for each and every subcarrier of the full spectrum in order to recover the transmitted signal.

The MIMO algorithms used in the LTE standard can be subdivided into four broad categories: receiver-combining, transmit-diversity, beamforming, and spatial-multiplexing. We will provide a short discussion of three of these techniques in this section.

6.3.1   Receiver-Combining Methods

Receiver-combining methods combine multiple versions of the transmitted signal at the receiver to improve performance. They have been used in 3G mobile standards and WiFi and WiMAX systems. Two types of combining method can be used at the receiver: Maximum Ratio Combining (MRC) and Selection Combining (SC) [2]. In MRC, we combine the multiple received signals (usually by averaging them) to find the most likely estimate of the transmitted signal. In SC, we forego the extensive complexity of MRC and use only the received signal with the highest SNR (Signal-to-Noise Ratio) to estimate the transmitted signal.

6.3.2   Transmit Diversity

In transmit diversity, redundant information is transmitted on different antennas at each subcarrier. In this mode, LTE does not increase the data rate but only makes the communications link more robust. Transmit diversity belongs to a class of multi-antenna techniques known as space–time coding. Space–time codes are capable of delivering a diversity order equal to the product of the number of receive and transmit antennas. SFBC, a technique closely related to Space–Time Block Coding (STBC), is the transmit-diversity technique used in the LTE standard.

6.3.3   Spatial Multiplexing

In spatial multiplexing, the system transmits independent (nonredundant) information on different antennas. This mode of MIMO can substantially boost the data rate of a given communications link as the data rate can increase linearly in proportion to the number of transmit antennas. The ability to transmit independent data streams in spatial multiplexing comes with a cost, however. Spatial multiplexing is susceptible to deficiencies in rank of the matrix representing the MIMO equation. Multiple techniques are introduced in LTE spatial multiplexing in order to minimize the probability of these rank deficiencies occurring and to harness its benefits.

6.4   Scope of MIMO Coverage

In this book we focus on signal processing related to the first four modes of MIMO transmission. Beamforming, used in mode 6, relates to multicast and is important for coordinated multipoint. Multi-user MIMO (MU-MIMO), used in modes 5 and 7–9, can be best understood as an extension of the single-user cases of modes 3 and 4. A detailed discussion of beamforming methods and MU-MIMO in both downlink and uplink deserves further study in a different volume.

6.5   MIMO Channels

MIMO channels specify the relationships between signals transmitted over multiple transmit antennas and signals received at multiple receive antennas. The number of connection links is equal to the product of the number of transmit antennas (*numTx*) and the number of receive antennas (*numRx*).

In a flat-fading scenario, the relationship between any given pair of transmit and receive antennas at any point in time is given by a scalar gain value known as the channel path gain. The collection of these path gains specifies the channel matrix *H*. The dimension of the channel matrix is equal to (*numTx*, *numRx*). A system of linear equations characterizes the relationship between the received signal at each receive antenna, the transmitted signal at each transmit antenna, and the channel matrix. Figure 6.2 illustrates this relationship between *X(n)* (the transmitted vector at sample time *n*), *Y(n)* (the received vector at sample time *n*), and *H(n)* (the channel matrix at sample time *n*) in a 2×2 MIMO channel characterized by a flat-fading response.

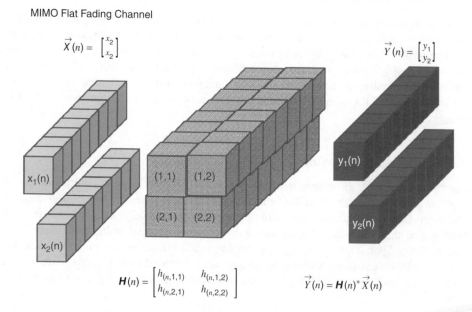

Figure 6.2   A 2×2 MIMO channel with a flat-fading response

The range for the time index n is equal to $n = 1, \ldots, nSamp$, where $nSamp$ is the number of transmitted symbols in each subframe per antenna. As a result, over a full subframe the transmitted signal has a dimension of $(nSamp, numTx)$, the received signal has a dimension of $(nSamp, numRx)$, and the channel matrix is a 3D matrix with dimensions of $(nSamp, numTx, numRx)$.

In a multipath fading scenario, the relationship between any given transmit and receive antenna at any point in time is characterized by the channel-path gain vector. So each received signal at any point in time depends on the present and past values of transmitted signals. This necessitates the introduction of one more parameter: the number of path delays L. To compute the received signals in a multipath case, the MIMO operations mentioned in the flat-fading scenario must be repeated for each value of the path-delay vector.

As a result, over a full subframe the transmitted signal has a dimension of $(nSamp, numTx)$, the received signal has a dimension of $(nSamp, numRx)$, but the channel matrix is a 4D matrix with dimensions of $(nSamp, L, numTx, numRx)$. Figure 6.3 illustrates this relationship between the transmitted signal $X(n)$, the received signal $Y(n)$, and the channel matrix $H(n,k)$ in a 2×2 MIMO channel characterized by a multipath fading response. Here the range for the time index n is equal to $n = 1, \ldots, nSamp$, where $nSamp$ is defined as before and the range for the path-delay index k is equal to $k = 1, \ldots, L$, where L is the number of path delays.

6.5.1 MATLAB® Implementation

We can use the *comm.MIMOChannel* System object to study the effects of multiple antennas and multiple propagation paths and to implement a MIMO channel model.

MIMO Multipath Fading Channel

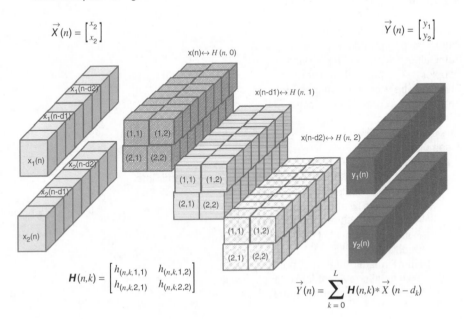

Figure 6.3 A 2×2 MIMO channel with a multipath fading response

The *comm.MIMOChannel* System object uses such parameters as number of transmit and receive antennas, delay profile, and Doppler shift to model the dynamics of a flat- or frequency-selective-fading MIMO channel.

The following MATLAB function shows a MIMO fading channel model that can handle frequency-flat- or selective-fading characteristics. This function takes as input a variable (x) that is organized as a 2D matrix. The first dimension of the matrix (*nSamp*) is the number of samples processed by each transmit antenna in a subframe. The second dimension is the number of transmit antennas (*numTx*). The function has two output variables. The first (y) is the filtered version of the input variable (x), processed by the fading channel. The first dimension of the first output signal is the same as the first dimension of the input signal (*nSamp*). The second dimension is equal to the number of receive antennas (*numRx*). The second output of the function is a multidimensional matrix (H) representing the channel matrix (otherwise known as path gains). The path gains operate on the input variable (x) to generate the output faded signal (y).

Algorithm

MATLAB function

```
function [y, yPg] = MIMOFadingChan(in, prmLTE, prmMdl)
% MIMOFadingChan
%#codegen
% Get simulation params
numTx       = prmLTE.numTx;
numRx       = prmLTE.numRx;
chanSRate   = prmLTE.chanSRate;
chanMdl     = prmMdl.chanMdl;
corrLvl     = prmMdl.corrLevel;
PathDelays  = prmMdl.PathDelays ;
PathGains   = prmMdl.PathGains ;
Doppler     = prmMdl.Doppler;
ChannelType = prmMdl.ChannelType ;
AntConfig   = prmMdl.AntConfig;
% Initialize objects
persistent chanObj;
if isempty(chanObj)
    if ChannelType ==1
        chanObj = comm.MIMOChannel('SampleRate', chanSRate, ...
            'MaximumDopplerShift', Doppler, ...
            'PathDelays', PathDelays,...
            'AveragePathGains', PathGains,...
            'RandomStream', 'mt19937ar with seed',...
            'Seed', 100,...
            'NumTransmitAntennas', numTx,...
            'TransmitCorrelationMatrix', eye(numTx),...
            'NumReceiveAntennas', numRx,...
```

```
                'ReceiveCorrelationMatrix', eye(numRx),...
                'PathGainsOutputPort', true,...
                'NormalizePathGains', false,...
                'NormalizeChannelOutputs', true);
        else
            chanObj = comm.LTEMIMOChannel('SampleRate', chanSRate, ...
                'Profile', chanMdl, ...
                'AntennaConfiguration', AntConfig, ...
                'CorrelationLevel', corrLvl,...
                'RandomStream', 'mt19937ar with seed',...
                'Seed', 100,...
                'PathGainsOutputPort', true);
        end
    end
    [y, yPg] = step(chanObj, in);
```

In this function we use two different System objects to perform MIMO channel processing. The *comm.MIMOChannel* System object is a generic model for MIMO channels. It takes such parameters as path delay, path gains, and Doppler shift to specify the model.

The *comm.LTEMIMOChannel* System object is specific to LTE channel modeling and is fully described in the next section. It takes a different set of parameters, such as antenna configurations and the correlation level between transmit antennas, to compute all the necessary channel-modeling operations. This function implements the MIMO fading profiles prescribed in the LTE standard [3].

6.5.2   LTE-Specific Channel Models

The 3GPP (Third Generation Partnership Project) Technical Recommendation (TR) 36.104 [3] specifies three different multipath fading channel models: the Extended Pedestrian A (EPA), Extended Vehicular A (EVA), and Extended Typical Urban (ETU). The channel-modeling functions used in this book explicitly take advantage of these models. We will not use the higher-mobility profiles as the closed-loop spatial-multiplexing mode is applicable to high-data-rate and low-mobility scenarios only. Together with the generic channel models described earlier, these models enable us to evaluate the performance of the transceiver in various reference channel conditions.

A multipath fading channel model is specified by the combination of delay profiles and a maximum Doppler frequency. The delay profiles of these channel models correspond to a low, medium, and high delay spread environment, respectively and a value of 5, 70, or 300 Hz will be used as the maximum Doppler shift. Table 6.1 illustrates the delay profile of each of the channel models expressed with excess tap delay values (in nanoseconds) and relative power (in decibels).

In a MIMO transmission scenario, the spatial correlations between the transmit antennas and the receiver antennas are important parameters that directly affect the overall performance.

Table 6.1 LTE channel models (EPA, EVA, ETU): delay profiles

Channel model	Excess tap delay (ns)	Relative power (dB)
Extended Pedestrian A (EPA)	[0 30 70 90 110 190 410]	[0 −1 −2 −3 −8 −17.2 −20.8]
Extended Vehicular A (EVA)	[0 30 150 310 370 710 1090 1730 2510]	[0 −1.5 −1.4 −3.6 −0.6 −9.1 −7 −12 −16.9]
Extended Typical Urban (ETU)	[0 50 120 200 230 500 1600 2300 5000]	[−1 −1 −1 0 0 0 −3 −5 −7]

MIMO works best under maximum-scattering and multipath fading environments. Therefore, it is desirable to minimize the correlation between various antenna ports in the transmitter or the receiver side. This will minimize the chance of rank deficiency in the MIMO channel matrices and boost the performance.

For example, in a 2×2 MIMO antenna configuration, the transmitter-side (eNodeB, enhanced Node Base station) spatial correlation matrix (M_{tx}) is expressed as a 2×2 matrix with diagonal elements equal to one and off-diagonal elements specified by a parameter (α) as $M_{tx} = \begin{bmatrix} 1 & \alpha \\ \alpha^* & 1 \end{bmatrix}$. Similarly, the receiver-side (UE, User Equipment) spatial correlation matrix (M_{rx}) is expressed as a 2×2 matrix specified by another parameter (β) as $M_{rx} = \begin{bmatrix} 1 & \beta \\ \beta^* & 1 \end{bmatrix}$. Note that if both parameters α and β are real-valued we do not need to perform the conjugation.

In a 4×4 antenna configuration, the spatial-correlation matrices of the transmitter and the receiver side are specified in identical ways as a function of either parameter α or parameter β. The transmitter-side (eNodeB) spatial correlation matrix (M_{tx}) is expressed with a 4×4 matrix as

$$M_{tx} = \begin{bmatrix} 1 & \alpha^{\frac{1}{9}} & \alpha^{\frac{4}{9}} & \alpha \\ \alpha^{\frac{1}{9}} & 1 & \alpha^{\frac{1}{9}} & \alpha^{\frac{4}{9}} \\ \alpha^{\frac{4}{9}} & \alpha^{\frac{1}{9}} & 1 & \alpha^{\frac{1}{9}} \\ \alpha & \alpha^{\frac{4}{9}} & \alpha^{\frac{1}{9}} & 1 \end{bmatrix}.$$

Three different correlation levels are defined in the LTE specification: low (actually no correlation), medium, and high. These correlation levels are reflected in the values of the parameters (α and β) specifying the correlation matrices, as illustrated in Table 6.2.

Table 6.2 LTE channel models: correlation levels and coefficients of the spatial-correlation matrices

LTE MIMO channel correlation levels	α	β
Low correlation	0	0
Medium correlation	0.3	0.9
High correlation	0.9	0.9

6.5.3   MATLAB Implementation

The System object *comm.LTEMIMOChannel* is specific to LTE channel modeling and implements the three types of channel model (EPA, EVA, and ETU) discussed in the previous section. It takes different sets of parameters, such as antenna configurations and the correlation level between transmit antennas, to compute all the necessary channel-modeling operations. The System object implements the MIMO fading profiles prescribed in LTE Recommendation 36.104 [3].

  Since this System object is implemented as a MATLAB-authored object, we can use the command *edit comm.LTEMIMOChannel* to examine the MATLAB code implementing its various functionalities. For example, the delay profiles of various LTE channel models are implemented with a few lines of MATLAB code in the *setDelayDopplerProfiles* function of the System object:

Algorithm

MATLAB code segment

```
function setDelayDopplerProfiles(obj)
    EPAPathDelays = [0 30 70 90 110 190 410]*1e-9;
    EPAPathGains  = [0 -1 -2 -3 -8 -17.2 -20.8];
    EVAPathDelays = [0 30 150 310 370 710 1090 1730 2510]*1e-9;
    EVAPathGains  = [0 -1.5 -1.4 -3.6 -0.6 -9.1 -7 -12 -16.9];
    ETUPathDelays = [0 50 120 200 230 500 1600 2300 5000]*1e-9;
    ETUPathGains  = [-1 -1 -1 0 0 0 -3 -5 -7];
    switch obj.Profile
    case 'EPA 5Hz'
        obj.PathDelays        = EPAPathDelays;
        obj.AveragePathGains   = EPAPathGains;
        obj.MaximumDopplerShift = 5;
    case 'EVA 5Hz'
        obj.PathDelays        = EVAPathDelays;
        obj.AveragePathGains   = EVAPathGains;
        obj.MaximumDopplerShift = 5;
    case 'EVA 70Hz'
        obj.PathDelays        = EVAPathDelays;
        obj.AveragePathGains   = EVAPathGains;
        obj.MaximumDopplerShift = 70;
    case 'ETU 70Hz'
        obj.PathDelays        = ETUPathDelays;
        obj.AveragePathGains   = ETUPathGains;
        obj.MaximumDopplerShift = 70;
    case 'ETU 300Hz'
        obj.PathDelays        = ETUPathDelays;
        obj.AveragePathGains   = ETUPathGains;
        obj.MaximumDopplerShift = 300;
    end
```

6.5.4   Initializing MIMO Channels

As we initialize the simulation, many properties that are either constant or reused in multiple functions are stored in various simulating parameter structures. In Chapter 4, we introduced a parameter structure called *prmLTEDLSCH*, which contains the properties needed to perform turbo coding and payload generation. In Chapter 5, we introduced a parameter structure called *prmLTEPDSCH*, which contains the properties needed to perform downlink shared-channel operations, including resource-grid mapping, OFDM signal generation, and MIMO operations. In this chapter, we introduce a parameter structure called *prmMdl*, which contains multiple properties related to specification of the MIMO fading channel and the criteria needed to stop the simulation.

The following MATLAB function initializes the *prmMdl* parameter structure. Depending on the values of nine parameters specified at the beginning of the simulation, the function sets a number of the structure's fields. For example, depending on the string specified as the *chanMdl* input argument, different values are set for the path delays, path gains, Doppler shift, and channel type. This determines whether a flat or frequency-selective fading is implemented and how the amount of mobility reflected by the Doppler-shift parameter affects the fading operations.

Algorithm

MATLAB function

```
function prmMdl = prmsMdl(chanSRate, chanMdl, numTx, numRx, ...
    corrLvl, chEstOn, snrdB, maxNumErrs, maxNumBits)
prmMdl.chanMdl = chanMdl;
prmMdl.AntConfig=char([48+numTx,'x',48+numRx]);
switch chanMdl
    case 'flat-low-mobility',
        prmMdl.PathDelays = 0*(1/chanSRate);
        prmMdl.PathGains  = 0;
        prmMdl.Doppler=0;
        prmMdl.ChannelType =1;
    case 'flat-high-mobility',
        prmMdl.PathDelays = 0*(1/chanSRate);
        prmMdl.PathGains  = 0;
        prmMdl.Doppler=70;
        prmMdl.ChannelType =1;
    case 'frequency-selective-low-mobility',
        prmMdl.PathDelays = [0 10 20 30 100]*(1/chanSRate);
        prmMdl.PathGains  = [0 -3 -6 -8 -17.2];
        prmMdl.Doppler=0;
        prmMdl.ChannelType =1;
    case 'frequency-selective-high-mobility',
        prmMdl.PathDelays = [0 10 20 30 100]*(1/chanSRate);
        prmMdl.PathGains  = [0 -3 -6 -8 -17.2];
        prmMdl.Doppler=70;
        prmMdl.ChannelType =1;
```

```
    case 'EPA 0Hz'
        prmMdl.PathDelays = [0 30 70 90 110 190 410]*1e-9;
        prmMdl.PathGains  = [0 -1 -2 -3 -8 -17.2 -20.8];
        prmMdl.Doppler=0;
        prmMdl.ChannelType =1;
    otherwise
        prmMdl.PathDelays = 0*(1/chanSRate);
        prmMdl.PathGains  = 0;
        prmMdl.Doppler=0;
        prmMdl.ChannelType =2;
end
prmMdl.corrLevel = corrLvl;
prmMdl.chEstOn = chEstOn;
prmMdl.snrdB=snrdB;
prmMdl.maxNumBits=maxNumBits;
prmMdl.maxNumErrs=maxNumErrs;
```

6.5.5 Adding AWGN

In Chapter 8, we introduced the *AWGNChannel* function, which adds white Gaussian noise to the signal. The following MATLAB code segment shows how channel modeling is performed by combining a fading channel with an AWGN (Additive White Gaussian Noise) channel. First, by calling the *MIMOFadingChan* function, we generate the faded version of the transmitted signal (*rxFade*) and the corresponding channel matrix (*chPathG*). Note that in the *MIMOFadingChan* function we specified path gains as being normalized. Despite this specification, since the MIMO fading channel computes the faded signal as a linear combination of multiple transmit antennas the output signal (*rxFade*) may not have a unity variance. To compute the noise variance needed to execute the *AWGNChannel* function, we must first compute the signal variance (*sigPow*) and derive the noise variance as the difference between the signal power and the SNR value in decibels.

Algorithm

MATLAB code segment

```
%% Channel
% MIMO Fading channel
[rxFade, chPathG] = MIMOFadingChan(txSig, prmLTEPDSCH, prmMdl);
% Add AWG noise
sigPow = 10*log10(var(rxFade));
nVar = 10.^(0.1.*(sigPow-snrdB));
rxSig =  AWGNChannel(rxFade, nVar);
```

Finally, through a decibel-to-linear transformation we compute the noise variance (*nVar*) as a vector of linear values. Since the second dimension of the faded output signal (*rxFade*) is equal to the number of receive antennas (*numRx*), the noise variance vector will have a dimension equal to the number of receive antennas. As we will see shortly, these noise-variance estimates are important parameters in equalization and demodulation procedures.

6.6 Common MIMO Features

Some of the functionalities introduced in the previous chapter for multicarrier transmission are common between it and the current chapter and need to be modified to accommodate multiple antennas. These functional components include resource-element mapping and demapping, channel-estimation methods, channel-response extraction, and equalization. On the other hand, some of the functionalities are unique to the MIMO implementation, including precoding, layer mapping, and the MIMO receiver. In this section we detail the modifications required for common functionalities and introduce the original MIMO operations.

6.6.1 MIMO Resource Grid Structure

The Cell-Specific Reference (CSR) signals play a critical role in both frequency-domain equalization (see Chapter 5) and MIMO receiver operations (to be described shortly). There is a fundamental difference, however, in the MIMO case, resulting from the multi-antenna requirements. When a CSR signal is transmitted on any antenna at any given subcarrier, all other antennas must transmit nothing (a zero-valued signal) at the same subcarrier. This requirement introduces a new set of components to be included within the resource grid, called spectral nulls.

Figure 6.4 shows the locations of CSR and spectral nulls within a typical resource block in cases where one, two, or four transmit antennas are used. The single-antenna case is illustrated at the top, showing that CSR signals are available in four OFDM symbols per subframe and that in every symbol there are two CSR samples available within each resource block. In this case there is no need for a spectral null, since only one antenna transmits any information. This configuration, for resource-element mapping and demapping, was implemented in the last chapter by the functions *REmapper_1Tx.m* and *REdemapper_1Tx.m*, respectively.

In the 2 × 2 MIMO configuration shown in the middle of Figure 6.4, we can see the addition of spectral nulls (zero-valued resource elements, marked by the letter x) in both antennas. Note also that the location of the spectral nulls in the resource block of one transmit antenna coincides exactly with the location of a CSR signal in the same resource block of the other one. In this 2 × 2 MIMO case, the density of the CSR signals is the same across multiple antennas. This means that in both antennas there are four OFDM symbols containing CSR signals and in each symbol there are two CSR signals per resource block.

In the 4 × 4 configuration, illustrated at the bottom of the figure, we can see two differences:

1. The CSR density in the first and second transmit antennas is different to that in the third and fourth.
2. The spectral nulls across all four transmit antennas have a higher overall density than in the two-antenna case.

The first and second transmit antennas have the same CSR signal density as found in the 2 × 2 MIMO configuration. This means there are four OFDM symbols containing CSR signals and

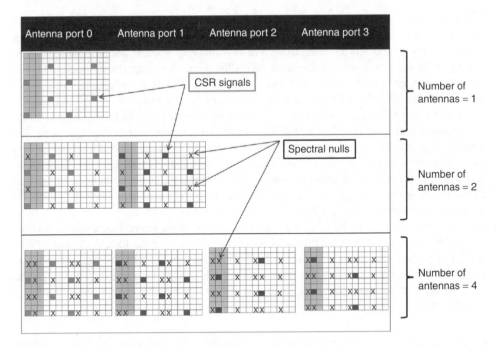

Figure 6.4 Cell-Specific Reference (CSR) signals and spectral nulls for one, two, and four antennas

that in each symbol and there are two CSR signals per resource block. In the third and fourth transmit antennas, there are only two OFDM symbols containing CSR signals, located on the first and eighth symbols, and there are two CSR signals per resource block. The location of the spectral nulls in any one transmit antenna coincides exactly with the location of all CSR signals in the other transmit antennas. As a consequence, the sum of CSR signals and spectral nulls is constant across different transmit antennas.

Details regarding the locations of CSR signals and spectral nulls form the basis for specifying resource-element mapping and demapping in the multi-antenna case. Next we will discuss the functions that implement this functionality: *REmapper_mTx.m* and *REdemapper_mTx.m*.

6.6.2   Resource-Element Mapping

In this section we detail the resource-element mapping for the MIMO transmission modes. As in single-antenna transmission, resource-element mapping is performed essentially by creating indices to the resource grid matrix and placing various information types within the grid. The types of signal that form the LTE downlink resource grid include the user data (PDSCH, Physical Downlink Shared Channel), CSR signals, Primary Synchronization Signal and Secondary Synchronization Signal (PSS, SSS), Physical Broadcast Channel (PBCH), and Physical Downlink Control Channel (PDCCH). The composition of the resource grid in the MIMO case is very similar to that for a single antenna, except that we need to include two more features. First, we must introduce the spectral nulls needed to mitigate interference among CSR signals during spectral estimation. Second, we must implement the special case of CSR placement in the 4×4 configuration, where the number of CSR symbols varies across multiple antennas.

The following MATLAB function shows the resource-element mapping. This function implements the mapping for the SISO (Single Input Single Output), SIMO (Single Input Multiple Output), and MIMO cases, using one, two, and four transmit antennas, respectively. The function takes as input the user data (*in*), CSR signal (*csr*), subframe index (*nS*), and PDSCH parameters, captured in a structure called prmLTEPDSCH. Depending on the availability of BCH (Broadcast Channel), SSS, PSS, and DCI (Downlink Control Information), the function may take on additional inputs. The output variable (*y*) is the resource grid matrix. The resource grid is a 3D matrix whose first dimension is the number of subcarriers, second dimension is equal to the number of OFDM symbols per subframe, and third dimension is the number of transmit antennas. The function is composed of three sections. In the first, depending on the number of transmit antennas (*numTx*), we initialize the indices for the user data (*idx_data*), the CSR signals (*idx_csr*), and the DCI (*idx_pdcch*). To compute indices for the user data we use the function *ExpungeFrom.m* to exclude the locations of all CSR indices. This way we exclude both the CSR and nulls in each transmit antenna. In the second section, we exclude from the user data and DCI indices the locations of the PSS, SSS, and PBCH, according to the value of the subframe index (*nS*). Finally, in the third section, we initialize the output buffer. By initializing the entire resource grid to zero we essentially place spectral nulls within it at locations where no other information is written. For each transmit antenna we fill up the resource grid using the indices generated in the first two sections.

Algorithm

MATLAB function

```
function y = REmapper_mTx(in, csr, nS, prmLTE, varargin)
%#codegen
switch nargin
   case 4, pdcch=[];pss=[];sss=[];bch=[];
   case 5, pdcch=varargin{1};pss=[];sss=[];bch=[];
   case 6, pdcch=varargin{1};pss=varargin{2};sss=[];bch=[];
   case 7, pdcch=varargin{1};pss=varargin{2};sss=varargin{3};bch=[];
   case 8, pdcch=varargin{1};pss=varargin{2};sss=varargin{3};bch=varargin{4};
   otherwise
      error('REMapper has 4 to 8 arguments!');
end
% NcellID = 0;                          % One of possible 504 values
% Get input params
numTx          = prmLTE.numTx;         % Number of transmit antennas
Nrb            = prmLTE.Nrb;
Nrb_sc         = prmLTE.Nrb_sc;        % 12 for normal mode
Ndl_symb       = prmLTE.Ndl_symb;      % 7   for normal mode
numContSymb    = prmLTE.contReg;       % either {1, 2, 3}
%% Specify resource grid location indices for CSR, PDCCH, PDSCH, PBCH, PSS, SSS
coder.varsize('idx_data');
lenOFDM = Nrb*Nrb_sc;
ContREs=numContSymb*lenOFDM;
idx_dci=1:ContREs;
```

```
lenGrid= lenOFDM * Ndl_symb*2;
idx_data  = ContREs+1:lenGrid;
%% 1st: Indices for CSR pilot symbols
idx_csr0  = 1:6:lenOFDM;           % More general starting point = 1+mod(NcellID, 6);
idx_csr4  = 4:6:lenOFDM;           % More general starting point = 1+mod(3+NcellID, 6);
% Depends on number of transmit antennas
switch numTx
  case 1
    idx_csr    = [idx_csr0, 4*lenOFDM+idx_csr4, 7*lenOFDM+idx_csr0, 11*lenOFDM
+idx_csr4];
    idx_data  = ExpungeFrom(idx_data,idx_csr);
    idx_pdcch = ExpungeFrom(idx_dci,idx_csr0);
    idx_ex     = 7.5* lenOFDM - 36 + (1:6:72);
    a=numel(idx_csr); IDX=[1, a];
  case 2
    idx_csr1   = [idx_csr0, 4*lenOFDM+idx_csr4, 7*lenOFDM+idx_csr0, 11*lenOFDM
+idx_csr4];
    idx_csr2   = [idx_csr4, 4*lenOFDM+idx_csr0, 7*lenOFDM+idx_csr4, 11*lenOFDM
+idx_csr0];
    idx_csr    = [idx_csr1, idx_csr2];
    % Exclude pilots and NULLs
    idx_data  = ExpungeFrom(idx_data,idx_csr1);
    idx_data  = ExpungeFrom(idx_data,idx_csr2);
    idx_pdcch = ExpungeFrom(idx_dci,idx_csr0);
    idx_pdcch = ExpungeFrom(idx_pdcch,idx_csr4);
    idx_ex     = 7.5* lenOFDM - 36 + (1:3:72);
    % Point to pilots only
    a=numel(idx_csr1); IDX=[1, a; a+1, 2*a];
  case 4
    idx_csr1   = [idx_csr0, 4*lenOFDM+idx_csr4, 7*lenOFDM+idx_csr0, 11*lenOFDM
+idx_csr4];
    idx_csr2   = [idx_csr4, 4*lenOFDM+idx_csr0, 7*lenOFDM+idx_csr4, 11*lenOFDM
+idx_csr0];
    idx_csr33  = [lenOFDM+idx_csr0, 8*lenOFDM+idx_csr4];
    idx_csr44  = [lenOFDM+idx_csr4, 8*lenOFDM+idx_csr0];
    idx_csr    = [idx_csr1, idx_csr2, idx_csr33, idx_csr44];
    % Exclude pilots and NULLs
    idx_data  = ExpungeFrom(idx_data,idx_csr1);
    idx_data  = ExpungeFrom(idx_data,idx_csr2);
    idx_data  = ExpungeFrom(idx_data,idx_csr33);
    idx_data  = ExpungeFrom(idx_data,idx_csr44);
    % From pdcch
    idx_pdcch = ExpungeFrom(idx_dci,idx_csr0);
    idx_pdcch = ExpungeFrom(idx_pdcch,idx_csr4);
    idx_pdcch = ExpungeFrom(idx_pdcch,lenOFDM+idx_csr0);
    idx_pdcch = ExpungeFrom(idx_pdcch,lenOFDM+idx_csr4);
    idx_ex     = [7.5* lenOFDM - 36 + (1:3:72), 8.5* lenOFDM - 36 + (1:3:72)];
    % Point to pilots only
    a=numel(idx_csr1); b=numel(idx_csr33);
```

```
      IDX =[1, a; a+1, 2*a; 2*a+1, 2*a+b; 2*a+b+1, 2*a+2*b];
   otherwise
      error('Number of transmit antennas must be {1, 2, or 4}');
end
%% 3rd: Indices for PDSCH and PDSCH data in OFDM symbols where pilots are present
%% Handle 3 types of subframes differently
switch nS
   %% 4th: Indices for BCH, PSS, SSS are only found in specific subframes 0 and 5
   % These symbols share the same 6 center sub-carrier locations (idx_ctr)
   % and differ in OFDM symbol number.
   case 0    % Subframe 0
      % PBCH, PSS, SSS are available + CSR, PDCCH, PDSCH
      idx_ctr = 0.5* lenOFDM - 36 + (1:72) ;
      idx_SSS  = 5* lenOFDM + idx_ctr;
      idx_PSS  = 6* lenOFDM + idx_ctr;
      idx_bch0=[7*lenOFDM + idx_ctr, 8*lenOFDM + idx_ctr, 9*lenOFDM + idx_ctr,
10*lenOFDM + idx_ctr];
      idx_bch = ExpungeFrom(idx_bch0,idx_ex);
      idx_data  = ExpungeFrom(idx_data,[idx_SSS, idx_PSS, idx_bch]);
   case 10  % Subframe 5
      % PSS, SSS are available + CSR, PDCCH, PDSCH
      % Primary and Secondary synchronization signals in OFDM symbols 5 and 6
      idx_ctr = 0.5* lenOFDM - 36 + (1:72) ;
      idx_SSS  = 5* lenOFDM + idx_ctr;
      idx_PSS  = 6* lenOFDM + idx_ctr;
      idx_data  = ExpungeFrom(idx_data,[idx_SSS, idx_PSS]);
   otherwise % other subframes
      % Nothing to do
end
% Initialize output buffer
y = complex(zeros(Nrb*Nrb_sc, Ndl_symb*2, numTx));
for m=1:numTx
   grid = complex(zeros(Nrb*Nrb_sc, Ndl_symb*2));
   grid(idx_data.')=in(:,m);                          % Insert user data
   Range=idx_csr(IDX(m,1):IDX(m,2)).';                 % How many pilots in this antenna
   csr_flat=packCsr(csr, m, numTx);                  % Pack correct number of CSR values
   grid(Range)= csr_flat(:);                          % Insert CSR pilot symbols
   if ~isempty(pdcch), grid(idx_pdcch)=pdcch(:,m);end
% Insert Physical Downlink Control Channel (PDCCH)
   if ~isempty(pss),    grid(idx_PSS)=pss(:,m);end
% Insert Primary Synchronization Signal (PSS)
   if ~isempty(sss),    grid(idx_SSS)=sss(:,m);end
% Insert Secondary Synchronization Signal (SSS)
   if ~isempty(bch),    grid(idx_bch)=bch(:,m);end % Insert Broadcast Channel data (BCH)
   y(:,:,m)=grid;
end
end
%% Helper function
function csr_flat=packCsr(csr, m, numTx)
```

```
    if ((numTx==4)&&(m>2))                  % Handle special case of 4Tx
        csr_flat=csr(:,[1,3],m);            % Extract pilots in this antenna
    else
        csr_flat=csr(:,:,m);
    end
end
```

6.6.3 Resource-Element Demapping

Resource-element demapping inverts the operations of resource-grid mapping. The following MATLAB function illustrates how the reference signal and data are extracted from the recovered resource grid at the receiver. The function has three input arguments: the received resource grid (*in*), the index of the subframe (*nS*), and the PDSCH parameter set. The function outputs extracted user data (*data*), the indices to the user data (*idx_data*), the CSR signals (*csr*), and optionally the DCI (*pdcch*), primary and secondary synchronization signals (*pss*, *sss*), and BCH signal (*bch*). As different subframes contain different content, the second input subframe index parameter (*nS*) enables the function to separate the correct data. The same algorithm used in the resource-mapping function is used here to generate indices in the demapping function. In the multi-antenna case, the resource-grid input is a 3D matrix. The first two dimensions specify the size of the resource grid for each receive antenna and the third dimension is the number of receive antennas. Like the resource-mapping function, resource demapping is performed in three sections. In the first two, we compute the indices localizing various components of the resource grid. These include indices for the user data (*idx_data*), the CSR signals (*idx_csr*), the DCI (*idx_pdcch*), primary and secondary synchronization signals (*idx_PSS*, *idx_SSS*), and the BCH signal (*idx_bch*). In the third section, we extract these data components from the resource grid for each receive antenna using the indices we generated in the first two sections.

Algorithm

MATLAB function

```
function [data, csr, idx_data, pdcch, pss, sss, bch] = REdemapper_mTx(in, nS, prmLTE)
%#codegen
% NcellID = 0;                           % One of possible 504 values
% Get input params
numTx        = prmLTE.numTx;            % number of receive antennas
numRx        = prmLTE.numRx;            % number of receive antennas
Nrb          = prmLTE.Nrb;             % either of {6,...,100 }
Nrb_sc       = prmLTE.Nrb_sc;          % 12 for normal mode
Ndl_symb     = prmLTE.Ndl_symb;        % 7   for normal mode
numContSymb  = prmLTE.contReg;         % either {1, 2, 3}
Npss         = prmLTE.numPSSRE;
Nsss         = prmLTE.numSSSRE;
Nbch         = prmLTE.numBCHRE;
%% Specify resource grid location indices for CSR, PDCCH, PDSCH, PBCH, PSS, SSS
```

```
coder.varsize('idx_data');
coder.varsize('idx_dataC');
lenOFDM = Nrb*Nrb_sc;
ContREs=numContSymb*lenOFDM;
idx_dci=1:ContREs;
lenGrid= lenOFDM * Ndl_symb*2;
idx_data  = ContREs+1:lenGrid;
%% 1st: Indices for CSR pilot symbols
idx_csr0  = 1:6:lenOFDM;              % More general starting point = 1+mod(NcellID, 6);
idx_csr4  = 4:6:lenOFDM;              % More general starting point = 1+mod(3+NcellID, 6);
% Depends on number of transmit antennas
switch numTx
  case 1
    idx_csr    = [idx_csr0, 4*lenOFDM+idx_csr4, 7*lenOFDM+idx_csr0, 11*lenOFDM
+idx_csr4];
    idx_data   = ExpungeFrom(idx_data,idx_csr);
    idx_pdcch  = ExpungeFrom(idx_dci,idx_csr0);
    idx_ex     = 7.5* lenOFDM - 36 + (1:6:72);
  case 2
    idx_csr1   = [idx_csr0, 4*lenOFDM+idx_csr4, 7*lenOFDM+idx_csr0, 11*lenOFDM
+idx_csr4];
    idx_csr2   = [idx_csr4, 4*lenOFDM+idx_csr0, 7*lenOFDM+idx_csr4, 11*lenOFDM
+idx_csr0];
    idx_csr    = [idx_csr1, idx_csr2];
    % Exclude pilots and NULLs
    idx_data   = ExpungeFrom(idx_data,idx_csr1);
    idx_data   = ExpungeFrom(idx_data,idx_csr2);
    idx_pdcch  = ExpungeFrom(idx_dci,idx_csr0);
    idx_pdcch  = ExpungeFrom(idx_pdcch,idx_csr4);
    idx_ex     = 7.5* lenOFDM - 36 + (1:3:72);
  case 4
    idx_csr1   = [idx_csr0, 4*lenOFDM+idx_csr4, 7*lenOFDM+idx_csr0, 11*lenOFDM
+idx_csr4];
    idx_csr2   = [idx_csr4, 4*lenOFDM+idx_csr0, 7*lenOFDM+idx_csr4, 11*lenOFDM
+idx_csr0];
    idx_csr33  = [lenOFDM+idx_csr0, 8*lenOFDM+idx_csr4];
    idx_csr44  = [lenOFDM+idx_csr4, 8*lenOFDM+idx_csr0];
    idx_csr    = [idx_csr1, idx_csr2, idx_csr33, idx_csr44];
    % Exclude pilots and NULLs
    idx_data   = ExpungeFrom(idx_data,idx_csr1);
    idx_data   = ExpungeFrom(idx_data,idx_csr2);
    idx_data   = ExpungeFrom(idx_data,idx_csr33);
    idx_data   = ExpungeFrom(idx_data,idx_csr44);
    % From pdcch
    idx_pdcch  = ExpungeFrom(idx_dci,idx_csr0);
    idx_pdcch  = ExpungeFrom(idx_pdcch,idx_csr4);
    idx_pdcch  = ExpungeFrom(idx_pdcch,lenOFDM+idx_csr0);
    idx_pdcch  = ExpungeFrom(idx_pdcch,lenOFDM+idx_csr4);
    idx_ex     = [7.5* lenOFDM - 36 + (1:3:72), 8.5* lenOFDM - 36 + (1:3:72)];
```

```
        otherwise
            error('Number of transmit antennas must be {1, 2, or 4}');
end
%% 3rd: Indices for PDSCH and PDSCH data in OFDM symbols where pilots are present
%% Handle 3 types of subframes differently
switch nS
    %% 4th: Indices for BCH, PSS, SSS are only found in specific subframes 0 and 5
    % These symbols share the same 6 center sub-carrier locations (idx_ctr)
    % and differ in OFDM symbol number.
    case 0    % Subframe 0
        % PBCH, PSS, SSS are available + CSR, PDCCH, PDSCH
        idx_ctr = 0.5* lenOFDM - 36 + (1:72) ;
        idx_SSS  = 5* lenOFDM + idx_ctr;
        idx_PSS  = 6* lenOFDM + idx_ctr;
        idx_bch0=[7*lenOFDM + idx_ctr, 8*lenOFDM + idx_ctr, 9*lenOFDM + idx_ctr,
10*lenOFDM + idx_ctr];
        idx_bch = ExpungeFrom(idx_bch0,idx_ex);
        idx_data  = ExpungeFrom(idx_data,[idx_SSS, idx_PSS, idx_bch]);
    case 10  % Subframe 5
        % PSS, SSS are available + CSR, PDCCH, PDSCH
        % Primary and Secondary synchronization signals in OFDM symbols 5 and 6
        idx_ctr = 0.5* lenOFDM - 36 + (1:72) ;
        idx_SSS  = 5* lenOFDM + idx_ctr;
        idx_PSS  = 6* lenOFDM + idx_ctr;
        idx_data  = ExpungeFrom(idx_data,[idx_SSS, idx_PSS]);
    otherwise % other subframes
        % Nothing to do
end
%% Write user data PDCCH, PBCH, PSS, SSS, CSR
pss=complex(zeros(Npss,numRx));
sss=complex(zeros(Nsss,numRx));
bch=complex(zeros(Nbch,numRx));
pdcch = complex(zeros(numel(idx_ pdcch),numRx));
data=complex(zeros(numel(idx_data),numRx));
idx_dataC=idx_data.';
for n=1:numRx
    grid=in(:,:,n);
    data(:,n)=grid(idx_dataC);              % Physical Downlink Shared Chan-
nel (PDSCH) = user data
    pdcch(:,n) = grid(idx_ pdcch.');        % Physical Downlink Control Channel (PDCCH)
    if nS==0
        pss(:,n)=grid(idx_PSS.');           % Primary Synchronization Signal (PSS)
        sss(:,n)=grid(idx_SSS.');           % Secondary Synchronization Signal (SSS)
        bch(:,n)=grid(idx_bch.');           % Broadcast Channel data (BCH)
    elseif nS==10
        pss(:,n)=grid(idx_PSS.');           % Primary Synchronization Signal (PSS)
        sss(:,n)=grid(idx_SSS.');           % Secondary Synchronization Signal (SSS)
    end
end
```

```
%% Cell-specific Reference Signal (CSR) = pilots
switch numTx
    case 1                                    % Case of 1 Tx
        csr=complex(zeros(2*Nrb,4,numRx));    % 4 symbols have CSR  per Subframe
        for n=1:numRx
            grid=in(:,:,n);
            csr(:,:,n)=reshape(grid(idx_csr'), 2*Nrb,4) ;
        end
    case 2                                    % Case of 2 Tx
        idx_0=(1:3:lenOFDM);                  % Total number of Nulls + CSR are constant
        idx_all=[idx_0,  4*lenOFDM+idx_0, 7*lenOFDM+idx_0,  11*lenOFDM+idx_0]';
        csr=complex(zeros(4*Nrb,4,numRx));    % 4 symbols have CSR+NULLs per Subframe
        for n=1:numRx
            grid=in(:,:,n);
            csr(:, :,n)=reshape(grid(idx_all), 4*Nrb,4) ;
        end
    case 4
        idx_0=(1:3:lenOFDM);                  % Total number of Nulls + CSR are constant
        idx_all=[idx_0,          lenOFDM+idx_0,    4*lenOFDM+idx_0, ...
            7*lenOFDM+idx_0,  8*lenOFDM+idx_0,  11*lenOFDM+idx_0]';
        csr=complex(zeros(4*Nrb,6,numRx));    % 4 symbols have CSR+NULLs  per Subframe
        for n=1:numRx
            grid=in(:,:,n);
            csr(:, :,n)=reshape(grid(idx_all), 4*Nrb,6) ;
        end
end
end
```

6.6.4 CSR-Based Channel Estimation

The system of linear equations characterizing a MIMO channel can be expressed as follows:

$$\vec{Y}(n) = \boldsymbol{H}(n) * \vec{X}(n) + \vec{n} \tag{6.1}$$

where at time index n and at any given subcarrier, $\vec{Y}(n)$ is the received signal, $\vec{X}(n)$ is the transmitted signal, $\boldsymbol{H}(n)$ is the channel matrix, and $\vec{n}$ represents the AWGN vector. When the receiver has obtained the received signal $\vec{Y}(n)$, we must compute an estimate for the channel matrix $\boldsymbol{H}(n)$ and the noise $\vec{n}$ in order to properly estimate the transmitted signal $\vec{X}(n)$. Assuming that an estimate of the channel AWGN is available, we focus in this section on ways of estimating the channel matrix.

Let us denote the number of transmit antennas by *numTx* and the number of receive antennas by *numRx*. The channel matrix has a dimension of (*numRx, numTx*). For each subcarrier and for each OFDM symbol, $numRx \times numTx$ values must be estimated for the channel matrix. As discussed in the last chapter, we use the CSR (pilot) signals for channel-matrix estimation. Let us see how multi-antenna transmission affects the channel estimation process. Considering, for example, a 2×2 configuration for the MIMO channel, the MIMO system of equation at a given time index can be expressed as:

$$\begin{bmatrix} y_1(n) \\ y_2(n) \end{bmatrix} = \begin{bmatrix} h_{1,1}(n) & h_{1,2}(n) \\ h_{2,1}(n) & h_{2,2}(n) \end{bmatrix} * \begin{bmatrix} x_1(n) \\ x_2(n) \end{bmatrix} + \begin{bmatrix} n_1 \\ n_2 \end{bmatrix} \tag{6.2}$$

Focusing on a single receive antenna, for example $y_1(n)$, the value of the received signal is a linear combination of values in two transmit antennas scaled by two channel gains:

$$y_1(n) = h_{1,1}(n) * x_1(n) + h_{1,2}(n) * x_2(n) + n_1 \tag{6.3}$$

Since multicarrier transmission allows us to perform channel estimation in the frequency domain, by taking a discrete Fourier transform of this expression we can express the relationship between the channel gains and received and transmitted signals as follows:

$$y_1(\omega) = h_{1,1}(\omega) * x_1(\omega) + h_{1,2}(\omega) * x_2(\omega) + nVar \tag{6.4}$$

where $y_1(\omega)$, for example, is the Fourier transform of the corresponding time-domain signal $y_1(n) \overset{FFT}{\longleftrightarrow} y_1(\omega)$ and $nVar$ is the noise variance of the AWGN channel at a given subcarrier. Note that variables $y_1(\omega)$, $x_1(\omega)$, and $x_2(\omega)$ are received and transmitted values at a given subcarrier and a given OFDM symbol in a transmitted and received resource grid, respectively.

If we choose known pilot (CSR) signals for variables $x_1(\omega)$ and $x_2(\omega)$, then by knowing the received variable $y_1(\omega)$ and ignoring the noise variance we can easily estimate channel-matrix variables $h_{1,1}(\omega)$ and $h_{1,2}(\omega)$. This is where the need for spectral nulls becomes apparent. At a given subcarrier and with a given OFDM symbol, when the value of $x_1(\omega)$ is equal to a reference signal at the same subcarrier the value of $x_2(\omega)$ is equal to zero, because this variable represents a spectral null. As a result, the previous equation can be modified to derive an expression for the channel matrix:

$$y_1(\omega) = h_{1,1}(\omega) * x_1(\omega)]_{\omega=subcarrier} + h_{1,2}(\omega) * x_2(\omega)]_{\omega=subcarrier}$$

$$y_1(\omega) = h_{1,1}(\omega) * x_1(\omega) + h_{1,2}(\omega) * 0.0$$

$$y_1(\omega) = h_{1,1}(\omega) * x_1(\omega) \tag{6.5}$$

This discussion shows that by exploiting the CSR signals and spectral nulls embedded within the resource grid, we can estimate the channel-matrix path gain value $h_{m,n}(\omega)$ as:

$$h_{m,n}(\omega) = \frac{y_n(\omega)}{x_m(\omega)} \tag{6.6}$$

where m is the index of the transmit antenna, with a range equal to $m = 1, \ldots, numTx$ and n is the index of the receive antenna, with a range equal to $m = 1, \ldots, numRx$. In the next section we see how in MATLAB we can use the transmitted and received CSR signals to implement this equation and estimate the channel matrix. Then, by expanding the channel matrix across the resource grid through interpolation, we arrive at an estimate of the channel-frequency response over the entire grid.

6.6.5   Channel-Estimation Function

The following MATLAB function illustrates how channel estimation is performed using the transmitted and received reference symbols, also referred to as pilots, at regular intervals within the OFDM time–frequency grid. The function has four input arguments: the parameters of the PDSCH captured in a structure (*prmLTE*), the received CSR signal (*Rx*), the transmitted reference CSR signal (*Ref*), and a parameter representing the channel-estimation mode (*Mode*). As its output, the function computes the channel-frequency response over the entire grid (*hD*).

Algorithm

MATLAB function

```
function hD = ChanEstimate_mTx(prmLTE, Rx, Ref, Mode)
%#codegen
Nrb      = prmLTE.Nrb;     % Number of resource blocks
Nrb_sc   = prmLTE.Nrb_sc;           % 12 for normal mode
Ndl_symb = prmLTE.Ndl_symb;    % 7   for normal mode
numTx    = prmLTE.numTx;
numRx    = prmLTE.numRx;
% Initialize output buffer
switch numTx
  case 1                                           % Case of 1 Tx
    hD = complex(zeros(Nrb*Nrb_sc, Ndl_symb*2,numRx));   % Initialize Output
    % size(Rx) = [2*Nrb,  4,numRx]  size(Ref) = [2*Nrb, 4]
    Edges=[0,3,0,3];
    for n=1:numRx
      Rec=Rx(:,:,n);
      hp= Rec./Ref;
      hD(:,:,n)=gridResponse(hp, Nrb, Nrb_sc, Ndl_symb, Edges,Mode);
    end
  case 2                                % Case of 2 Tx
    hD = complex(zeros(Nrb*Nrb_sc, Ndl_symb*2,numTx, numRx));
    % size(Rx) = [4*Nrb,  4,numRx]  size(Ref) = [2*Nrb, 4, numTx]
    for n=1:numRx
      Rec=Rx(:,:,n);
      for m=1:numTx
        [R,Edges]=getBoundaries2(m, Rec);
        T=Ref(:,:,m);
        hp= R./T;
        hD(:,:,m,n)=gridResponse(hp, Nrb, Nrb_sc, Ndl_symb, Edges,Mode);
      end
    end
  case 4
    hD = complex(zeros(Nrb*Nrb_sc, Ndl_symb*2,numTx, numRx));
    % size(Rx) = [4*Nrb,  4,numRx]  size(Ref) = [2*Nrb, 4, numTx]
    for n=1:numRx
      Rec=Rx(:,:,n);
```

```
        for m=1:numTx
           [R,idx3, Edges]=getBoundaries4(m, Rec);
           T=Ref(:,idx3,m);
           hp= R./T;
           hD(:,:,m,n)=gridResponse(hp, Nrb, Nrb_sc, Ndl_symb, Edges,Mode);
        end
     end
end
end
%% Helper function
function [R,idx3, Edges]=getBoundaries4(m, Rec)
coder.varsize('Edges');coder.varsize('idx3');
numPN=size(Rec,1);
idx_0=(1:2:numPN);
idx_1=(2:2:numPN);
Edges=[0,3,0,3];
idx3=1:4;
switch m
  case 1
     index=[idx_0,  2*numPN+idx_1, 3*numPN+idx_0,  5*numPN+idx_1]';
     Edges=[0,3,0,3];   idx3=1:4;
  case 2
     index=[idx_1,  2*numPN+idx_0, 3*numPN+idx_1,  5*numPN+idx_0]';
     Edges=[3,0,3,0];   idx3=1:4;
  case 3
     index=[numPN+idx_0,  4*numPN+idx_1]';
     Edges=[0,3];       idx3=[1 3];
  case 4
     index=[numPN+idx_1,  4*numPN+idx_0]';
     Edges=[3,0];       idx3=[1 3];
end
R=reshape(Rec(index),numPN/2,numel(Edges));
end
%% Helper function
function [R, Edges]=getBoundaries2(m, Rec)
numPN=size(Rec,1);
idx_0=(1:2:numPN);
idx_1=(2:2:numPN);
Edges=[0,3,0,3];
switch m
  case 1
     index=[idx_0,  numPN+idx_1, 2*numPN+idx_0, 3*numPN+idx_1]';
     Edges=[0,3,0,3];
  case 2
     index=[idx_1,  numPN+idx_0, 2*numPN+idx_1, 3*numPN+idx_0]';
     Edges=[3,0,3,0];
end
R=reshape(Rec(index),numPN/2,4);
end
```

The function performs channel estimation in two steps. First, it computes the channel matrix over elements of the resource grid aligned with the reference signal. This is accomplished by accessing all combinations of the transmitted reference signal (T) and received reference signal (R) and using the elementwise division operator in MATLAB to compute the channel-matrix elements. In the second step, we call the *gridResponse* function to expand the channel-matrix estimates over the entire grid from those computed based only on CSR values. The type of interpolation or averaging of values that makes expansion possible is specified by the input argument (*Mode*).Next we will look at various channel-expansion operations.

6.6.6   Channel-Estimate Expansion

The following MATLAB function shows three algorithms that can expand the channel matrices computed only over CSR signals to generate the function output (y), channel-frequency responses over the entire resource grid. The function takes as input the following arguments: a limited set of channel responses computed over pilots (hp) and parameters related to the dimensions of the resource grid, including the number of resource blocks (Nrb), number of subcarriers in each resource block (Nrb_sc), and number of OFDM symbols per slot (*Ndl_symb*), as well as two others: a vector that specifies the location of the CSR signal relative to the edge of the resource block (*Edges*) and the algorithm chosen to expand the response to the entire grid (*Mode*).

Algorithm

MATLAB function

```
function y=gridResponse(hp, Nrb, Nrb_sc, Ndl_symb, Edges,Mode)
%#codegen
switch Mode
   case 1
      y=gridResponse_interpolate(hp, Nrb, Nrb_sc, Ndl_symb, Edges);
   case 2
      y=gridResponse_averageSlot(hp, Nrb, Nrb_sc, Ndl_symb, Edges);
   case 3
      y=gridResponse_averageSubframe(hp,  Ndl_symb, Edges);
   otherwise
      error('Choose the right Mode in function ChanEstimate.');
end
end
```

The following MATLAB function (*gridResponse_interpolate.m*) executes if the value chosen for the *Mode* argument in the *gridResponse.m* function is 1. It performs an expansion algorithm based on frequency-and-time-domain interpolation. This algorithm involves interpolation between subcarriers in the frequency domain in OFDM symbols that contain CSR signals. Having computed the channel response for all subcarriers on these symbols, the function then interpolates in time to find the channel response across the whole resource grid.

The difference between this algorithm and the one used in the single-antenna case is the separate treatment of two-antenna and four-antenna cases. Note that the number of OFDM symbols containing CSR signals in the third and fourth antennas in the four-antenna case is only two. The interpolation between OFDM symbols must take this detail into account.

Algorithm

MATLAB function

```
function hD=gridResponse_interpolate(hp, Nrb, Nrb_sc, Ndl_symb, Edges)
% Average over the two same Freq subcarriers, and then interpolate between
% them - get all estimates and then repeat over all columns (symbols).
% The interpolation assmues NCellID = 0.
% Time average two pilots over the slots, then interpolate (F)
% between the 4 averaged values, repeat for all symbols in sframe
Separation=6;
hD = complex(zeros(Nrb*Nrb_sc, Ndl_symb*2));
N=numel(Edges);
% Compute channel response over all resource elements of OFDM symbols
switch N
    case 2
        Symbols=[2, 9];
        % Interpolate between subcarriers
        for n=1:N
            E=Edges(n);Edge=[E, 5-E];
            y = InterpolateCsr(hp(:,n),  Separation, Edge);
            hD(:,Symbols(n))=y;
        end
        % Interpolate between OFDM symbols
        for m=[1,3:8,10:14]
            alpha=(1/7)*(m-2);
            beta=1-alpha;
            hD(:,m)   = beta*hD(:,2) + alpha*hD(:, 9);
        end
    case 4
        Symbols=[1, 5, 8, 12];
        % Interpolate between subcarriers
        for n=1:N
            E=Edges(n);Edge=[E, 5-E];
            y = InterpolateCsr(hp(:,n),  Separation, Edge);
            hD(:,Symbols(n))=y;
        end
        % Interpolate between OFDM symbols
        for m=[2, 3, 4, 6, 7]
            alpha=0.25*(m-1);
            beta=1-alpha;
            hD(:,m)   = beta*hD(:,1) + alpha*hD(:, 5);
            hD(:,m+7) =beta*hD(:,8) + alpha*hD(:,12);
        end
```

```
      otherwise
         error('Wrong Edges parameter for function gridResponse.');
   end
```

The following MATLAB function (*gridResponse_averageSlot.m*) executes if the value of the *Mode* argument in the *gridResponse.m* function is set to 2. It performs an expansion algorithm based on frequency-domain interpolation and averaging in time among OFDM symbols in each slot. The operations of this algorithm depend on whether one or two OFDM symbols containing CSR signals are found in a given slot. If there are two OFDM symbols containing CSR signals, the algorithm combines the CSR signals from the first two OFDM symbols. In this case, instead of a separation of six subcarriers between CSR signals, we have a separation of three subcarriers. If there is only one OFDM symbol that contains CSR signals per slot (e.g., in a four-antenna case, in the third and fourth antennas), no CSR combination is performed and the separation between CSR values remains six subcarriers. As a next step, the function interpolates the values along the frequency axis based on the separation value determined previously. Finally, it applies the same channel response to all the OFDM symbols of a given slot and repeats the operations for the next slot in order to compute the channel response of the whole resource grid.

Algorithm

MATLAB function

```
function hD=gridResponse_averageSlot(hp, Nrb, Nrb_sc, Ndl_symb, Edges)
% Average over the two same Freq subcarriers, and then interpolate between
% them - get all estimates and then repeat over all columns (symbols).
% The interpolation assmues NCellID = 0.
% Time average two pilots over the slots, then interpolate (F)
% between the 4 averaged values, repeat for all symbols in sframe
Separation=3;
hD = complex(zeros(Nrb*Nrb_sc, Ndl_symb*2));
N=numel(Edges);
% Compute channel response over all resource elements of OFDM symbols
switch N
   case 2
      % Interpolate between subcarriers
      Index=1:Ndl_symb;
      for n=1:N
      E=Edges(n);Edge=[E, 5-E];
      y = InterpolateCsr(hp(:,n),  2* Separation, Edge);
      % Repeat between OFDM symbols in each slot
      yR=y(:,ones(1,Ndl_symb));
      hD(:,Index)=yR;
      Index=Index+Ndl_symb;
      end
   case 4
```

```
        Edge=[0 2];
        h1_a_mat = [hp(:,1),hp(:,2)].';
        h1_a = h1_a_mat(:);
        h2_a_mat = [hp(:,3),hp(:,4)].';
        h2_a = h2_a_mat(:);
        hp_a=[h1_a,h2_a];
        Index=1:Ndl_symb;
        for n=1:size(hp_a,2)
        y = InterpolateCsr(hp_a(:,n),  Separation, Edge);
        % Repeat between OFDM symbols in each slot
        yR=y(:,ones(1,Ndl_symb));
        hD(:,Index)=yR;
        Index=Index+Ndl_symb;
        end
    otherwise
        error('Wrong Edges parameter for function gridResponse.');
end
```

Finally, the following MATLAB function (*gridResponse_averageSubframe.m*) executes if the value of the *Mode* argument in the *gridResponse.m* function is set to 3. It performs an expansion algorithm based on frequency-domain interpolation and averaging in time among OFDM symbols in the entire subframe. The operations of this algorithm depend on whether or not there are two or four OFDM symbols containing CSR signals found in a given subframe. If there are four, the algorithm averages the values first in the first and third OFDM symbols and then in the second and fourth symbols, and then combines these average vectors. In this case, instead of a separation of six subcarriers between CSR signals we now have a separation of three subcarriers. If there is only one OFDM symbol per slot that contains CSR signals, the algorithm combines the two OFDM symbols, resulting in a separation of three subcarriers between combined CSR signals. As the next step, the function interpolates the values along the frequency axis based on a separation value of 3 in all cases. Finally, it applies the same channel response to all the OFDM symbols of the subframe as the channel response of the whole resource grid.

Algorithm

MATLAB function

```
function hD=gridResponse_averageSubframe(hp, Ndl_symb, Edges)
% Average over the two same Freq subcarriers, and then interpolate between
% them - get all estimates and then repeat over all columns (symbols).
% The interpolation assmues NCellID = 0.
% Time average two pilots over the slots, then interpolate (F)
% between the 4 averaged values, repeat for all symbols in sframe
Separation=3;
N=numel(Edges);
Edge=[0 2];
```

```
% Compute channel response over all resource elements of OFDM symbols
switch N
    case 2
        h1_a_mat = hp.';
        h1_a = h1_a_mat(:);
        % Interpolate between subcarriers
        y = InterpolateCsr(h1_a,  Separation, Edge);
        % Repeat between OFDM symbols
        hD=y(:,ones(1,Ndl_symb*2));
    case 4
        h1_a1 = mean([hp(:, 1), hp(:, 3)],2);
        h1_a2 = mean([hp(:, 2), hp(:, 4)],2);
        h1_a_mat = [h1_a1 h1_a2].';
        h1_a = h1_a_mat(:);
        % Interpolate between subcarriers
        y = InterpolateCsr(h1_a,  Separation, Edge);
        % Repeat between OFDM symbols
        hD=y(:,ones(1,Ndl_symb*2));
    otherwise
        error('Wrong Edges parameter for function gridResponse.');
end
```

The three algorithms mentioned here provide different dynamic behaviors within each subframe. Note that in OFDM transmission with a normal cyclic prefix, each slot contains seven OFDM symbols and each subframe contains fourteen. The first algorithm results in a channel estimate where the response within a subframe is dynamic and changes from one OFDM symbol to the next. The second algorithm results in a constant channel response for the first seven OFDM symbols (first slot) and in a different constant response for the next seven OFDM symbols (second slot). The third algorithm is the least dynamic implementation, with a single response applying to all OFDM symbols of a subframe.

6.6.7   Ideal Channel Estimation

So far we have discussed algorithms that rely on the pilots (CSR signals) to provide a channel-response estimate. These algorithms are realistic implementations and can be incorporated as part of a real system. In this section, we present what we call an "ideal channel estimator." This type of ideal algorithm relies on exact knowledge of the channel matrix or the path gain values that the MIMO channel model provides. Since the second output of the *MIMOFadingChan.m* function is the multidimensional channel matrix representing the path gains, the ideal channel estimator can use these path gains to compute the best estimate of the channel-frequency response for the entire resource grid. Note that because of the way it is formulated, the ideal channel estimator cannot be implemented as part of a real system. It can only be used during simulation as a yardstick or as the best "upper-bound" solution to the problem of channel estimation.

The function *IdChEst.m* implements an ideal channel estimator. It takes as input the parameters of the PDSCH captured in a structure (*prmLTEPDSCH*), the channel model parameter structure (*prmMdl*), and the channel matrix (*chPathG*), which is the second output of the *MIMOFadingChan.m* function. As its output, the function computes the channel-frequency response over the entire grid (*H*).

Algorithm

MATLAB function

```
function H = IdChEst(prmLTEPDSCH, prmMdl, chPathG)
% Ideal channel estimation for LTE subframes
%
%   Given the system parameters and the MIMO channel path Gains, provide
%   the ideal channel estimates for the RE corresponding to the data.
%   Limitation - will work for path delays that are multiple of channel sample
%   time and largest pathDelay < size of FFT
%   Implementation based on FFT of channel impulse response
persistent hFFT;
if isempty(hFFT)
  hFFT = dsp.FFT;
end
% get parameters
numDataTones = prmLTEPDSCH.Nrb*12; % Nrb_sc = 12
N               = prmLTEPDSCH.N;
cpLen0          = prmLTEPDSCH.cpLen0;
cpLenR          = prmLTEPDSCH.cpLenR;
Ndl_symb        = prmLTE.Ndl_symb;      % 7   for normal mode
slotLen         = (N*Ndl_symb + cpLen0 + cpLenR*6);
% Get path delays
pathDelays = prmMdl.PathDelays;
% Delays, in terms of number of channel samples, +1 for indexing
sampIdx = round(pathDelays/(1/prmLTEPDSCH.chanSRate)) + 1;
[~, numPaths, numTx, numRx] = size(chPathG);
% Initialize output
H = complex(zeros(numDataTones, 2*Ndl_symb, numTx, numRx));
for i= 1:numTx
  for j = 1:numRx
    link_PathG = chPathG(:, :, i, j);
    % Split this per OFDM symbol
    g = complex(zeros(2*Ndl_symb, numPaths));
    for n = 1:2 % over two slots
      % First OFDM symbol
      Index=(n-1)*slotLen + (1:(N+cpLen0));
      g((n-1)*Ndl_symb+1, :) = mean(link_PathG(Index, :), 1);
      % Next 6 OFDM symbols
      for k = 1:6
        Index=(n-1)*slotLen+cpLen0+k*N+(k-1)*cpLenR + (1:(N+cpLenR));
        g((n-1)*Ndl_symb+k+1, :) = mean(link_PathG(Index, :), 1);
      end
    end
```

```
      hImp = complex(zeros(2*Ndl_symb, N));
      % assign pathGains at impulse response sample locations
      hImp(:, sampIdx) = g;
      % FFT of impulse response
      h = step(hFFT, hImp.');
      % Reorder, remove DC, Unpack channel gains
      h = [h(N/2+1:N, :); h(1:N/2, :)];
      H(:, :, i, j) = [h(N/2-numDataTones/2+1:N/2, :); h(N/2+2:N/2+1+numDataTones/2, :)];
   end
end
```

This function essentially computes the channel frequency by applying a Fast Fourier Transform (FFT) to the channel impulse response. It is based on averaging over the entire subframe, so the same channel response is applied to all 14 OFDM symbols of a subframe. The operations of this function can be summarized as follows: (i) for any given transmit antenna and receive antenna, the channel path gains are extracted for all samples in time; (ii) the cyclic prefix samples are excluded; (iii) an average value is taken over the non-cyclic-prefix samples; (iv) a single impulse response vector (*hImp*) is initialized; (v) the non-zero samples of the impulse response are found by rounding the normalized path-delay values to the nearest integer; (vi) the impulse-response vector is updated by placing the average path gains in non-zero samples; (vii) an FFT is applied to the impulse response; (viii) the channel-response values are reordered and unpacked in oirder to compute the channel response over the entire resource grid.

6.6.8   Channel-Response Extraction

The received resource grid of each receive antenna contains multiple types of data, including user data, CSR and spectral-null signals, DCI, synchronization signals, and BCH signals. In order to focus on equalizing and recovering the user data, we must extract from the estimated channel response those elements that align with user data. The following MATLAB function (*ExtChResponse*.m) employs the PDSCH parameter structure (*prmLTEPDSCH*) and the user-data indices (*idx_data*) to extract from the full grid (*chEst*) the channel response values that align with the user data (*hD*). Note that when this function is called, the user-data indices (*idx_data*) have already been already computed as the third output of the resource-demapper function (*REdemapper_mTx*).

Algorithm

MATLAB function

```
function hD=ExtChResponse(chEst, idx_data, prmLTE)
%#codegen
numTx = prmLTE.numTx;
numRx = prmLTE.numRx;
if (numTx==1)
   hD=complex(zeros(numel(idx_data),numRx));
```

```
    for n=1:numRx
        tmp=chEst(:,:,n);
        hD(:,n)=tmp(idx_data);
    end
else
    hD=complex(zeros(numel(idx_data),numTx,numRx));
    for n=1:numRx
        for m=1:numTx
            tmp=chEst(:,:,m,n);
            hD(:,m,n)=tmp(idx_data);
        end
    end
end
```

6.7 Specific MIMO Features

In the following sections we will introduce functionalities that are unique to the MIMO implementation. These include precoding, layer mapping, and MIMO receiver. These operations will be markedly different depending on whether a transmit-diversity or a spatial-multiplexing technique is used. By adding these specific features to the common MIMO features – that is, resource grid computations, channel estimation, and OFDM specific features related to OFDM signal generation – we can completely specify the PDSCH operation. In this chapter we will feature PDSCH operations for modes 2, 3, and 4 of MIMO transmission in the LTE standard.

6.7.1 Transmit Diversity

Transmit diversity uses multiple antennas at the transmitter to exploit diversity gains and improve the link quality. There are two transmit-diversity schemes specified by LTE: one is a 2×2 SFBC technique and the other is a 4×4 technique. Both techniques feature full-rate codes and offer increased performance via their diversity as compared to single-antenna transmissions.

6.7.1.1 MIMO Operations in Transmit Diversity

The LTE standard specifies MIMO operations as a combination of layer mapping and precoding. In transmit-diversity mode, layer mapping and precoding are combined as a single encoding operation. The transmit-diversity encoder subdivides the modulated symbols into pairs and through diversity coding places transformed versions of modulated pairs on different transmit antennas. As the samples on each transmit antenna are derived from the original modulated stream, layer mapping is also implicit and precoding can be considered the result of various conjugations and negations. The number of layers is defined as the number of transmit antennas with independent and nonrelated data. Since samples on different antennas essentially reflect the same modulated data, the number of layers in transmit diversity is equal to one.

Two Antenna Ports

When using two transmit antennas, transmit diversity in LTE is based on SFBC. SFBC is closely related to the more familiar STBC. Transmit diversity using STBC has been deployed in various 3GPP and WiMAX standards. We will now provide a short overview of the STBC and SFBC techniques and show how SFBC can be derived from STBC through a simple transformation.

STBC can be regarded as a multi-antenna modulation and mapping technique that provides full diversity and results in simple encoders and decoders. One of the simplest forms of STBC is an Alamouti code defined for a two-antenna transmission. In STBC with Alamouti code, as illustrated in Figure 6.5, pairs of modulated symbols (s_1, s_2) are mapped on the first and second antenna ports in the initial sample time. In the following sample time, the symbols are swapped and conjugated $(-s_2{}^*, s_1{}^*)$ and mapped to the first and second antenna ports. Note that the two consecutive vectors in time are orthogonal.

In SFBC, as illustrated in Figure 6.6, pairs of consecutive modulated symbols (s_1, s_2) map directly on to consecutive samples in time on the first antenna port. On the second port, the swapped and transformed symbols $(-s_2{}^*, s_1{}^*)$ are mapped consecutively in time such that the consecutive vectors on different antennas are orthogonal.

We can produce the SFBC output symbols through a simple transformation followed by STBC using the Alamouti code. As illustrated in Figure 6.7, we first transform every second modulated symbol such that it is both negated and conjugated and then apply STBC with an Alamouti code. The result is the SFBC output for the pair of modulated inputs. This approach leverages the availability of efficient implementations for STBC and the Alamouti code and is considered advantageous as an example of software reuse.

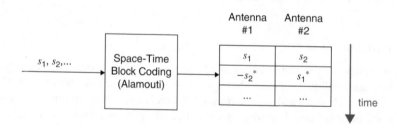

Figure 6.5   Space–time block coding: Alamouti code

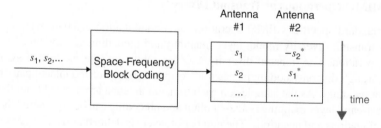

Figure 6.6   Space–frequency block coding

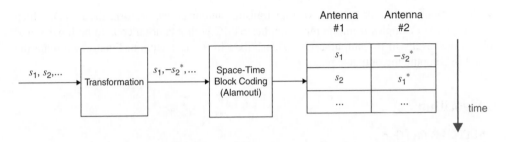

Figure 6.7   SFBC as a combination of a transformation and STBC

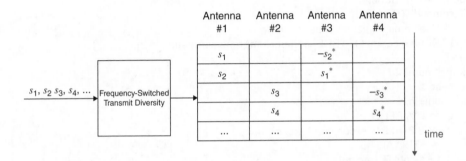

Figure 6.8   SFBC combined with Frequency-Switched Transmit Diversity (FSTD)

Four Antenna Ports

When using four transmit antennas, LTE combines SFBC with a Frequency-Switched Transmit Diversity (FSTD) technique. In this case, we perform transmit-diversity encoding on four consecutive modulated symbols at a time. First we apply SFBC to the first pair of modulated symbols (s_1, s_2) and place the results in first two samples in time and on the first and third transmit antennas. Then we apply SFBC on the third and fourth modulated symbols (s_3, s_4) and place the results in the third and fourth samples in time and on the second and fourth transmit antennas. Figure 6.8 illustrates the four-antenna transmit-diversity operations.

6.7.1.2   Transmit-Diversity Encoder Function

The following MATLAB function implements the transmit-diversity encoder for both two- and four-antenna configurations. The function takes as inputs the signal composed of modulated symbols (*in*) and the number of transmit antennas (*numTx*). The function output (*out*) is a 2D matrix. The first dimension is equal to the number of modulated symbols; that is, the size of the first input signal (*in*). The second dimension is equal to the number of transmit antennas (*numTx*). Operations performed for the two- and four-antenna cases include the following. First we transform the input signal by replacing every even-numbered element with its negative conjugate value. If we have two transmit antennas, we then perform STBC with Alamouti code. For the case of four transmit antennas, we perform the FSTD, which selects pairs of samples from the input, applies STBC with Alamouti code to both pairs, and places the results in the outputbuffer, as described in the last section. Finally, we scale the result to compute the output signal.

Algorithm

MATLAB function

```
function out = TDEncode(in, numTx)
%   Both SFBC and SFBC with FSTD
persistent hTDEnc;
if isempty(hTDEnc)
    % Use same object for either scheme
    hTDEnc = comm.OSTBCEncoder('NumTransmitAntennas', 2);
end
switch numTx
    case 1
        out=in;
    case 2 % SFBC
        in((2:2:end).') = -conj(in((2:2:end).'));
        % STBC Alamouti
        y= step(hTDEnc, in);
        % Scale
        out = y/sqrt(2);
    case 4
        inLen=size(in,1);
        y = complex(zeros(inLen, 4));
        in((2:2:end).') = -conj(in((2:2:end).'));
        idx12 = ([1:4:inLen; 2:4:inLen]); idx12 = idx12(:);
        idx34 = ([3:4:inLen; 4:4:inLen]); idx34 = idx34(:);
        y(idx12, [1 3]) = step(hTDEnc, in(idx12));
        y(idx34, [2 4]) = step(hTDEnc, in(idx34));
        out = y/sqrt(2);
end
```

Note that in order to perform STBC with the Alamouti code we take advantage of the *comm.OSTBCEncoder* System object from the Communications System Toolbox. As we will show in Chapter 9, using this System object results in a more efficient implementation of the STBC operation.

6.7.1.3 Transmit-Diversity Receiver Operations

To find the best estimates of the transmitted modulated symbols, we must perform transmit-diversity combining at the receiver. Transmit-diversity combining can be regarded as the inverse of transmit-diversity encoding.

Let us consider a 2×2 MIMO channel. A MIMO system of linear equations computes the received signals $\begin{bmatrix} y_1(n) \\ y_2(n) \end{bmatrix}$ at two receive antennas in each time index (n) as a function of the transmitted signals $\begin{bmatrix} x_1(n) \\ x_2(n) \end{bmatrix}$ and the MIMO channel matrix $\begin{bmatrix} h_{1,1}(n) & h_{1,2}(n) \\ h_{2,1}(n) & h_{2,2}(n) \end{bmatrix}$ in the same time:

$$\begin{bmatrix} y_1(n) \\ y_2(n) \end{bmatrix} = \begin{bmatrix} h_{1,1}(n) & h_{1,2}(n) \\ h_{2,1}(n) & h_{2,2}(n) \end{bmatrix} * \begin{bmatrix} x_1(n) \\ x_2(n) \end{bmatrix} \tag{6.7}$$

In the next time index ($n + 1$), the equation is expressed as:

$$\begin{bmatrix} y_1(n+1) \\ y_2(n+1) \end{bmatrix} = \begin{bmatrix} h_{1,1}(n+1) & h_{1,2}(n+1) \\ h_{2,1}(n+1) & h_{2,2}(n+1) \end{bmatrix} * \begin{bmatrix} x_1(n+1) \\ x_2(n+1) \end{bmatrix} \tag{6.8}$$

In both cases of two and four antennas, transmit-diversity encoding operations process pairs of consecutive modulated symbols. Let us consider a pair of consecutive received samples at the first receive antenna $\begin{bmatrix} y_1(n) \\ y_1(n+1) \end{bmatrix}$ and develop transmit-diversity equations with the assumption that STBC with Alamouti code has been used in the MIMO transmitter. The results can then be repeated for any pair of received signals at any receive-antenna port. The equation for the pair of consecutive received samples in the first receive antenna is expressed as:

$$\begin{bmatrix} y_1(n) \\ y_1(n+1) \end{bmatrix} = \begin{bmatrix} h_{1,1}(n) * x_1(n) + h_{1,2}(n) * x_2(n) \\ h_{1,1}(n+1) * x_1(n+1) + h_{1,2}(n+1) * x_2(n+1) \end{bmatrix} \tag{6.9}$$

Recall that the transmit-diversity encoder applies STBC with Alamouti code to pairs of modulated inputs symbols (s_1, s_2) and maps them into a 2×2 transmitted signal as:

$$\begin{bmatrix} x_1(n) & x_2(n) \\ x_1(n+1) & x_2(n+1) \end{bmatrix} = \begin{bmatrix} s_1 & s_2 \\ -s_2^* & s_1^* \end{bmatrix} \tag{6.10}$$

As a result, the equation for the pair of received signal in a 2×2 transmit-diversity case can be expressed as:

$$\begin{bmatrix} y_1(n) \\ y_1(n+1) \end{bmatrix} = \begin{bmatrix} h_{1,1}(n) * s_1 + h_{1,2}(n) * s_2 \\ -h_{1,1}(n+1) * s_2^* + h_{1,2}(n+1) * s_1^* \end{bmatrix} \tag{6.11}$$

Now, if we assume that the channel gains in two consecutive samples in time are similar to each other (i.e., $h_{1,1}(n) \approx h_{1,1}(n+1) = h_{1,1}$ and $h_{1,2}(n) \approx h_{1,2}(n+1) = h_{1,2}$) and fix the value of time index n (i.e., $y_1(n) = y_1$ and $y_1(n+1) = y_2$), we can further simply the equations as:

$$\begin{bmatrix} y_1 \\ y_2 \end{bmatrix} = \begin{bmatrix} h_{1,1} * s_1 + h_{1,2} * s_2 \\ -h_{1,1} * s_2^* + h_{1,2} * s_1^* \end{bmatrix} \tag{6.12}$$

Conjugating both sides of the second equation can lead to further simplification:

$$\begin{bmatrix} y_1 \\ y_2^* \end{bmatrix} = \begin{bmatrix} h_{1,1} * s_1 + h_{1,2} * s_2 \\ -h_{1,1}^* * s_2 + h_{1,2}^* * s_1 \end{bmatrix} = \begin{bmatrix} h_{1,1} & h_{1,2} \\ h_{1,2}^* & -h_{1,1}^* \end{bmatrix} * \begin{bmatrix} s_1 \\ s_2 \end{bmatrix} \tag{6.13}$$

By essentially inverting the matrix $H = \begin{bmatrix} h_{1,1} & h_{1,2} \\ h_{1,2}^* & -h_{1,1}^* \end{bmatrix}$, we can solve for the best estimates of the modulated transmitted symbols $\begin{bmatrix} \hat{s}_1 \\ \hat{s}_2 \end{bmatrix}$ as a function of received symbols $\begin{bmatrix} y_1 \\ y_2^* \end{bmatrix}$.

$$\begin{bmatrix} \hat{s}_1 \\ \hat{s}_2 \end{bmatrix} = \begin{bmatrix} h_{1,1} & h_{1,2} \\ h_{1,2}^* & -h_{1,1}^* \end{bmatrix}^{-1} \begin{bmatrix} y_1 \\ y_2^* \end{bmatrix}$$

$$\begin{bmatrix} \hat{s}_1 \\ \hat{s}_2 \end{bmatrix} = \frac{\begin{bmatrix} h_{1,1}^* & h_{1,2} \\ h_{1,2}^* & -h_{1,1} \end{bmatrix} \begin{bmatrix} y_1 \\ y_2^* \end{bmatrix}}{(h_{1,1} * h_{1,1}^* + h_{1,2} * h_{1,2}^*)} \tag{6.14}$$

This equation expresses an estimate of the transmitted symbols $\begin{bmatrix} \hat{s}_1 \\ \hat{s}_2 \end{bmatrix}$ at a given receiver antenna. To compute the best overall estimate of the transmitted symbols, a MRC algorithm is used. The MRC algorithm combines all the estimates computed at various receivers, as described next.

At each receiver (denoted by index n), let us call the estimate $\vec{s}_n = \begin{bmatrix} \hat{s}_1 \\ \hat{s}_2 \end{bmatrix}_n$, the channel matrix,

$H_n = \begin{bmatrix} h_{1,1}^* & h_{1,2} \\ h_{1,2}^* & -h_{1,1} \end{bmatrix}_n$ the received symbols $\vec{y}_n = \begin{bmatrix} y_1 \\ y_2^* \end{bmatrix}_n$ and the norm (the energy estimate) of

the channel matrix $E_n = (h_{1,1} * h_{1,1}^* + h_{1,2} * h_{1,2}^*)_n$.
The Equation 6.14 can then be re-written as

$$\vec{s}_n = \frac{1}{E_n} H_n \vec{y}_n \tag{6.15}$$

The MCR algorithm computes the overall estimate $(\hat{s})$ as a weighted sum of the individual estimates $(\vec{s}_n)$ across N receive antennas, where $1 < n < N$. Each individual estimate, at receiver n, is weighted by a gain factor α_n, that is

$$\hat{s} = \sum_{n=1}^{N} \alpha_n \vec{s}_n \tag{6.16}$$

The gain factor is defined as the ratio of a given channel matrix norm (E_n) over the sum of all channel matrix norms, that is

$$\alpha_n = \frac{E_n}{\sum_{k=1}^{N} E_k} \tag{6.17}$$

By combining Equations 6.15–6.17 and simplifying the formulation, we arrive at the maximum-ratio combing expression for the best overall estimate of the transmitted symbols:

$$\hat{s} = \sum_{n=1}^{N} \alpha_n \vec{s}_n = \sum_{n=1}^{N} \frac{E_n}{\sum_{k=1}^{N} E_k} \cdot \frac{1}{E_n} H_n \vec{y}_n = \frac{\sum_{n=1}^{N} H_n \vec{y}_n}{\sum_{k=1}^{N} E_k} \tag{6.18}$$

The following MATLAB function implements transmit-diversity combining for the 2×2 Alamouti code. The function has two inputs: (i) the received symbols (u), with dimensions of (LEN, 2), and (ii) the estimated channel matrix, with dimensions of (LEN, 2, 2). The function subdivides the received symbols in consecutive pairs in time and at each receive antenna performs an ML combining estimate as outlined in this section.

Algorithm

MATLAB function

```
function s = Alamouti_Combiner1(u,H)
%#codegen
% STBC_DEC STBC Combiner
%   Outputs the recovered symbol vector
LEN=size(u,1);
Nr=size(u,2);
BlkSize=2;
NoBlks=LEN/BlkSize;
% Initialize outputs
h=complex(zeros(1,2));
s=complex(zeros(LEN,1));
% Alamouti code for 2 Tx
indexU=(1:BlkSize);
for m=1:NoBlks
   t_hat=complex(zeros(BlkSize,1));
   h_norm=0.0;
   for n=1:Nr
      h(:)=H(2*m-1,:,n);
      h_norm=h_norm+real(h*h');
      r=u(indexU,n);
      r(2)=conj(r(2));
      shat=[conj(h(1)), h(2); conj(h(2)), -h(1)]*r;
      t_hat=t_hat+shat;
   end
   s(indexU)=t_hat/h_norm; % Maximum-likelihood combining
   indexU=indexU+BlkSize;
end
end
```

This function is an explicit and descriptive formulation of ML combining for the 2×2 Alamouti code. However, the runtime performance of this function is not optimal. As we will see in Chapter 9, a vectorized MATLAB function performs much better in runtime. As a result, we will use the *comm.OSTBCCombiner* System object of the Communications System Toolbox, which is optimized for performance. The following MATLAB function uses the *comm.OSTBCCombiner* System object to implement transmit-diversity combining. It requires just six lines of MATLAB code to achieve the same functionality as the previous function.

Algorithm

MATLAB function

```
function s = Alamouti_CombinerS(u,H)
%#codegen
% STBC_DEC STBC Combiner
persistent hTDDec
if isempty(hTDDec)
   hTDDec= comm.OSTBCCombiner(...
      'NumTransmitAntennas',2,'NumReceiveAntennas',2);
end
s = step(hTDDec, u, H);
```

6.7.1.4 Transmit-Diversity Combiner Function

The following MATLAB function implements the transmit-diversity combiner for both two- and four-antenna configurations. The function takes as inputs: (i) the 2D received signal (*in*), (ii) the 3D channel-estimate signal (*chEst*), (iii) the number of transmit antennas (*numTx*), and (iv) the number of receive antennas (*numRx*). The function output (*y*) is an ML estimate of the transmitted modulated symbols. The number of samples in the output vector (*y*) is equal to the number of transmitted modulated symbols (*inLen*); that is, the first dimension of input signals (*in* and *chEst*). The second dimension of input signals (*in* and *chEst*) is equal to the number of transmit antennas (*numTx*). The third dimension of the channel-estimate signal (*chEst*) is the number of receive antennas (*numRx*).

The operations performed for the two- and four-antenna cases are the inverse of those in transmit-diversity encoding. We first scale the input signal. If we have two transmit antennas, we then perform STBC combining. In the case of four transmit antennas, we perform FSTD combining. First we rearrange the 3D channel-estimate matrix (*chEst*) to form a new matrix (*H*) with dimensions equal to (*inLen, 2, 4*). Then we perform STBC combining on matrix *H*; that is, we repeat Alamouti code combining for transmit antennas (1, 3) and (2, 4) separately. Finally we replace every even-numbered element of the combiner output with its negative and conjugate value to return to SFBC and compute the output signal.

Algorithm

MATLAB function

```
function y = TDCombine(in, chEst, numTx, numRx)
% LTE transmit diversity combining
%   SFBC and SFBC with FSTD.
inLen = size(in, 1);
Index=(2:2:inLen)';
switch numTx
    case 1
        y=in;
    case 2   % For 2TX - SFBC
        in = sqrt(2) * in; % Scale
        y = Alamouti_CombinerS(in,chEst);
        % ST to SF transformation.
        % Apply blockwise correction for 2nd symbol combining
        y(Index) = -conj(y(Index));
    case 4   % For 4Tx - SFBC with FSTD
        in = sqrt(2) * in; % Scale
        H = complex(zeros(inLen, 2, numRx));
        idx12 = ([1:4:inLen; 2:4:inLen]); idx12 = idx12(:);
        idx34 = ([3:4:inLen; 4:4:inLen]); idx34 = idx34(:);
        H(idx12, :, :) = chEst(idx12, [1 3], :);
        H(idx34, :, :) = chEst(idx34, [2 4], :);
        y = Alamouti_CombinerS(in, H);
        % ST to SF transformation.
        % Apply blockwise correction for 2nd symbol combining
        y(Index) = -conj(y(Index));
end
```

6.7.2   *Transceiver Setup Functions*

Before we look at models of the various MIMO transmission modes, we will present in this section the testbench, initialization, and visualization functions. These types of function are common among all simulations and help verify the performance of each transceiver model.

6.7.2.1   Initialization Functions

The following initialization function (*commlteMIMO_initialize*) sets simulation parameters. This function is used for all MIMO modes, including transmit diversity and spatial multiplexing. The first input argument (*txMode*) determines which MIMO mode is used: a value of 2 signals a transmit-diversity mode, a value of 3 an open-loop spatial-multiplexing mode, and a value of 4 a closed-loop spatial-multiplexing mode. In order to set *prmLTEPDSCH*, *prmLTEDLSCH*, and *prmMdl* parameter structures, this function calls three functions: *prmsPDSCH*, *prmsDLSCH*, and *prmsMdl*, respectively.

Algorithm

MATLAB function

```
function [prmLTEPDSCH, prmLTEDLSCH, prmMdl] = commlteMIMO_initialize(txMode, ...
chanBW, contReg, modType, Eqmode,numTx, numRx,cRate,maxIter, fullDecode,
chanMdl, corrLvl, ...
    chEstOn, snrdB, maxNumErrs, maxNumBits)
% Create the parameter structures
% PDSCH parameters
CheckAntennaConfig(numTx, numRx);
prmLTEPDSCH = prmsPDSCH(txMode, chanBW, contReg, modType,numTx, numRx);
prmLTEPDSCH.Eqmode=Eqmode;
prmLTEPDSCH.modType=modType;
[SymbolMap, Constellation]=ModulatorDetail(modType);
prmLTEPDSCH.SymbolMap=SymbolMap;
prmLTEPDSCH.Constellation=Constellation;
% DLSCH parameters
prmLTEDLSCH = prmsDLSCH(cRate,maxIter, fullDecode, prmLTEPDSCH);
% Channel parameters
chanSRate  = prmLTEPDSCH.chanSRate;
 prmMdl = prmsMdl(chanSRate,  chanMdl, numTx, numRx, ...
    corrLvl, chEstOn, snrdB, maxNumErrs, maxNumBits);
```

The functions *prmsDLSCH* and *prmsMdl* are unchanged from those described in this and the previous chapter. The function *prmLTEPDSCH* is however modified to handle all MIMO cases. Depending on the transmission mode, number of antennas, channel bandwidth, and modulation mode used, this function sets all necessary parameters for many functions in PDSCH processing.

Algorithm

MATLAB function

```
function p = prmsPDSCH(txMode, chanBW, contReg, modType, numTx, numRx,
numCodeWords)
%% PDSCH parameters
switch chanBW
    case 1     % 1.4 MHz
       BW = 1.4e6; N = 128; cpLen0 = 10; cpLenR = 9;
       Nrb = 6; chanSRate = 1.92e6;
    case 2     % 3 MHz
       BW = 3e6; N = 256; cpLen0 = 20; cpLenR = 18;
       Nrb = 15; chanSRate = 3.84e6;
```

```
  case 3      % 5 MHz
    BW = 5e6; N = 512; cpLen0 = 40; cpLenR = 36;
    Nrb = 25; chanSRate = 7.68e6;
  case 4      % 10 MHz
    BW = 10e6; N = 1024; cpLen0 = 80; cpLenR = 72;
    Nrb = 50; chanSRate = 15.36e6;
  case 5      % 15 MHz
    BW = 15e6; N = 1536; cpLen0 = 120; cpLenR = 108;
    Nrb = 75; chanSRate = 23.04e6;
  case 6      % 20 MHz
    BW = 20e6; N = 2048; cpLen0 = 160; cpLenR = 144;
    Nrb = 100; chanSRate = 30.72e6;
end
p.BW = BW;              % Channel bandwidth
p.N = N;               % NFFT
p.cpLen0 = cpLen0;         % Cyclic prefix length for 1st symbol
p.cpLenR = cpLenR;          % Cyclic prefix length for remaining
p.Nrb = Nrb;               % Number of resource blocks
p.chanSRate = chanSRate;    % Channel sampling rate
p.contReg = contReg;
switch txMode
  case 1 % SISO transmission
    p.numTx = numTx;
    p.numRx = numRx;
    numCSRRE_RB = 2*2*2; % CSR, RE per OFDMsym/slot/subframe per RB
    p.numLayers = 1;
    p.numCodeWords = 1;
  case 2 % Transmit diversity
    p.numTx = numTx;
    p.numRx = numRx;
    switch numTx
      case 1
        numCSRRE_RB = 2*2*2; % CSR, RE per OFDMsym/slot/subframe per RB
      case 2    % 2xnumRx
        % RE - resource element, RB - resource block
        numCSRRE_RB = 4*2*2; % CSR, RE per OFDMsym/slot/subframe per RB
      case 4      % 4xnumRx
        numCSRRE_RB = 4*3*2; % CSR, RE per OFDMsym/slot/subframe per RB
    end
    p.numLayers = 1;
    p.numCodeWords = 1; % for transmit diversity
  case 3 % CDD Spatial multiplexing
    p.numTx = numTx;
    p.numRx = numRx;
    switch numTx
      case 1
        numCSRRE_RB = 2*2*2; % CSR, RE per OFDMsym/slot/subframe per RB
      case 2      % 2x2
        % RE - resource element, RB - resource block
```

```
                numCSRRE_RB = 4*2*2; % CSR, RE per OFDMsym/slot/subframe per RB
          case 4     % 4x4
                numCSRRE_RB = 4*3*2; % CSR, RE per OFDMsym/slot/subframe per RB
        end
        p.numLayers = min([p.numTx, p.numRx]);
        p.numCodeWords = 1; % for spatial multiplexing
      case 4 % Spatial multiplexing
        p.numTx = numTx;
        p.numRx = numRx;
        switch numTx
          case 1
                numCSRRE_RB = 2*2*2; % CSR, RE per OFDMsym/slot/subframe per RB
          case 2     % 2x2
                % RE - resource element, RB - resource block
                numCSRRE_RB = 4*2*2; % CSR, RE per OFDMsym/slot/subframe per RB
          case 4     % 4x4
                numCSRRE_RB = 4*3*2; % CSR, RE per OFDMsym/slot/subframe per RB
        end
        p.numLayers = min([p.numTx, p.numRx]);
        p.numCodeWords = numCodeWords; % for spatial multiplexing
end
% For Normal cyclic prefix, FDD mode
p.deltaF = 15e3;    % subcarrier spacing
p.Nrb_sc = 12;      % no. of subcarriers per resource block
p.Ndl_symb = 7;     % no. of OFDM symbols in a slot
%% Modeling a subframe worth of data (=> 2 slots)
numResources = (p.Nrb*p.Nrb_sc)*(p.Ndl_symb*2);
numCSRRE = numCSRRE_RB * p.Nrb;          % CSR, RE per
OFDMsym/slot/subframe per RB
% Actual PDSCH bits calculation - accounting for PDCCH, PBCH, PSS, SSS
switch p.numTx
    % numRE in control region - minus the CSR
    case 1
      numContRE = (10 + 12*(p.contReg-1))*p.Nrb;
      numBCHRE = 60+72+72+72; % removing the CSR present in 1st symbol
    case 2
      numContRE = (8 + 12*(p.contReg-1))*p.Nrb;
      numBCHRE = 48+72+72+72; % removing the CSR present in 1st symbol
    case 4
      numContRE = (8 + (p.contReg>1)*(8+ 12*(p.contReg-2)))*Nrb;
      numBCHRE = 48+48+72+72; % removing the CSR present in 1,2 symbol
end
numSSSRE=72;
numPSSRE=72;
numDataRE=zeros(3,1);
% Account for BCH, PSS, SSS and PDCCH for subframe 0
numDataRE(1)=numResources-numCSRRE-numContRE-numSSSRE
- numPSSRE-numBCHRE;
% Account for PSS, SSS and PDCCH for subframe 5
```

```
numDataRE(2)=numResources-numCSRRE-numContRE-numSSSRE - numPSSRE;
% Account for PDCCH only in all other subframes
numDataRE(3)=numResources-numCSRRE-numContRE;
% Maximum data resources - with no extra overheads (only CSR + data)
p.numResources=numResources;
p.numCSRResources =  numCSRRE;
p.numDataResources = p.numResources - p.numCSRResources;
p.numContRE = numContRE;
p.numBCHRE = numBCHRE;
p.numSSSRE=numSSSRE;
p.numPSSRE=numPSSRE;
p.numDataRE=numDataRE;
% Modulation types , bits per symbol, number of layers per codeword
Qm = 2 * modType;
p.Qm = Qm;
p.numLayPerCW = p.numLayers/p.numCodeWords;
% Maximum data bits - with no extra overheads (only CSR + data)
p.numDataBits = p.numDataResources*Qm*p.numLayPerCW;
numPDSCHBits =numDataRE*Qm*p.numLayPerCW;
p.numPDSCHBits = numPDSCHBits;
p.maxG = max(numPDSCHBits);
```

The *CheckAntennaConfig* function is called within *commlteMIMO_initialize*. It ensures that a valid antenna configuration is selected for simulation. In this book we limit our antenna configurations to four single-antenna cases (1×1, 1×2, 1×3, and 1×4), one two-antenna configuration (2×2), and one four-antenna configuration (4×4).

Algorithm

MATLAB function

```
function CheckAntennaConfig(numTx, numRx)
MyConfig=[numTx,numRx];
Allowed=[1,1;1,2;1,3;1,4;2,2;4,4];
tmp=MyConfig(ones(size(Allowed,1),1),:);
err=sum(abs(tmp-Allowed),2);
if isempty(find(~err,1))
   Status=0;
else
   Status=1;
end
if ~Status
   disp('Wrong antenna configuration! Allowable configurations are:');
   disp(Allowed);
   error('Please change number of Tx and/or Rx antennas!');
end
```

The *ModulatorDetail* function is also called within *commlteMIMO_initialize*. Depending on the modulation mode, the function provides the constellation and symbol mapping used in the visualization function and in one of the MIMO receiver functions, known as the Sphere Decoder (SD).

Algorithm

MATLAB function

```
function [SymMap, Constellation]=ModulatorDetail(Mode)
%% Initialization
persistent QPSK QAM16 QAM64
if isempty(QPSK)
   QPSK       = comm.PSKModulator(4, 'BitInput', true, ...
      'PhaseOffset', pi/4, 'SymbolMapping', 'Custom', ...
      'CustomSymbolMapping', [0 2 3 1]);
   QAM16     = comm.RectangularQAMModulator(16, 'BitInput',true,...
      'NormalizationMethod','Average power',...
      'SymbolMapping', 'Custom', ...
      'CustomSymbolMapping', [11 10 14 15 9 8 12 13 1 0 4 5 3 2 6 7]);
   QAM64     = comm.RectangularQAMModulator(64, 'BitInput',true,...
      'NormalizationMethod','Average power',...
      'SymbolMapping', 'Custom', ...
      'CustomSymbolMapping', [47 46 42 43 59 58 62 63 45 44 40 41 ...
      57 56 60 61 37 36 32 33 49 48 52 53 39 38 34 35 51 50 54 55 7 ...
      6 2 3 19 18 22 23 5 4 0 1 17 16 20 21 13 12 8 9 25 24 28 29 15 ...
      14 10 11 27 26 30 31]);
end
%% Processing
switch Mode
   case 1
      Constellation=constellation(QPSK);
      SymMap = QPSK.CustomSymbolMapping;
   case 2
      Constellation=constellation(QAM16);
      SymMap = QAM16.CustomSymbolMapping;
   case 3
      Constellation=constellation(QAM64);
      SymMap = QAM64.CustomSymbolMapping;
   otherwise
      error('Invalid Modulation Mode. Use {1,2, or 3}');
end
```

6.7.2.2   Visualization Functions

In this chapter we have updated the *zVisualize* function, which enables us to directly observe the effects of fading on transmitted symbols before and after MIMO receiver processing.

Algorithm

MATLAB function

```
function zVisualize(prmLTE, txSig, rxSig, yRec, dataRx, csr, nS)
% Constellation Scopes & Spectral Analyzers
zVisConstell(prmLTE, yRec, dataRx, nS);
zVisSpectrum(prmLTE, txSig, rxSig, yRec, csr, nS);
```

The function performs two tasks. First, it shows the constellation diagram of the user data at the receiver before and after equalization by calling the function *zVisConstell*, which shows constellation diagrams for data transmitted over multiple transmit antennas. Depending on the number of transmit antennas used, it creates and configures multiple Constellation Diagram System objects from the Communications System Toolbox.

Algorithm

MATLAB function

```
function zVisConstell(prmLTE, yRec, dataRx, nS)
% Constellation Scopes
switch prmLTE.numTx
   case 1
      zVisConstell_1(prmLTE, yRec, dataRx, nS);
   case 2
      zVisConstell_2(prmLTE, yRec, dataRx, nS);
   case 4
      zVisConstell_4(prmLTE, yRec, dataRx, nS);
end
end
%% Case of numTx =1
function zVisConstell_1(prmLTE, yRec, dataRx, nS)
persistent h1 h2
if isempty(h1)
   h1 = comm.ConstellationDiagram('SymbolsToDisplay',...
      prmLTE.numDataResources, 'ReferenceConstellation', prmLTE.
Constellation,...
      'YLimits', [-2 2], 'XLimits', [-2 2], 'Position', ...
      figposition([5 60 20 25]), 'Name', 'Before Equalizer');
   h2 = comm.ConstellationDiagram('SymbolsToDisplay',...
      prmLTE.numDataResources, 'ReferenceConstellation', prmLTE.
Constellation,...
      'YLimits', [-2 2], 'XLimits', [-2 2], 'Position', ...
      figposition([6 61 20 25]), 'Name', 'After Equalizer');
end
% Update Constellation Scope
if (nS~=0 && nS~=10)
```

```
    step(h1, dataRx(:,1));
    step(h2, yRec(:,1));
end
end
%% Case of numTx =2
function zVisConstell_2(prmLTE, yRec, dataRx, nS)
persistent h11 h21 h12 h22
if isempty(h11)
    h11 = comm.ConstellationDiagram('SymbolsToDisplay',...
        prmLTE.numDataResources, 'ReferenceConstellation', prmLTE.Constellation,...
        'YLimits', [-2 2], 'XLimits', [-2 2], 'Position', ...
        figposition([5 60 20 25]), 'Name', 'Before Equalizer');
    h21 = comm.ConstellationDiagram('SymbolsToDisplay',...
        prmLTE.numDataResources, 'ReferenceConstellation', prmLTE.Constellation,...
        'YLimits', [-2 2], 'XLimits', [-2 2], 'Position', ...
        figposition([6 61 20 25]), 'Name', 'After Equalizer');
    h12 = clone(h11);
    h22 = clone(h21);
end
yRecM = sqrt(2) *TDEncode( yRec, 2);
% Update Constellation Scope
if (nS~=0 && nS~=10)
    step(h11, dataRx(:,1));
    step(h21, yRecM(:,1));
    step(h12, dataRx(:,2));
    step(h22, yRecM(:,2));
end
end
%% Case of numTx =4
function zVisConstell_4(prmLTE, yRec, dataRx, nS)
persistent ha1 hb1 ha2 hb2 ha3 hb3 ha4 hb4
if isempty(ha1)
    ha1 = comm.ConstellationDiagram('SymbolsToDisplay',...
        prmLTE.numDataResources, 'ReferenceConstellation', prmLTE.Constellation,...
        'YLimits', [-2 2], 'XLimits', [-2 2], 'Position', ...
        figposition([5 60 20 25]), 'Name', 'Before Equalizer');
    hb1 = comm.ConstellationDiagram('SymbolsToDisplay',...
        prmLTE.numDataResources, 'ReferenceConstellation', prmLTE.Constellation,...
        'YLimits', [-2 2], 'XLimits', [-2 2], 'Position', ...
        figposition([6 61 20 25]), 'Name', 'After Equalizer');
    ha2 = clone(ha1);
    hb2 = clone(hb1);
    ha3 = clone(ha1);
    hb3 = clone(hb1);
    ha4 = clone(ha1);
    hb4 = clone(hb1);
end
yRecM = sqrt(2) *TDEncode( yRec, 4);
% Update Constellation Scope
```

```
if (nS~=0 && nS~=10)
   step(ha1, dataRx(:,1));
   step(hb1, yRecM(:,1));
   step(ha2, dataRx(:,2));
   step(hb2, yRecM(:,2));
   step(ha3, dataRx(:,3));
   step(hb3, yRecM(:,3));
   step(ha4, dataRx(:,4));
   step(hb4, yRecM(:,4));
end
end
```

Second, the *zVisualize* function illustrates the spectra of the transmitted signal and of the received signal both before and after equalization, by calling the function *zVisSpectrum*, which shows the magnitude spectrum of data transmitted over multiple transmit antennas. Depending on the number of transmit antennas used, it creates and configures multiple Spectrum Analyzer System objects from the DSP System Toolbox.

Algorithm

MATLAB function

```
function zVisSpectrum(prmLTE, txSig, rxSig, yRec, csr, nS)
% Spectral Analyzers
switch prmLTE.numTx
   case 1
      zVisSpectrum_1(prmLTE, txSig, rxSig, yRec, csr, nS);
   case 2
      zVisSpectrum_2(prmLTE, txSig, rxSig, yRec, csr, nS);
   case 4
      zVisSpectrum_4(prmLTE, txSig, rxSig, yRec, csr, nS);
end
end
%% Case of numTx = 1
function zVisSpectrum_1(prmLTE, txSig, rxSig, yRec, csr, nS)
persistent hSpecAnalyzer
if isempty(hSpecAnalyzer)
   hSpecAnalyzer = dsp.SpectrumAnalyzer('SampleRate', prmLTE.chanSRate, ...
      'SpectrumType', 'Power density', 'PowerUnits', 'dBW', ...
      'RBWSource', 'Property', 'RBW', 15000,...
      'FrequencySpan', 'Span and center frequency',...
      'Span', prmLTE.BW, 'CenterFrequency', 0,...
      'FFTLengthSource', 'Property', 'FFTLength', prmLTE.N,...
      'Title', 'Transmitted & Received Signal Spectrum', 'YLimits', [-110 -60],...
      'YLabel', 'PSD');
end
```

```
alamoutiRx = TDEncode(yRec, prmLTE.numTx);
yRecGrid = REmapper_mTx(alamoutiRx, csr, nS, prmLTE);
yRecGridSig = lteOFDMTx(yRecGrid, prmLTE);
step(hSpecAnalyzer, ...
   [SymbSpec(txSig(:,1), prmLTE), SymbSpec(rxSig(:,1), prmLTE),
SymbSpec(yRecGridSig(:,1), prmLTE)]);
end
%% Case of numTx = 2
function zVisSpectrum_2(prmLTE, txSig, rxSig, yRec, csr, nS)
persistent hSpec1 hSpec2
if isempty(hSpec1)
   hSpec1 = dsp.SpectrumAnalyzer('SampleRate', prmLTE.chanSRate, ...
      'SpectrumType', 'Power density', 'PowerUnits', 'dBW', ...
      'RBWSource', 'Property', 'RBW', 15000,...
      'FrequencySpan', 'Span and center frequency',...
      'Span', prmLTE.BW, 'CenterFrequency', 0,...
      'FFTLengthSource', 'Property', 'FFTLength', prmLTE.N,...
      'Title', 'Transmitted & Received Signal Spectrum', 'YLimits', [-110 -60],...
      'YLabel', 'PSD');
   hSpec2 = clone(hSpec1);
end
alamoutiRx = TDEncode(yRec, prmLTE.numTx);
yRecGrid = REmapper_mTx(alamoutiRx, csr, nS, prmLTE);
yRecGridSig = lteOFDMTx(yRecGrid, prmLTE);
step(hSpec1, ...
   [SymbSpec(txSig(:,1), prmLTE), SymbSpec(rxSig(:,1), prmLTE),
SymbSpec(yRecGridSig(:,1), prmLTE)]);
step(hSpec2, ...
   [SymbSpec(txSig(:,2), prmLTE), SymbSpec(rxSig(:,2), prmLTE),
SymbSpec(yRecGridSig(:,2), prmLTE)]);
end
%% Case of numTx = 4
function zVisSpectrum_4(prmLTE, txSig, rxSig, yRec, csr, nS)
persistent hSpec1 hSpec2 hSpec3 hSpec4
if isempty(hSpec1)
   hSpec1 = dsp.SpectrumAnalyzer('SampleRate', prmLTE.chanSRate, ...
      'SpectrumType', 'Power density', 'PowerUnits', 'dBW', ...
      'RBWSource', 'Property', 'RBW', 15000,...
      'FrequencySpan', 'Span and center frequency',...
      'Span', prmLTE.BW, 'CenterFrequency', 0,...
      'FFTLengthSource', 'Property', 'FFTLength', prmLTE.N,...
      'Title', 'Transmitted & Received Signal Spectrum', 'YLimits', [-110 -60],...
      'YLabel', 'PSD');
   hSpec2 = clone(hSpec1);
   hSpec3 = clone(hSpec1);
   hSpec4 = clone(hSpec1);
end
alamoutiRx = TDEncode(yRec, prmLTE.numTx);
yRecGrid = REmapper_mTx(alamoutiRx, csr, nS, prmLTE);
```

```
yRecGridSig = lteOFDMTx(yRecGrid, prmLTE);
step(hSpec1, ...
   [SymbSpec(txSig(:,1), prmLTE), SymbSpec(rxSig(:,1), prmLTE),
SymbSpec(yRecGridSig(:,1), prmLTE)]);
step(hSpec2, ...
   [SymbSpec(txSig(:,2), prmLTE), SymbSpec(rxSig(:,2), prmLTE),
SymbSpec(yRecGridSig(:,2), prmLTE)]);
step(hSpec3, ...
   [SymbSpec(txSig(:,3), prmLTE), SymbSpec(rxSig(:,3), prmLTE),
SymbSpec(yRecGridSig(:,3), prmLTE)]);
step(hSpec4, ...
   [SymbSpec(txSig(:,4), prmLTE), SymbSpec(rxSig(:,4), prmLTE),
SymbSpec(yRecGridSig(:,4), prmLTE)]);
end
%% Helper function
function y = SymbSpec(in, prmLTE)
N = prmLTE.N;
cpLenR = prmLTE.cpLen0;
y = complex(zeros(N+cpLenR, 1));
% Use the first Tx/Rx antenna of the input for the display
y(:,1) = in(end-(N+cpLenR)+1:end, 1);
end
```

6.7.3   Downlink Transmission Mode 2

The following MATLAB function shows a transceiver model for transmit diversity mode 2 of the LTE standard. It includes both the two- and the four-transmit-antenna configurations. In essence, both the 2×2 and the 4×4 schemes specified by LTE are full-rate codes and both offer increased performance benefits, due to their diversity when compared to single-antenna transmissions. The key components highlighted in this example include:

- Generation of payload data for a single subframe (a transport block).
- **DLSCH processing:** Transport-block CRC (Cyclic Redundancy Check) attachment, code-block segmentation and CRC attachment, turbo coding based on a $1/3$-rate code, rate matching, and codeblock concatenation to generate a codeword input to PDSCH.
- **PDSCH transmitter processing:** Bit-level scrambling, data modulation, layer mapping, and precoding for two and four antennas with transmit diversity encoding, plus resource-element mapping and OFDM signal generation.
- **Channel modeling:** A MIMO fading channel followed by an AWGN channel.
- **PDSCH receiver processing:** An OFDM signal receiver generating the resource grid, resource element demapping to separate the CSR signal from the user data, channel estimation, SFBC-based combining using channel estimates and soft-decision demodulation and descrambling, and DLSCH decoding.

Algorithm

MATLAB function

```
function [dataIn, dataOut, modOut, rxSig, dataRx, yRec, csr_ref]...
   = commIteMIMO_TD_step(nS, snrdB, prmLTEDLSCH, prmLTEPDSCH, prmMdl)
%% TX
%  Generate payload
dataIn = genPayload(nS,  prmLTEDLSCH.TBLenVec);
% Transport block CRC generation
tbCrcOut1 =CRCgenerator(dataIn);
% Channel coding includes - CB segmentation, turbo coding, rate matching,
% bit selection, CB concatenation - per codeword
[data, Kplus1, C1] = IteTbChannelCoding(tbCrcOut1, nS, prmLTEDLSCH, prmLTEPDSCH);
%Scramble codeword
scramOut = IteScramble(data, nS, 0, prmLTEPDSCH.maxG);
% Modulate
modOut = Modulator(scramOut, prmLTEPDSCH.modType);
% TD with SFBC
numTx=prmLTEPDSCH.numTx;
alamouti = TDEncode(modOut(:,1),numTx);
% Generate Cell-Specific Reference (CSR) signals
csr = CSRgenerator(nS, numTx);
csr_ref=complex(zeros(2*prmLTEPDSCH.Nrb, 4, numTx));
for m=1:numTx
   csr_pre=csr(1:2*prmLTEPDSCH.Nrb,:,:,m);
   csr_ref(:,:,m)=reshape(csr_pre,2*prmLTEPDSCH.Nrb,4);
end
% Resource grid filling
txGrid = REmapper_mTx(alamouti, csr_ref, nS, prmLTEPDSCH);
% OFDM transmitter
txSig = OFDMTx(txGrid, prmLTEPDSCH);
%% Channel : MIMO Fading channel
[rxFade, chPathG] = MIMOFadingChan(txSig, prmLTEPDSCH, prmMdl);
% Add AWG noise
nVar = 10.^(0.1.*(-snrdB));
rxSig =  AWGNChannel(rxFade, nVar);
%% RX
% OFDM Rx
rxGrid = OFDMRx(rxSig, prmLTEPDSCH);
% updated for numLayers -> numTx
[dataRx, csrRx, idx_data] = REdemapper_mTx(rxGrid, nS, prmLTEPDSCH);
% MIMO channel estimation
if prmMdl.chEstOn
   chEst = ChanEstimate_mTx(prmLTEPDSCH, csrRx,  csr_ref, prmMdl.chEstOn);
   hD    = ExtChResponse(chEst, idx_data, prmLTEPDSCH);
```

```
else
   idealChEst = IdChEst(prmLTEPDSCH, prmMdl, chPathG);
   hD = ExtChResponse(idealChEst, idx_data, prmLTEPDSCH);
end
% Frequency-domain equalizer
if (numTx==1)
   % Based on Maximum-Combining Ratio (MCR)
   yRec = Equalizer_simo(dataRx, hD, nVar, prmLTEPDSCH.Eqmode);
else
   % Based on Transmit Diversity  with SFBC combiner
   yRec = TDCombine(dataRx, hD, prmLTEPDSCH.numTx, prmLTEPDSCH.numRx);
end
% Demodulate
demodOut = DemodulatorSoft(yRec, prmLTEPDSCH.modType, nVar);
% Descramble received codeword
rxCW = lteDescramble(demodOut, nS, 0, prmLTEPDSCH.maxG);
% Channel decoding includes - CB segmentation, turbo decoding, rate dematching
[decTbData1, ˜,˜] = lteTbChannelDecoding(nS, rxCW, Kplus1, C1,  prmLTEDLSCH,
prmLTEPDSCH);
% Transport block CRC detection
[dataOut, ˜] = CRCdetector(decTbData1);
end
```

6.7.3.1   Structure of the Transceiver Model

The following MATLAB script is the testbench that calls the MIMO transceiver function *commlteMIMO*. First it calls the initialization function (*commlteMIMO_initialize*) to set all relevant parameter structures (*prmLTEDLSCH*, *prmLTEPDSCH*, *prmMdl*). Then it uses a while loop to perform subframe processing by calling the MIMO transceiver function *commlteMIMO_TD_step*. Finally, it computes the BER and calls the visualization function to illustrate the channel response and modulation constellation before and after equalization.

Algorithm

MATLAB function

```
% Script for MIMO LTE (mode 2)
%
% Single codeword transmission only,
%
clear all
clear functions
%% Set simulation parameters & initialize parameter structures
commlteMIMO_ params;
[prmLTEPDSCH, prmLTEDLSCH, prmMdl] = commlteMIMO_initialize(txMode, ...
chanBW, contReg, modType, Eqmode,numTx, numRx,cRate,maxIter, fullDecode,
chanMdl, corrLvl, chEstOn, snrdB, maxNumErrs, maxNumBits);
```

```
clear txMode chanBW contReg modType Eqmode numTx numRx cRate maxIter
fullDecode chanMdl corrLvl chEstOn snrdB maxNumErrs maxNumBits
%%
disp('Simulating the LTE Mode 2: Multiple Tx & Rx antrennas with transmit diversity');
zReport_data_rate(prmLTEPDSCH, prmLTEDLSCH);
hPBer = comm.ErrorRate;
snrdB=prmMdl.snrdB;
maxNumErrs=prmMdl.maxNumErrs;
maxNumBits=prmMdl.maxNumBits;
%% Simulation loop
nS = 0; % Slot number, one of [0:2:18]
Measures = zeros(3,1); %initialize BER output
while (( Measures(2)< maxNumErrs) && (Measures(3) < maxNumBits))
   [dataIn, dataOut, txSig, rxSig, dataRx, yRec, csr] = ...
      commlteMIMO_TD_step(nS, snrdB, prmLTEDLSCH, prmLTEPDSCH, prmMdl);
   % Calculate  bit errors
   Measures = step(hPBer, dataIn, dataOut);
    % Visualize constellations and spectrum
   if visualsOn, zVisualize( prmLTEPDSCH, txSig, rxSig, yRec, dataRx, csr, nS);end;
   % Update subframe number
   nS = nS + 2; if nS > 19, nS = mod(nS, 20); end;
end
disp(Measures);
```

6.7.3.2   Verifying Transceiver Performance

By executing the MATLAB script of the MIMO transceiver model (*commlteMIMO*), we can look at various signals to assess the performance of the system. The parameters used in simulation are summarized in the following MATLAB script (*commlteMIMO_params*). This set of parameters specifies a transceiver model using the transmit-diversity MIMO mode, with the number of transmit and receive antennas equal to two, a channel bandwidth of 10 MHz (with 1 OFDM symbol per subframe carrying the DCI), a 16QAM (Quadrature Amplitude Modulation) modulation type (with ⅓-rate turbo coding with early termination enabled), the maximum number of iterations set to 6), and a frequency-selective MIMO channel with a Doppler shift of 70 Hz (estimating channel response based on the interpolation method and using a transmit-diversity combiner as a MIMO receiver). In this simulation, 10 million bits of user data are processed, the SNR of the AWGN channel is set to 16 dB, and the visualization function is turned on.

Algorithm

MATLAB function

```
% PDSCH
txMode     = 2;  % Transmission mode one of {1, 2, 4}
numTx      = 2;   % Number of transmit antennas
```

```
numRx       = 2;    % Number of receive antennas
chanBW      = 4;    % Index to chanel bandwidth used [1,....6]
contReg     = 1;    % No. of OFDM symbols dedictaed to control information [1,....,3]
modType     = 2;    % Modulation type [1, 2, 3] for ['QPSK,'16QAM','64QAM']
% DLSCH
cRate       = 1/3; % Rate matching target coding rate
maxIter     = 6;    % Maximum number of turbo decoding terations
fullDecode  = 0;    % Whether "full" or "early stopping" turbo decoding is performed
% Channel model
chanMdl     = 'frequency-selective-high-mobility';
corrLvl     = 'Medium';
% Simulation parameters
Eqmode      = 2;    % Type of equalizer used [1,2] for ['ZF', 'MMSE']
chEstOn     = 1;    % One of [0,1,2,3] for 'Ideal estimator','Interpolation',
Slot average','Subframe average'
snrdB       = 16;  % Signal to Noise ratio
maxNumErrs = 5e5; % Maximum number of errors found before simulation stops
maxNumBits = 5e5; % Maximum number of bits processed before simulation stops
visualsOn   = 1;    % Whether to visualize channel response and constellations
```

Figure 6.9 shows the constellation diagrams before (first row) and after (second row) equalization of user data obtained from each of the two receive antennas in a subframe. It shows that the equalizer can compensate for the effect of a fading channel to result in a constellation that more closely resembles that of the 16QAM modulator.

Figure 6.10 illustrates the spectra of user data obtained from each of the two receive antennas in a subframe. It shows the transmitted signal and the received signal before and after equalization. The received signal before equalization (showing the effects of frequency-selective fading) is effectively equalized by the transmit diversity (showing a more frequency-flat nature), which closely resembles the transmitted signal spectrum.

6.7.3.3 BER Measurements

In order to verify the BER performance of the transceiver, we create a testbench called *commlteMIMO_test_timing_ber*. This testbench first initializes the LTE system parameters and then iterates through a range of SNR values and calls the *commlteMIMO_fcn* function in the loop in order to compute the corresponding BER values.

Algorithm

MATLAB script: commlteMIMO_test_timing_ber

```
% Script for MIMO LTE (mode 2)
%
% Single codeword transmission only,
%
```

Figure 6.9 LTE model: MIMO transmit-diversity constellation diagram of user data before and after equalization

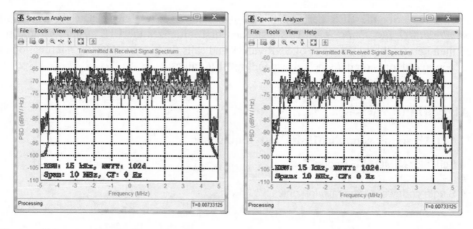

Figure 6.10 LTE MIMO transmit-diversity model: spectra of the transmitted signal and of the received signal before and after equalization

```
clear all
clear functions
%% Set simulation parameters & initialize parameter structures
commlteMIMO_ params;
maxNumErrs=5e7;
maxNumBits=5e7;
[prmLTEPDSCH, prmLTEDLSCH, prmMdl] = commlteMIMO_initialize(txMode, ...
chanBW, contReg, modType, Eqmode,numTx, numRx,cRate,maxIter, fullDecode,
chanMdl, corrLvl, chEstOn, snrdB, maxNumErrs, maxNumBits);
clear txMode chanBW contReg modType Eqmode numTx numRx cRate maxIter
fullDecode chanMdl corrLvl chEstOn snrdB maxNumErrs maxNumBits
%%
disp('Simulating the LTE Mode 2: Multiple Tx & Rx antrennas with transmit diversity');
zReport_data_rate(prmLTEPDSCH, prmLTEDLSCH);
%%
MaxIter=8;
snr_vector=getSnrVector(prmLTEPDSCH.modType, MaxIter);
ber_vector=zeros(size(snr_vector));
tic;
for n=1:MaxIter
    fprintf(1,'Iteration %2d out of %2d:  Processing %10d bits. SNR = %3d\n', ...
        n, MaxIter, prmMdl.maxNumBits, snr_vector(n));
    [ber, ~] = commlteMIMO_ fcn(snr_vector(n), prmLTEPDSCH, prmLTEDLSCH, prmMdl);
    ber_vector(n)=ber;
end;
toc;
semilogy(snr_vector, ber_vector);
title('BER - commlteMIMO TD');xlabel('SNR (dB)');ylabel('ber');grid;
```

Figure 6.11 shows the BER of the transceiver as a function of the SNR after processing of 50 million bits of user data in each of the eight iterations.

6.7.4 Spatial Multiplexing

Spatial multiplexing is a multiple-antenna technique that allows MIMO wireless systems to obtain high spectral efficiencies by dividing the bit stream into multiple substreams. Because these substreams are independently modulated, spatial multiplexing can accommodate higher data rates than comparable space–time or space–frequency block codes. However, this absence of redundancy in the transmitted signal makes spatial multiplexing susceptible to deficiencies in the rank of the matrix characterizing the MIMO equation. Channel-estimation inaccuracies when computing the MIMO channel matrix can severely limit performance gains. As a result, the LTE standard introduces various mechanisms, including adaptive precoding and layer mapping based on rank estimation, to make the implementation more robust in the presence of various channel impairments.

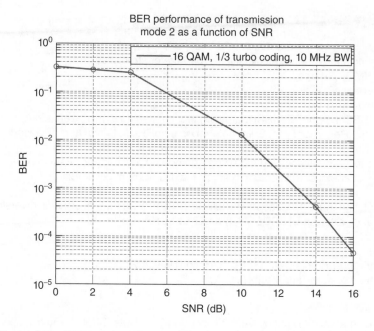

Figure 6.11   BER results: LTE mode 2, transmit diversity, 2×2 MIMO channel

In this section we will discuss details regarding the spatial-multiplexing approach to MIMO transmission in the LTE standard. These include the way in which it implements precoding and layer mapping, whcih eventually lead to generation of OFDM signals for simultaneous transmission over multiple antennas. Finally, by examining the receiver operations, including various MIMO equalization methodologies, we will study the performance of the system under various conditions.

6.7.4.1   Motivation for Precoding

The spectral-efficiency benefits associated with MIMO processing hinge on the availability of a rich scattering environment. A MIMO channel with a high degree of scattering enables independent multipath links to be made from each transmit antenna to each receive antenna. As a result, the matrix of channel gains connecting each pair of transmit and receive antennas pairs will have a full rank and the resulting MIMO equation will be solvable.

In a typical MIMO transmission, however, the assumption regarding a high level of scattering cannot be guaranteed. As a result, in order to design a practical system, steps must be taken to reduce the probability of channel matrices with reduced ranks occuring. Precoding is one of the most effective approaches taken by the LTE standard to combating the rank-deficiency problem. In this section we will elaborate on the nature of channel-matrix rank deficiencies, introduce a precoding formulation, provide a beamforming interpretation for precoding, and introduce different types of precoding used in the LTE standard. Then we will show the MAT-LAB functions that efficiently implement these operations.

6.7.4.2 Rank-Deficiency Problem

Spatial multiplexing solves the following system of linear equations, which expresses the received signal (Y) as a modified version of the transmitted signal (X) transformed linearly by the MIMO channel matrix (H) plus an added white noise (n):

$$Y = H X + n \tag{6.19}$$

For example, for a 4×4 MIMO configuration the received vector $\overrightarrow{Y}$ can be expressed as follows:

$$\overrightarrow{Y} = \begin{bmatrix} y_1 \\ y_2 \\ y_3 \\ y_4 \end{bmatrix} = \begin{bmatrix} h_{1,1} & \cdots & h_{1,4} \\ \vdots & \ddots & \vdots \\ h_{4,1} & \cdots & h_{4,4} \end{bmatrix} \begin{bmatrix} x_1 \\ x_2 \\ x_3 \\ x_4 \end{bmatrix} + \begin{bmatrix} n_1 \\ n_2 \\ n_3 \\ n_4 \end{bmatrix} \tag{6.20}$$

When the paths connecting transmit antennas to receive antennas become similar, multiple rows or columns of the channel matrix H can become linearly dependent; for example, in the following matrix the first two rows are identical.

$$H = \begin{bmatrix} h_{1,1} & h_{1,2} & h_{1,3} & h_{1,4} \\ h_{1,1} & h_{1,2} & h_{1,3} & h_{1,4} \\ h_{3,1} & h_{3,2} & h_{3,3} & h_{3,4} \\ h_{4,1} & h_{4,2} & h_{4,3} & h_{4,4} \end{bmatrix} \tag{6.21}$$

In this scenario, the rank of the channel matrix (the number of linearly independent equations) is three, whereas the dimension of the matrix is four. This system of linear equations is singular and has no inverses. As a result, the MIMO system of equation represented by this type of linearly dependent matrix cannot be uniquely solved.

6.7.4.3 Formulation of Precoding

Precoding techniques have been developed to solve the problem of rank deficiency. The optimal precoder can be determined by exploiting the singular-value decomposition of the channel matrix. Singular-value decomposition expresses the channel matrix as:

$$H = UDV \tag{6.22}$$

where V is a square matrix whose size is equal to the rank of the channel matrix, D is a diagonal matrix with diagonal elements composed of singular values of the channel matrix, and U is a square matrix whose size is equal to the number of receiver antennas. As developed in References [4] and [5], one of the theoretically optimal precoders can be defined as a column-permuted version of matrix V. This precoder operates only on transmitter antennas with sufficient rank and guarantees that the resulting MIMO equation can be solved.

Such an optimal precoder cannot be practically implemented, since it requires complete knowledge of the channel matrix at the transmitter. As the channel matrix can only be estimated at the receiver, communicating this information to the transmitter would require an excessive amount of bandwidth. The LTE chooses a more practical approach, based on choosing among a finite set of predetermined precoding matrices. Through a process similar to vector quantization, we can choose the best precoder at both the transmitter and the receiver.

At the transmitter, precoding performs a matrix multiplication on modulated symbols after layer mapping. As a result, the MIMO equation with the precoder is expressed as:

$$Y = HV X + n \tag{6.23}$$

where V is the precoding matrix. At the receiver, following the MIMO receiver operations, we apply the received signal with the inverse of the same precoding matrix V as used in the transmitter. LTE defines the precoding matrices as Hermitian matrices, which means that the precoder matrix is composed of a set of orthonormal vectors. This implies that the inverse of a precoder matrix is simply equal to its Hermitian transpose. It also results in efficient implementation of precoding, since transposing a matrix is much less computationally expensive than performing a matrix inversion.

6.7.4.4 Precoder-Matrix Codebooks

The finite sets of precoder matrices used in the LTE standard are known as the precoder codebook. Table 6.3 shows the precoder codebooks for two transmit antennas.

The precoding operation essentially spreads the input signal and reduces the probability of error by combating rank-deficiency problems. The efficacy of precoding in reducing the probability of rank deficiencies can be explained by interpreting the precoder matrix columns as beamforming vectors. In the case of single-layer transmission, for example, choosing each codebook index results in a multiplication of the transmitted signal X with different beamforming vectors. This multiplication is essentially a transformation that rotates the transmitted signal in various directions. Since precoder vectors are orthonormal, the direction of rotation results in phase differences of $\left\{ 0, \pi, \frac{\pi}{2}, -\frac{\pi}{2} \right\}$. Large phase differences make it more likely that different streams will take different multipath trajectories before arriving at any receive antenna. This in turn reduces the possibility of channel matrices with linearly dependent rows or columns occurring and increases the chance of there being full-rank channel matrices.

Table 6.3   Precoding matrices for two transmit antennas in LTE spatial multiplexing

Codebook index	Number of layers	
	1	2
0	$\frac{1}{\sqrt{2}} \begin{bmatrix} 1 \\ 1 \end{bmatrix}$	$\frac{1}{\sqrt{2}} \begin{bmatrix} 1 & 0 \\ 0 & 1 \end{bmatrix}$
1	$\frac{1}{\sqrt{2}} \begin{bmatrix} 1 \\ -1 \end{bmatrix}$	$\frac{1}{\sqrt{2}} \begin{bmatrix} 1 & 1 \\ 1 & -1 \end{bmatrix}$
2	$\frac{1}{\sqrt{2}} \begin{bmatrix} 1 \\ j \end{bmatrix}$	$\frac{1}{\sqrt{2}} \begin{bmatrix} 1 & 1 \\ j & -j \end{bmatrix}$
3	$\frac{1}{\sqrt{2}} \begin{bmatrix} 1 \\ -j \end{bmatrix}$	–

The same interpretation applies to the two-antenna and four-antenna precoder matrices. For example, in the case of two transmit antennas, when we multiply the two modulated substreams by any of the precoder matrices, $\frac{1}{\sqrt{2}} \begin{bmatrix} 1 & 1 \\ j & -j \end{bmatrix}$ for example, each substream is steered like a beamformer by each of the precoder matrix column vectors. Since these vectors are orthonormal, they can represent rotation operations in different N-dimensional directions [6]. When viewed as a beamformer, precoding enhances the chance of the transmitted streams following different multipaths, since it can force each substream to take different directions, as specified by the angle of rotation. This explains why spatial-multiplexing systems that use precoding have been shown to provide dramatic performance gains over unprecoded systems.

6.7.4.5   Types of Precoding

Precoding can be performed within an open- or a closed-loop MIMO processing context. Open-loop precoding is used in the third MIMO transmission mode and closed-loop precoding in the fourth. In open-loop precoding, the transmitter and receiver use a predefined set of precoding matrix indices and periodically rotate through them without any need for codebook index transmission. Precoding with closed-loop feedback prompts the receiver to choose the precoder matrix from a finite codebook and then to convey the selected matrix to the transmitter using a limited number of bits. Precoder-matrix codebook selection and closed-loop precoder-matrix feedback are discussed in Chapter 7.

6.7.5   MIMO Operations in Spatial Multiplexing

Among the nine modes of transmissions in the LTE standard, six are based on spatial multiplexing. In spatial multiplexing, layer mapping and precoding are distinct and explicit operations. As the samples on each transmit antenna are independent of each other, the original modulated stream will be mapped to various substreams to be transmitted on each transmit antenna. Since different samples are transmitted on different antennas, spatial multiplexing has the potential to boost data rates in proportion to the number of transmit antennas. The MIMO receiver operation is performed at the receiver to recover the best estimate of the modulated symbols from the received signal. The estimation processes featured in this book are based on the following three algorithms: Zero Forcing (ZF), Minimum Mean Square Error (MMSE), and Sphere Decoder (SD). Next, we will discuss in detail layer mapping, precoding, and MIMO receiver operations.

6.7.5.1   Layer Mapping

Layer mapping divides a single data stream into substreams destined for different antennas. The following MATLAB function shows how the modulated data stream from one or two codewords is mapped to layers (antenna ports) defined by the LTE standard. At this stage we assume a full rank transmission. As a result, the number of layers is equal to the number of transmit antennas. This function takes as input the modulated streams of the first (*in1*) and second (*in2*) codewords and the parameter structure of the PDSCH (*prmLTEPDSCH*). Depending on the number of codewords and the number of layers, the function reorganizes the input symbol stream to generate the output signal (*out*). The output signal is a 2D matrix whose second dimension is equal to the number of layers.

Algorithm

MATLAB function

```
function out = LayerMapper(in1, in2, prmLTEPDSCH)
% Layer mapper for spatial multiplexing.
%
%#codegen
% Assumes the incoming codewords are of the same length.
q = prmLTEPDSCH.numCodeWords;          % Number of codewords
v = prmLTEPDSCH.numLayers;            % Number of layers
inLen1 = size(in1, 1);
inLen2 = size(in2, 1);
switch q
    case 1 % Single codeword
        % for numLayers = 1,2,3,4
        out = reshape(in1, v, inLen1/v).';
    case 2 % Two codewords
        switch v
            case 2
                out = complex(zeros(inLen1, v));
                out(:,1) = in1;
                out(:,2) = in2;
            case 4
                out = complex(zeros(inLen1/2, v));
                out(:,1:2) = reshape(in1, 2, inLen1/2).';
                out(:,3:4) = reshape(in2, 2, inLen2/2).';
            case 6
                out = complex(zeros(inLen1/3, v));
                out(:,1:3) = reshape(in1, 3, inLen1/3).';
                out(:,4:6) = reshape(in2, 3, inLen2/3).';
            case 8
                out = complex(zeros(inLen1/4, v));
                out(:,1:4) = reshape(in1, 4, inLen1/4).';
                out(:,5:8) = reshape(in2, 4, inLen2/4).';
            otherwise
                assert(false, 'This mode is not implemented yet.');
        end
end
```

6.7.5.2 Precoding

Precoding performs linear transformations on the data of each substream to improve the overall receiver performance. The following MATLAB function shows how the multi-antenna data substreams that follow layer mapping are precoded prior to resource element mapping and generation of the resource grid. The function takes as input the modulated symbols organized in layers (*in*), the precoder matrix index (*cbIdx*), and the PDSCH parameter structure (*prmLTEPDSCH*). First we compute the orthonormal precoder matrix (*Wn*) by calling the *SpatialMuxPrecoder* function. Then we compute the precoded output (*out*) by multiplying the precoder matrix with vectors of input, selected by taking samples from all transmit antennas at a given sample time.

Algorithm

MATLAB function

```
function [out, Wn] = SpatialMuxPrecoder(in, prmLTEPDSCH, cbIdx)
% Precoder for PDSCH spatial multiplexing
%#codegen
% Assumes the incoming codewords are of the same length
v = prmLTEPDSCH.numLayers;          % Number of layers
numTx = prmLTEPDSCH.numTx;            % Number of Tx antennas
% Compute the precoding matrix
Wn = PrecoderMatrix(cbIdx, numTx, v);
% Initialize the output
out = complex(zeros(size(in)));
inLen = size(in, 1);
% Apply the relevant precoding matrix to the symbol over all layers
for n = 1:inLen
   temp = Wn * (in(n, :).');
   out(n, :) = temp.';
end
```

The *PrecoderMatrix* function computes the precoder matrix (*Wn*) from the values stored in a codebook. The codebook values are defined in [7]. The function takes as input the precoder index (*cbIdx*), the number of transmit antennas (*numTx*), and the number of layers (*v*). Regardless of whether an open- or a closed-loop precoding technique is used, at each subframe a common codebook index is selected at both the transmitter and the receiver. In this chapter, we choose a constant value of 1 as our codebook index. For a two-transmit-antenna configuration, valid values for the codebook index are from 0 to 3, and for a four-antenna configuration, from 0 to 15. Note that for a two-antenna transmission in which the number of layers is also two, only codebook indices of 1 and 2 are valid. Note also that for a four-antenna configuration, the precoder matrix is computed from 1×4 codebook vectors and a matrix operation that results in an orthonormal precoder matrix for any given index.

Algorithm

MATLAB function

```
function Wn = PrecoderMatrix(cbIdx, numTx, v)
% LTE Precoder for PDSCH spatial multiplexing.
%#codegen
%  v       = Number of layers
```

```
%  numTx  = Number of Tx antennas
switch numTx
  case 2
    Wn = complex(ones(numTx, v));
    switch v
      case 1
        a=(1/sqrt(2));
        codebook = [a,a; a,-a; a, 1j*a; a, -1j*a];
        Wn = codebook(cbIdx+1,:);
      case 2
        if cbIdx==1
          Wn = (1/2)*[1 1; 1 -1];
        elseif cdIdx==2
          Wn = (1/2)*[1 1; 1j -1j];
        else
          error('Not used. Please try with a different index.');
        end
    end
  case 4
    un = complex(ones(numTx, 1));
    switch cbIdx
      case 0, un = [1 -1 -1 -1].';
      case 1, un = [1 -1j 1 1j].';
      case 2, un = [1 1 -1 1].';
      case 3, un = [1 1j 1 -1j].';
      case 4, un = [1 (-1-1j)/sqrt(2) -1j (1-1j)/sqrt(2)].';
      case 5, un = [1 (1-1j)/sqrt(2) 1j (-1-1j)/sqrt(2)].';
      case 6, un = [1 (1+1j)/sqrt(2) -1j (-1+1j)/sqrt(2)].';
      case 7, un = [1 (-1+1j)/sqrt(2) 1j (1+1j)/sqrt(2)].';
      case 8, un = [1 -1 1 1].';
      case 9, un = [1 -1j -1 -1j].';
      case 10, un = [1 1 1 -1].';
      case 11, un = [1 1j -1 1j].';
      case 12, un = [1 -1 -1 1].';
      case 13, un = [1 -1 1 -1].';
      case 14, un = [1 1 -1 -1].';
      case 15, un = [1 1 1 1].';
    end
    Wn = eye(4) - 2*(un*un')./(un'*un);
    switch cbIdx % order columns, for numLayers=4 only
      case {2, 3, 14}
        Wn = Wn(:, [3 2 1 4]);
      case {6, 7, 10, 11, 13}
        Wn = Wn(:, [1 3 2 4]);
    end
    Wn = Wn./sqrt(v);
end
```

6.7.5.3 MIMO Receiver

The MIMO receiver inverts the combination of precoding and MIMO channel operations to recover the best estimate of the modulated symbols. As a result of the MIMO channel modeling, at each time index n the vector of received signals $\vec{Y}(n)$ can be modeled as a linear combination of transmitted signals from transmit antennas $\vec{X}(n)$ scaled by the channel matrix $H(n)$, with an added vector of white Gaussian noise $\vec{n}(n)$. In this book we will model the channel matrices only as 2×2 or 4×4 square matrices in order to simplify the discussion. The results can be easily generalized to non-square matrices where matrix inverse operations can be replaced by pseudo-inverse operations.

For example, for a 4×4 MIMO configuration, in any subframe and at any point in time, the received vector $\vec{Y}$ can be expressed as:

$$
\vec{Y} = \begin{bmatrix} y_1 \\ y_2 \\ y_3 \\ y_4 \end{bmatrix} = \begin{bmatrix} h_{1,1} & \cdots & h_{1,4} \\ \vdots & \ddots & \vdots \\ h_{4,1} & \cdots & h_{4,4} \end{bmatrix} \begin{bmatrix} x_1 \\ x_2 \\ x_3 \\ x_4 \end{bmatrix} + \begin{bmatrix} n_1 \\ n_2 \\ n_3 \\ n_4 \end{bmatrix} \tag{6.24}
$$

The objective of the MIMO receiver operation is to solve for best estimates of the modulated transmitted symbols $\begin{bmatrix} x_1 \\ x_2 \\ x_3 \\ x_4 \end{bmatrix}$ as a function of received symbols $\begin{bmatrix} y_1 \\ y_2 \\ y_3 \\ y_4 \end{bmatrix}$. Since the AWGN is a stochastic process, actual values of the noise vector $\begin{bmatrix} n_1 \\ n_2 \\ n_3 \\ n_4 \end{bmatrix}$ are not exactly known. We can only estimate the noise variance in each receive antenna. As a result, the effect of the AWGN is already included in the received vector.

Defining the received signal as follows, $\begin{bmatrix} r_1 \\ r_2 \\ r_3 \\ r_4 \end{bmatrix} = \begin{bmatrix} y_1 \\ y_2 \\ y_3 \\ y_4 \end{bmatrix} - \begin{bmatrix} n_1 \\ n_2 \\ n_3 \\ n_4 \end{bmatrix}$, we can rewrite Equation 6.24 as:

$$
\begin{bmatrix} r_1 \\ r_2 \\ r_3 \\ r_4 \end{bmatrix} = \begin{bmatrix} h_{1,1} & \cdots & h_{1,4} \\ \vdots & \ddots & \vdots \\ h_{4,1} & \cdots & h_{4,4} \end{bmatrix} \begin{bmatrix} x_1 \\ x_2 \\ x_3 \\ x_4 \end{bmatrix} \tag{6.25}
$$

In this section we present the three most popular methods of MIMO equalizer design, which produce the best estimate of the modulated transmitted symbols $\begin{bmatrix} x_1 \\ x_2 \\ x_3 \\ x_4 \end{bmatrix}$:

- **ZF equalizer:** We apply the inverse of the channel matrix $H = \begin{bmatrix} h_{1,1} & \cdots & h_{1,4} \\ \vdots & \ddots & \vdots \\ h_{4,1} & \cdots & h_{4,4} \end{bmatrix}$ to both sides

  of the equation. As we will see shortly, ZF equalizers can augment the effect of uncorrelated noise on the equalization process, especially in a low-SNR transmission environment.

- **MMSE equalizer:** We minimize the mean square error between the transmitted vector $\begin{bmatrix} x_1 \\ x_2 \\ x_3 \\ x_4 \end{bmatrix}$

  and its estimate $\begin{bmatrix} x_1 \\ x_2 \\ x_3 \\ x_4 \end{bmatrix}$. This approach takes into account the effect of AWGN and offsets the

  inverse matrix with the noise variance. MMSE equalizers have been shown to outperform ZF equalizers in terms of reconstruction error.

- **SD equalizer:** Our objective to find the Maximum-Likelihood solution for Equation 6.25. The SD algorithm needs to know the modulation scheme used on all of the transmit antennas. It combines MIMO equalization and soft-decision demodulation and maximizes the *a posteriori* probability measure to output the Log-Likelihood Ratios (LLRs) of the transmitted bits most likely to be involved in generating the received signal $\begin{bmatrix} r_1 \\ r_2 \\ r_3 \\ r_4 \end{bmatrix}$ at each sample time.

The following function implements the MIMO receiver operation, taking as input the received signal (*in*), the channel matrix (*chEst*), the PDSCH parameter structure (*prmLTE*), the noise-variance vector (*nVar*), and the precoder matrix (*Wn*). Depending on the equalization mode specified (*prmLTE.Eqmode*), either of the functions implementing a ZF, MMSE, or SD receiver can be called to generate the output signal (*y*).

We will now discuss each of the equalizer methodologies. Each provides a unique way of inverting layer mapping, precoding, and MIMO channel operations. The ZF and MMSE techniques help arrive at an estimate of the transmitted modulated symbols. In the case of SD, the output is not actually an estimate of the modulated symbols but rather of the bits that if modulated would generate the symbols.

Algorithm

MATLAB function

```
function y = MIMOReceiver(in, chEst, prmLTE, nVar, Wn)
%#codegen
switch prmLTE.Eqmode
    case 1 % ZF receiver
        y = MIMOReceiver_ZF(in, chEst, Wn);
    case 2 % MMSE receiver
```

```
    y = MIMOReceiver_MMSE(in, chEst, nVar, Wn);
  case 3 % Sphere Decoder
    y = MIMOReceiver_SphereDecoder(in, chEst, prmLTE, nVar, Wn);
  otherwise
    error('Function MIMOReceiver: ZF, MMSE, Sphere decoder are only
supported MIMO detectors');
end
```

ZF Receiver

The following MATLAB function shows a MIMO receiver that employs a ZF receiver to undo the effects of the MIMO channel and combat the interference from multi-antenna transmission. The function takes as input the received signal (*in*), the 2D channel matrix (*chEst*), and the precoder matrix used in this subframe (*Wn*). The function generates as its output (*y*) the estimated modulated symbols in this subframe based on the ZF equalization method. In a ZF approach, we simply invert the channel matrix and multiply the received signal by the inverse matrix. Since the vector of transmitted signals is also subject to precoding in the transmitter, in the MIMO receiver we need to multiply the equalized vector by the inverse of the precoder matrix.

Algorithm

MATLAB function

```
function y = MIMOReceiver_ZF(in, chEst, Wn)
%#codegen
% MIMO Receiver:
%   Based on received channel estimates, process the data elements
%   to equalize the MIMO channel. Uses the ZF detector.
% Get params
numData = size(in, 1);
y = complex(zeros(size(in)));
iWn = inv(Wn);
%% ZF receiver
for n = 1:numData
  h = squeeze(chEst(n, :, :)); % numTx x numRx
  h = h.';                % numRx x numTx
  Q = inv(h);
  x = Q * in(n, :).';%#ok
  tmp = iWn * x; %#ok
  y(n, :) = tmp.';
end
```

MMSE Receiver

The objective of the MMSE equalizer is to minimize the power of the error signal $e(n)$, defined as the difference between the equalized signal $X(n)$ and the original transmitted modulated signal $X(n)$. Let us define G as the optimum equalizer that transforms the received signal $Y(n)$ into the equalized signal. The error signal can then be expressed as:

$$e(n) = \widehat{X}(n) - X(n) = GY(n) - X(n) \tag{6.26}$$

Now, combining this expression with the definition of the received signal $Y(n)$ as the transformed version of the transmitted signal $X(n)$ by the MIMO channel matrix H:

$$Y(n) = H\,X(n) + n(n) \tag{6.27}$$

Assuming square matrices for both the channel matrix H and the equalizer matrix G, we obtain the following expression for the error signal:

$$e(n) = G\,Y(n) - X(n) = G(H\,X(n) + n(n)) - X(n) = (GH - I)\,X(n) + G\,n(n) \tag{6.28}$$

Modeling this expression as a Wiener filtering problem that minimizes the expected value of the error signal, we find the MMSE optimal equalizer to be:

$$G_{mmse} = H^H(HH^H + \sigma_n^2 I_n)^{-1} \tag{6.29}$$

where H^H represents the Hermitian of the channel matrix H, σ_n^2 represents the channel noise variance, and I_n represents the identity matrix of the same size as the number of transmit antennas.

The following MATLAB function shows a MIMO receiver based on an MMSE equalizer. The function takes as input the received signal (in), the 2D channel matrix ($chEst$), and the precoder matrix used in this subframe (Wn). The function generates as its output (y) the estimated modulated symbols in this subframe based on the MMSE equalization method. For each vector of received signal at sample time n, we compute the equalizer matrix (Q) based on the optimal MMSE equalizer formula and multiply the received vector by the equalizer matrix. To undo the precoding operation, we also need to multiply the equalized vector by the inverse of the precoder matrix.

Algorithm

MATLAB function

```
function y = MIMOReceiver_MMSE(in, chEst, nVar, Wn)
%#codegen
% MIMO Receiver:
%   Based on received channel estimates, process the data elements
%   to equalize the MIMO channel. Uses the MMSE detector.
% Get params
numLayers = size(Wn,1);
% noisFac = numLayers*diag(nVar);
noisFac = diag(nVar);
numData = size(in, 1);
y = complex(zeros(size(in)));
iWn = inv(Wn);
%% MMSE receiver
```

```
for n = 1:numData
    h = chEst(n, :, :);                    % numTx x numRx
    h = reshape(h(:), numLayers, numLayers).';  % numRx x numTx
    Q = (h'*h + noisFac)\h';
    x = Q * in(n, :).';
    tmp = iWn * x;
    y(n, :) = tmp.';
end
```

SD Receiver

In SD, the objective is to find the ML solution for the MIMO equation. Given the MIMO channel modeling equation at a given time sample:

$$Y = H X + n \tag{6.30}$$

the SD finds the ML estimate for the transmitted modulated symbols $\widehat{X}_{ML}$, such that:

$$\widehat{X}_{ML} = \arg \min \|Y - H X\|^2 \tag{6.31}$$

where $X \in \Omega$ and Ω is the complex-valued constellation from which the elements of X are chosen. The SD algorithm makes use of knowledge concerning the modulation scheme and the actual constellation and symbol mapping used in the modulator. It combines MIMO equalization and soft-decision demodulation and maximizes the *a posteriori* probability measure to produce its output. The output of an SD is the LLRs of the transmitted bits most likely to be involved in generation of the received signal. The *comm.SphereDecoder* System object of the Communications System Toolbox implements an SD algorithm. The ML receiver used in the System object is implemented in a reduced-complexity form by means of a Soft Sphere Decoder (SSD).

The following MATLAB function shows a MIMO receiver implemented with an SD. The function takes as input the received signal (*in*), the 3D channel matrix (*chEst*), the PDSCH parameter structure (*prmLTE*), the noise variance vector (*nVar*), and the precoder matrix used in this subframe (*Wn*). It generates as its output (*y*) the estimated modulated symbols in this subframe based on the Sphere Decoder (SD) equalization method. First, we transform the channel matrices of each sample time by the inverse of the precoder matrix. Then , we use the *comm.SphereDecoder* System object to implement maximum-likelihood (ML) Sphere Decoding operation.

Algorithm

MATLAB function

```
function [y, bittable] = MIMOReceiver_SphereDecoder(in, chEst, prmLTE, nVar, Wn)
%#codegen
% MIMO Receiver:
```

```
%   Based on received channel estimates, process the data elements
%   to equalize the MIMO channel. Uses the Sphere detector.
% Soft-Sphere Decoder
symMap=prmLTE.SymbolMap;
numBits=prmLTE.Qm;
constell=prmLTE.Constellation;
bittable = de2bi(symMap, numBits, 'left-msb');
iWn=Wn.';
nVar1=(-1/mean(nVar));
persistent SphereDec
if isempty(SphereDec)
    % Soft-Sphere Decoder
    SphereDec = comm.SphereDecoder('Constellation', constell,...
        'BitTable', bittable, 'DecisionType', 'Soft');
end
% SSD receiver
temp = complex(zeros(size(chEst)));
% Account for precoding
for n = 1:size(chEst,1)
    temp(n, :, :) = iWn * squeeze(chEst(n, :, :));
end
hD = temp;
y = nVar1 * step(SphereDec, in, hD);
```

6.7.6   Downlink Transmission Mode 4

In this section, we will focus on what is in my view one of the most innovative MIMO modes in the LTE standard, responsible for its highest data rates: mode 4. This mode employs spatial multiplexing with precoding and closed-loop channel feedback. In low-mobility scenarios, a closed-loop feedback of the channel quality can lead to performance improvements. We will actually perform the closed-loop feedback operations in the receiver in Chapter 7. In this chapter, we use a constant precoder matrix index as a stepping stone to implementation of the closed-loop adaptive precoding featured in the next chapter.

We will build two variants of this mode:

1. **Single-codeword case:** Only one codeword is generated at the DLSCH and processed by the PDSCH.
2. **Two-codeword case:** Two distinct codewords are generated at the DLSCH and multiplexed by the layer-mapping operation for precoding, resource-element mapping, and eventual OFDM transmission.

6.7.6.1   Single-Codeword Case

The following MATLAB function shows a transmitter, receiver, and channel model for the fourth mode of the LTE standard, featuring single-codeword spatial multiplexing. Using multiple antennas at both the transmitter and the receiver, we showcase both 2×2 and 4×4

MIMO antenna configurations. The key components highlighted in the example include the following:

- Generation of payload data for a single subframe (a transport block)
- DLSCH processing, as described earlier
- PDSCH transmitter processing, including bit-level scrambling, data modulation, layer mapping, and precoding for two or four antennas, as well as precoding for spatial multiplexing, resource-element mapping, and OFDM signal generation
- Channel modeling, including a MIMO fading channel followed by an AWGN channel
- PDSCH receiver processing, including an OFDM signal receiver to generate the resource grid, resource-element demapping to separate the CSR signal from the user data, channel estimation, MIMO receiver and layer demapping, soft-decision demodulation, descrambling, and DLSCH decoding.

Algorithm

MATLAB script

```
function [dataIn, dataOut, txSig, rxSig, dataRx, yRec, csr_ref]...
    = commlteMIMO_SM_step(nS, snrdB, prmLTEDLSCH, prmLTEPDSCH, prmMdl)
%% TX
persistent hPBer1
if isempty(hPBer1), hPBer1=comm.ErrorRate; end;
%  Generate payload
dataIn = genPayload(nS,  prmLTEDLSCH.TBLenVec);
% Transport block CRC generation
tbCrcOut1 =CRCgenerator(dataIn);
% Channel coding includes - CB segmentation, turbo coding, rate matching,
% bit selection, CB concatenation - per codeword
[data, Kplus1, C1] = lteTbChannelCoding(tbCrcOut1, nS, prmLTEDLSCH, prmLTEPDSCH);
%Scramble codeword
scramOut = lteScramble(data, nS, 0, prmLTEPDSCH.maxG);
% Modulate
modOut = Modulator(scramOut, prmLTEPDSCH.modType);
% Map modulated symbols  to layers
numTx=prmLTEPDSCH.numTx;
LayerMapOut = LayerMapper(modOut, [], prmLTEPDSCH);
usedCbIdx = prmMdl.cbIdx;
% Precoding
[PrecodeOut, Wn] = lteSpatialMuxPrecoder(LayerMapOut, prmLTEPDSCH, usedCbIdx);
% Generate Cell-Specific Reference (CSR) signals
csr = CSRgenerator(nS, numTx);
csr_ref=complex(zeros(2*prmLTEPDSCH.Nrb, 4, numTx));
for m=1:numTx
    csr_pre=csr(1:2*prmLTEPDSCH.Nrb,:,:,m);
    csr_ref(:,:,m)=reshape(csr_pre,2*prmLTEPDSCH.Nrb,4);
end
```

```
% Resource grid filling
txGrid = REmapper_mTx(PrecodeOut, csr_ref, nS, prmLTEPDSCH);
% OFDM transmitter
txSig = OFDMTx(txGrid, prmLTEPDSCH);
%% Channel
% MIMO Fading channel
[rxFade, chPathG] = MIMOFadingChan(txSig, prmLTEPDSCH, prmMdl);
% Add AWG noise
sigPow = 10*log10(var(rxFade));
nVar = 10.^(0.1.*(sigPow-snrdB));
rxSig = AWGNChannel(rxFade, nVar);
%% RX
% OFDM Rx
rxGrid = OFDMRx(rxSig, prmLTEPDSCH);
% updated for numLayers -> numTx
[dataRx, csrRx, idx_data] = REdemapper_mTx(rxGrid, nS, prmLTEPDSCH);
% MIMO channel estimation
if prmMdl.chEstOn
    chEst = ChanEstimate_mTx(prmLTEPDSCH, csrRx, csr_ref, prmMdl.chEstOn);
    hD   = ExtChResponse(chEst, idx_data, prmLTEPDSCH);
else
    idealChEst = IdChEst(prmLTEPDSCH, prmMdl, chPathG);
    hD = ExtChResponse(idealChEst, idx_data, prmLTEPDSCH);
end
% Frequency-domain equalizer
if (numTx==1)
    % Based on Maximum-Combining Ratio (MCR)
    yRec = Equalizer_simo(dataRx, hD, nVar, prmLTEPDSCH.Eqmode);
else
    % Based on Spatial Multiplexing
    yRec = MIMOReceiver(dataRx, hD, prmLTEPDSCH, nVar, Wn);
end
% Demap received codeword(s)
[cwOut, ~] = LayerDemapper(yRec, prmLTEPDSCH);
if prmLTEPDSCH.Eqmode < 3
    % Demodulate
    demodOut = DemodulatorSoft(cwOut, prmLTEPDSCH.modType, mean(nVar));
else
    demodOut = cwOut;
end
    % Descramble received codeword
    rxCW = lteDescramble(demodOut, nS, 0, prmLTEPDSCH.maxG);
% Channel decoding includes - CB segmentation, turbo decoding, rate dematching
[decTbData1, ~,~] = lteTbChannelDecoding(nS, rxCW, Kplus1, C1, prmLTEDLSCH,
prmLTEPDSCH);
% Transport block CRC detection
[dataOut, ~] = CRCdetector(decTbData1);
end
```

Structure of the Transceiver Model

The following MATLAB script is the testbench that calls the MIMO transceiver function *commlteMIMO*. First it calls the initialization function (*commlteMIMO_initialize*) to set all relevant parameter structures (*prmLTEDLSCH, prmLTEPDSCH, prmMdl*). Then it uses a while loop to perform subframe processing by calling the MIMO transceiver function *commlteMIMO_SM_step*. Finally, it computes the BER and calls the visualization function to illustrate the channel response and modulation constellation before and after equalization.

Algorithm

MATLAB script

```
% Script for MIMO LTE (mode 4)
%
% Single codeword transmission
%
clear all
clear functions
%% Set simulation parameters & initialize parameter structures
commlteMIMO_ params;
[prmLTEPDSCH, prmLTEDLSCH, prmMdl] = commlteMIMO_initialize(txMode, ...
chanBW, contReg, modType, Eqmode,numTx, numRx,cRate,maxIter, fullDecode,
chanMdl, corrLvl, ...
    chEstOn, numCodeWords, enPMIfback, cbIdx, snrdB, maxNumErrs, maxNumBits);
clear txMode chanBW contReg modType Eqmode numTx numRx cRate maxIter
fullDecode chanMdl corrLvl chEstOn numCodeWords enPMIfback cbIdx snrdB
maxNumErrs maxNumBits
%%
disp('Simulating the LTE Mode 3: Multiple Tx & Rx antrennas with Spatial Multiplexing');
zReport_data_rate(prmLTEPDSCH, prmLTEDLSCH);
hPBer = comm.ErrorRate;
snrdB=prmMdl.snrdB;
maxNumErrs=prmMdl.maxNumErrs;
maxNumBits=prmMdl.maxNumBits;
%% Simulation loop
tic;
nS = 0; % Slot number, one of [0:2:18]
Measures = zeros(3,1); %initialize BER output
while (( Measures(2)< maxNumErrs) && (Measures(3) < maxNumBits))
  [dataIn, dataOut, txSig, rxSig, dataRx, yRec, csr] = ...
    commlteMIMO_SM_step(nS, snrdB, prmLTEDLSCH, prmLTEPDSCH, prmMdl);
  % Calculate  bit errors
  Measures = step(hPBer, dataIn, dataOut);
   % Visualize constellations and spectrum
  if (visualsOn && prmLTEPDSCH.Eqmode~=3)
    zVisualize( prmLTEPDSCH, txSig, rxSig, yRec, dataRx, csr, nS);
  end;
```

```
    % Update subframe number
    nS = nS + 2; if nS > 19, nS = mod(nS, 20); end;
end
disp(Measures);
toc;
```

Verifying Transceiver Performance

By executing the MATLAB script of the MIMO transceiver model (*commlteMIMO*) we can examine various signals to assess the performance of the system. The parameters used in simulation are summarized in the following MATLAB script (*commlteMIMO_params*). This parameter set copies the common MIMO parameters used in Section 6.7.1. The parameters that are different reflect the use of spatial multiplexing MIMO mode 4 with a single code-word, turning off of the precoder matrix feedback, and use of the MMSE equalizer for MIMO receiver processing. In this simulation, 1 million bits of user data are processed, the SNR of the AWGN channel is set to 16 dB, and the visualization function is turned on.

Algorithm

MATLAB script

```
% PDSCH
txMode        = 4;   % Transmission mode one of {1, 2, 4}
numTx         = 2;   % Number of transmit antennas
numRx         = 2;   % Number of receive antennas
chanBW        = 4;       % [1,2,3,4,5,6] maps to [1.4, 3, 5, 10, 15, 20]MHz
contReg       = 1;       % {1,2,3} for >=10MHz, {2,3,4} for <10Mhz
modType       = 2;       % [1,2,3] maps to ['QPSK','16QAM','64QAM']
% DLSCH
cRate         = 1/3; % Rate matching target coding rate
maxIter       = 6;   % Maximum number of turbo decoding terations
fullDecode    = 0;   % Whether "full" or "early stopping" turbo decoding is performed
% Channel model
chanMdl       = 'frequency-selective-high-mobility';
% one of {'flat-low-mobility', 'flat-high-mobility','frequency-selective-low-mobility',
% 'frequency-selective-high-mobility', 'EPA 0Hz', 'EPA 5Hz', 'EVA 5Hz', 'EVA 70Hz'}
corrLvl       = 'Medium';
% Simulation parameters
Eqmode        = 2;   % Type of equalizer used [1,2,3] for ['ZF', 'MMSE','Sphere Decoder']
chEstOn       = 1;       % use channel estimation or ideal channel
snrdB         = 16;  % Signal to Noise Ratio in dB
maxNumErrs    = 1e6; % Maximum number of errors found before simulation stops
maxNumBits    = 1e6; % Maximum number of bits processed before simulation stops
visualsOn     = 1;       % Whether to visualize channel response and constellations
numCodeWords  = 1; % Number of codewords in PDSCH
enPMIfback    = 0;   % Enable/Disable Precoder Matrix Indicator (PMI) feedback
cbIdx         = 1;       % Initialize PMI index
```

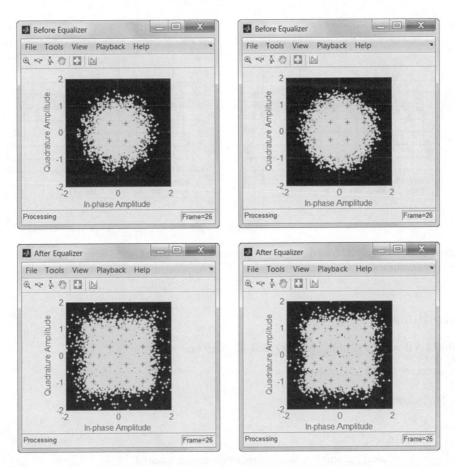

Figure 6.12 LTE model: MIMO spatial-multiplexing constellation diagram of user data before and after equalization

Figure 6.12 shows the constellation diagrams before (first row) and after (second row) equalization of user data obtained from each of the two receive antennas in a subframe. It shows that the equalizer can compensate for the effect of a fading channel to result in a constellation that more closely resembles that of the 16QAM modulator.

Figure 6.13 illustrates the spectra of user data obtained from each of the two receive antennas in a subframe. It shows the transmitted signal and the received signal before and after equalization. The received signal before equalization (showing the effects of frequency-selective fading) is effectively equalized by the closed-loop spatial multiplexing (showing a more frequency-flat nature), which closely resembles the transmitted signal spectrum.

BER Measurements

In order to verify the BER performance of the transceiver, we create a testbench called *commlteMIMO_test_timing_ber*, which first initializes the LTE system parameters and then iterates through a range of SNR values and calls the *commlteMIMO_fcn* function in the loop in order to compute the corresponding BER values.

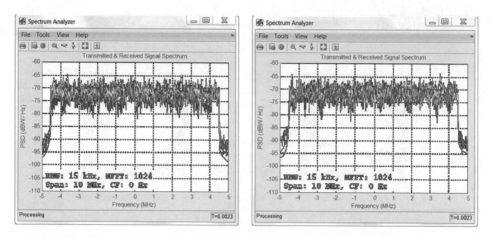

Figure 6.13 LTE MIMO spatial-multiplexing spectra of transmitted and of the received signal before and after equalization

Algorithm

MATLAB script: commlteMIMO_test_timing_ber

```
% Script for MIMO LTE (mode 4)
%
% Single codeword transmission only
%
clear all
clear functions
%% Set simulation parameters & initialize parameter structures
commlteMIMO_ params;
maxNumErrs=5e7;
maxNumBits=5e7;
[prmLTEPDSCH, prmLTEDLSCH, prmMdl] = commlteMIMO_initialize(txMode, ...
chanBW, contReg, modType, Eqmode,numTx, numRx,cRate,maxIter, fullDecode,
chanMdl, corrLvl, ...
   chEstOn, numCodeWords, enPMIfback, cbIdx, snrdB, maxNumErrs, maxNumBits);
clear txMode chanBW contReg modType Eqmode numTx numRx cRate maxIter
fullDecode chanMdl corrLvl chEstOn numCodeWords enPMIfback cbIdx snrdB
maxNumErrs maxNumBits
%%
disp('Simulating the LTE Mode 3: Multiple Tx & Rx antrennas with Spatial Multiplexing');
zReport_data_rate(prmLTEPDSCH, prmLTEDLSCH);
%% Geerate code and setup parallelism
disp('Generating code for commlteMIMO_ fcn.m ...');
arg1=coder.Constant(prmLTEPDSCH);
arg2=coder.Constant( prmLTEDLSCH);
arg3=coder.Constant(prmMdl);
```

```
codegen commlteMIMO_ fcn -args {16, arg1, arg2, arg3} -report
disp('Done.');
parallel_setup;
%%
MaxIter=8;
snr_vector=getSnrVector(prmLTEPDSCH.modType, MaxIter);
ber_vector=zeros(size(snr_vector));
maxNumBits=prmMdl.maxNumBits;
tic;
parfor n=1:MaxIter
    fprintf(1,'Iteration %2d out of %2d:  Processing %10d bits. SNR = %3d\n', ...
        n, MaxIter, maxNumBits, snr_vector(n));
    [ber, ˜] = commlteMIMO_ fcn_mex(snr_vector(n), prmLTEPDSCH, prmLTEDLSCH,
prmMdl);
    ber_vector(n)=ber;
end;
toc;
semilogy(snr_vector, ber_vector);
title('BER - commlteMIMO SM');xlabel('SNR (dB)');ylabel('ber');grid;
```

Figure 6.14 shows the BER of the transceiver as a function of the SNR values after processing of 50 million bits of user data in each of eight iterations.

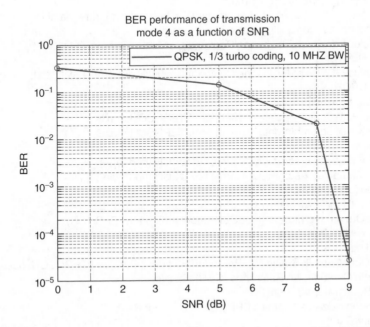

Figure 6.14   BER results: LTE mode 4 spatial-multiplexing single-codeword (2 × 2) MIMO channel

6.7.6.2 Two-Codeword Case

The following MATLAB function shows a transmitter, receiver, and channel model for mode 4 of the LTE standard featuring two-codeword spatial multiplexing. The structure of the transceiver model is very similar to that in the single-codeword case, except that we create and process a pair of data bits and repeat operations such as CRC generation, DLSCH processing, scrambling, and modulation on data pairs. The layer-mapping operation transforms the data into layers, and from then until layer demapping everything is similar to the single-antenna case. After that, demodulation, descrambling, CRC detection, and transport-block-channel decoding (the inverse of DLSCH processing) also occur as pairs of operations.

Algorithm

MATLAB function

```
function [dataIn, dataOut, txSig, rxSig, dataRx, yRec, csr_ref]...
    = commlteMIMO_SM2_step(nS, snrdB, prmLTEDLSCH, prmLTEPDSCH, prmMdl)
%% TX
persistent hPBer1
if isempty(hPBer1), hPBer1=comm.ErrorRate; end;
%  Generate payload
dataIn1 = genPayload(nS,  prmLTEDLSCH.TBLenVec);
dataIn2 = genPayload(nS,  prmLTEDLSCH.TBLenVec);
dataIn=[dataIn1;dataIn2];
% Transport block CRC generation
tbCrcOut1 =CRCgenerator(dataIn1);
tbCrcOut2 =CRCgenerator(dataIn2);
% Channel coding includes - CB segmentation, turbo coding, rate matching,
% bit selection, CB concatenation - per codeword
[data1, Kplus1, C1] = lteTbChannelCoding(tbCrcOut1, nS, prmLTEDLSCH,
prmLTEPDSCH);
[data2, Kplus2, C2] = lteTbChannelCoding(tbCrcOut2, nS, prmLTEDLSCH,
prmLTEPDSCH);
%Scramble codeword
scramOut1 = lteScramble(data1, nS, 0, prmLTEPDSCH.maxG);
scramOut2 = lteScramble(data2, nS, 0, prmLTEPDSCH.maxG);
% Modulate
modOut1 = Modulator(scramOut1, prmLTEPDSCH.modType);
modOut2 = Modulator(scramOut2, prmLTEPDSCH.modType);
% Map modulated symbols  to layers
numTx=prmLTEPDSCH.numTx;
LayerMapOut = LayerMapper(modOut1, modOut2, prmLTEPDSCH);
usedCbIdx = prmMdl.cbIdx;
% Precoding
[PrecodeOut, Wn] = lteSpatialMuxPrecoder(LayerMapOut, prmLTEPDSCH, usedCbIdx);
% Generate Cell-Specific Reference (CSR) signals
csr = CSRgenerator(nS, numTx);
csr_ref=complex(zeros(2*prmLTEPDSCH.Nrb, 4, numTx));
for m=1:numTx
```

```
   csr_pre=csr(1:2*prmLTEPDSCH.Nrb,:,:,m);
   csr_ref(:,:,m)=reshape(csr_pre,2*prmLTEPDSCH.Nrb,4);
end
% Resource grid filling
txGrid = REmapper_mTx(PrecodeOut, csr_ref, nS, prmLTEPDSCH);
% OFDM transmitter
txSig = OFDMTx(txGrid, prmLTEPDSCH);
%% Channel
% MIMO Fading channel
[rxFade, chPathG] = MIMOFadingChan(txSig, prmLTEPDSCH, prmMdl);
% Add AWG noise
sigPow = 10*log10(var(rxFade));
nVar = 10.^(0.1.*(sigPow-snrdB));
rxSig =  AWGNChannel(rxFade, nVar);
%% RX
% OFDM Rx
rxGrid = OFDMRx(rxSig, prmLTEPDSCH);
% updated for numLayers -> numTx
[dataRx, csrRx, idx_data] = REdemapper_mTx(rxGrid, nS, prmLTEPDSCH);
% MIMO channel estimation
if prmMdl.chEstOn
   chEst = ChanEstimate_mTx(prmLTEPDSCH, csrRx,  csr_ref, prmMdl.chEstOn);
   hD     = ExtChResponse(chEst, idx_data, prmLTEPDSCH);
else
   idealChEst = IdChEst(prmLTEPDSCH, prmMdl, chPathG);
   hD = ExtChResponse(idealChEst, idx_data, prmLTEPDSCH);
end
% Frequency-domain equalizer
if (numTx==1)
   % Based on Maximum-Combining Ratio (MCR)
   yRec = Equalizer_simo(dataRx, hD, nVar, prmLTEPDSCH.Eqmode);
else
   % Based on Spatial Multiplexing
   yRec = MIMOReceiver(dataRx, hD, prmLTEPDSCH, nVar, Wn);
end
% Demap received codeword(s)
[cwOut1, cwOut2] = LayerDemapper(yRec, prmLTEPDSCH);
if prmLTEPDSCH.Eqmode < 3
   % Demodulate
   demodOut1 = DemodulatorSoft(cwOut1, prmLTEPDSCH.modType, mean(nVar));
   demodOut2 = DemodulatorSoft(cwOut2, prmLTEPDSCH.modType, mean(nVar));
else
   demodOut1 = cwOut1;
   demodOut2 = cwOut2;
end
% Descramble received codeword
rxCW1 = lteDescramble(demodOut1, nS, 0, prmLTEPDSCH.maxG);
rxCW2 = lteDescramble(demodOut2, nS, 0, prmLTEPDSCH.maxG);
% Channel decoding includes - CB segmentation, turbo decoding, rate dematching
```

```
[decTbData1, ~,~] = lteTbChannelDecoding(nS, rxCW1, Kplus1, C1, prmLTEDLSCH,
prmLTEPDSCH);
[decTbData2, ~,~] = lteTbChannelDecoding(nS, rxCW2, Kplus2, C2, prmLTEDLSCH,
prmLTEPDSCH);
% Transport block CRC detection
[dataOut1, ~] = CRCdetector(decTbData1);
[dataOut2, ~] = CRCdetector(decTbData2);
dataOut=[dataOut1;dataOut2];
end
```

Structure of the Transceiver Model

The following MATLAB script is the testbench that calls the MIMO transceiver function *commlteMIMO*. First it calls the initialization function (*commlteMIMO_initialize*) to set all the relevant parameter structures (*prmLTEDLSCH*, *prmLTEPDSCH*, *prmMdl*). Then it uses a while loop to perform subframe processing by calling the MIMO transceiver function *commlteMIMO_SM2_step*. Finally, it computes the BER and calls the visualization function to illustrate the channel response and modulation constellation before and after equalization.

Algorithm

MATLAB script

```
% Script for MIMO LTE (mode 4)
%
% Two codeword transmission
%
clear all
clear functions
%% Set simulation parameters & initialize parameter structures
commlteMIMO_ params;
[prmLTEPDSCH, prmLTEDLSCH, prmMdl] = commlteMIMO_initialize(txMode, ...
chanBW, contReg, modType, Eqmode,numTx, numRx,cRate,maxIter, fullDecode,
chanMdl, corrLvl, ...
    chEstOn, numCodeWords, enPMIfback, cbIdx, snrdB, maxNumErrs, maxNumBits);
clear txMode chanBW contReg modType Eqmode numTx numRx cRate maxIter
fullDecode chanMdl corrLvl chEstOn numCodeWords enPMIfback cbIdx snrdB
maxNumErrs maxNumBits
%%
disp('Simulating the LTE Mode 3: Multiple Tx & Rx antrennas with Spatial Multiplexing');
zReport_data_rate(prmLTEPDSCH, prmLTEDLSCH);
hPBer = comm.ErrorRate;
snrdB=prmMdl.snrdB;
maxNumErrs=prmMdl.maxNumErrs;
maxNumBits=prmMdl.maxNumBits;
%% Simulation loop
tic;
```

```
nS = 0; % Slot number, one of [0:2:18]
Measures = zeros(3,1); %initialize BER output
while (( Measures(2)< maxNumErrs) && (Measures(3) < maxNumBits))
  [dataIn, dataOut, txSig, rxSig, dataRx, yRec, csr] = ...
    commlteMIMO_SM2_step(nS, snrdB, prmLTEDLSCH, prmLTEPDSCH, prmMdl);
  % Calculate  bit errors
  Measures = step(hPBer, dataIn, dataOut);
   % Visualize constellations and spectrum
  if (visualsOn && prmLTEPDSCH.Eqmode¯=3)
    zVisualize( prmLTEPDSCH, txSig, rxSig, yRec, dataRx, csr, nS);
  end;
  % Update subframe number
  nS = nS + 2; if nS > 19, nS = mod(nS, 20); end;
end
disp(Measures);
toc;
```

Verifying Transceiver Performance

By executing the MATLAB script of the MIMO transceiver model (*commlteMIMO*) we can examine various signals in order to assess the performance of the system. The parameters used in simulation are summarized in the following MATLAB script (*commlteMIMO_params*). This parameter set copies the common MIMO parameters used in Section 6.7.1. The parameters that are different reflect the use of spatial-multiplexing MIMO mode 4 with two codewords, turning off of the precoder matrix feedback, and the use of the MMSE equalizer for MIMO receiver processing. In this simulation, 1 million bits of user data are processed, the SNR of the AWGN channel is set to 16 dB, and the visualization function is turned on.

Algorithm

MATLAB script

```
% PDSCH
txMode       = 4;  % Transmission mode one of {1, 2, 4}
numTx        = 2;   % Number of transmit antennas
numRx        = 2;   % Number of receive antennas
chanBW       = 4;      % [1,2,3,4,5,6] maps to [1.4, 3, 5, 10, 15, 20]MHz
contReg      = 1;      % {1,2,3} for >=10MHz, {2,3,4} for <10Mhz
modType      = 2;      % [1,2,3] maps to ['QPSK','16QAM','64QAM']
% DLSCH
cRate        = 1/3; % Rate matching target coding rate
maxIter      = 6;    % Maximum number of turbo decoding terations
fullDecode   = 0;   % Whether "full" or "early stopping" turbo decoding is performed
% Channel model
chanMdl      = 'frequency-selective-high-mobility';
% one of {'flat-low-mobility', 'flat-high-mobility','frequency-selective-low-mobility',
% 'frequency-selective-high-mobility', 'EPA 0Hz', 'EPA 5Hz', 'EVA 5Hz', 'EVA 70Hz'}
```

```
corrLvl          = 'Medium';
% Simulation parameters
Eqmode           = 2;    % Type of equalizer used [1,2,3] for ['ZF', 'MMSE','Sphere Decoder']
chEstOn          = 1;        % use channel estimation or ideal channel
snrdB            = 16;   % Signal to Noise Ratio in dB
maxNumErrs       = 1e6; % Maximum number of errors found before simulation stops
maxNumBits       = 1e6; % Maximum number of bits processed before simulation stops
visualsOn        = 1;    % Whether to visualize channel response and constellations
numCodeWords     = 2; % Number of codewords in PDSCH
enPMIfback       = 0;    % Enable/Disable Precoder Matrix Indicator (PMI) feedback
cbIdx            = 1;    % Initialize PMI index
```

Figure 6.15 shows the constellation diagrams before (first row) and after (second row) equalization of the user data obtained from each of the two receive antennas in a given subframe.

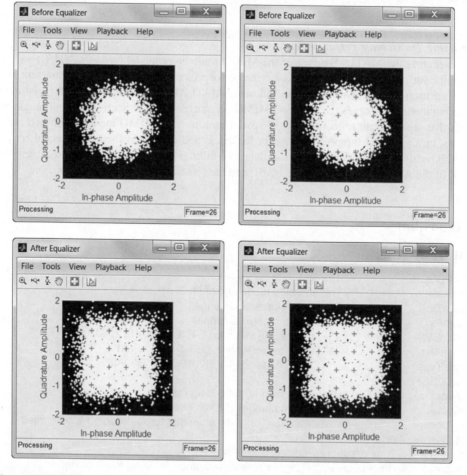

Figure 6.15 LTE model: MIMO spatial-multiplexing two-codeword constellation diagram before and after equalization

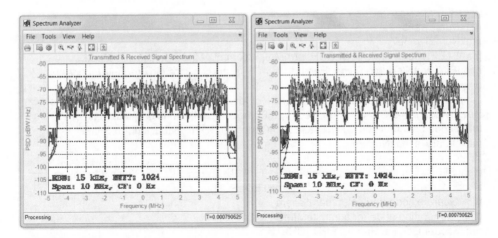

Figure 6.16   MIMO spatial-multiplexing two-codewords spectra of the transmited signal and of the received signal before and after equalization

It shows that the equalizer can compensate for the effect of a fading channel to result in a constellation that more closely resembles that of the 16QAM modulator.

Figure 6.16 illustrates the spectra of user data obtained from each of the two receive antennas in a subframe. It shows the transmitted signal and the received signal before and after equalization. The received signal before equalization is effectively equalized in the 2-codeword spatial multiplexing case which closely resembles the transmitted signal spectrum.

BER Measurements
In order to verify the BER performance of the transceiver, we create a testbench called *commlteMIMO_test_timing_ber*, which first initializes the LTE system parameters and then iterates through a range of SNR values and calls the *commlteMIMO_fcn* function in the loop to compute the corresponding BER values. The results obtained are very similar to the single-codeword results illustrated in Figure 6.14.

Algorithm

MATLAB script: commlteMIMO_test_timing_ber

```
% Script for MIMO LTE (mode 4)
%
% Single codeword transmission only
%
clear all
clear functions
%% Set simulation parameters & initialize parameter structures
commlteMIMO_ params;
```

```
maxNumErrs=5e7;
maxNumBits=5e7;
[prmLTEPDSCH, prmLTEDLSCH, prmMdl] = commlteMIMO_initialize(txMode, ...
chanBW, contReg, modType, Eqmode,numTx, numRx,cRate,maxIter, fullDecode,
chanMdl, corrLvl, ...
    chEstOn, numCodeWords, enPMIfback, cbIdx, snrdB, maxNumErrs, maxNumBits);
clear txMode chanBW contReg modType Eqmode numTx numRx cRate maxIter
fullDecode chanMdl corrLvl chEstOn numCodeWords enPMIfback cbIdx snrdB
maxNumErrs maxNumBits
%%
disp('Simulating the LTE Mode 3: Multiple Tx & Rx antrennas with Spatial Multiplexing');
zReport_data_rate(prmLTEPDSCH, prmLTEDLSCH);
%% Geerate code and setup parallelism
disp('Generating code for commlteMIMO_ fcn.m ...');
arg1=coder.Constant(prmLTEPDSCH);
arg2=coder.Constant( prmLTEDLSCH);
arg3=coder.Constant(prmMdl);
codegen commlteMIMO_ fcn -args {16, arg1, arg2, arg3} -report
disp('Done.');
parallel_setup;
%%
MaxIter=8;
snr_vector=getSnrVector(prmLTEPDSCH.modType, MaxIter);
ber_vector=zeros(size(snr_vector));
maxNumBits=prmMdl.maxNumBits;
tic;
parfor n=1:MaxIter
    fprintf(1,'Iteration %2d out of %2d:  Processing %10d bits. SNR = %3d\n', ...
        n, MaxIter, maxNumBits, snr_vector(n));
    [ber, ~] = commlteMIMO_ fcn_mex(snr_vector(n), prmLTEPDSCH, prmLTEDLSCH,
prmMdl);
    ber_vector(n)=ber;
end;
toc;
semilogy(snr_vector, ber_vector);
title('BER - commlteMIMO SM');xlabel('SNR (dB)');ylabel('ber');grid;
```

6.7.7   Open-Loop Spatial Multiplexing

The following MATLAB function shows the spatial-multiplexing-with-large-CDD (Cyclic
Delay Diversity) algorithm that implements MIMO transmission mode 3 of the LTE standard. Open-loop precoding is designed for transmission in high-mobility scenarios and does
not rely on Precoder Matrix Indicator (PMI) by the mobile terminal. When the mobile terminal
moves rapidly, an open-loop approach works best, since the channel-state feedback from a previous subframe cannot accurately predict the channel quality in the current one. As a result, in
open-loop spatial multiplexing no explicit information regarding the precoder matrix is transmitted from the base station to the mobile terminal. Instead, the precoder matrix is selected in a

deterministic way that can be computed synchronously in both the transmitter and the receiver in every subframe.

6.7.7.1 Open-Loop Precoding

In open-loop precoding, the transmitter and receiver do not communicate the choice of code-book indices using a feedback loop. Instead, a predefined set of precoding matrix indices is used, whcih is periodically updated in the transmitter and the receiver and is synchronized with each transmitted sample.

The transmission rank for open-loop precoding can also vary, from a full rank down to a minimum of two layers. When rank estimation results in a single layer, we switch from spatial multiplexing to transmit diversity. For a mode 3 transmission, when the rank takes on a value of 1, SFBC is used for two antennas and combined SFBC/FSTD for four.

The *PrecoderMatrix* function computes the precoder matrix (W), the diagonal matrix (D), and the matrix of eigenvectors (U) for each sample of the input signal. The function takes as input the sample index (n) and the number of layers (v). The codebook values are defined in Reference [7].

Algorithm

MATLAB function

```
function [W, D, U] = PrecoderMatrix(n, v)
% LTE Precoder for PDSCH spatial multiplexing.
%#codegen
idx=mod(n-1,4);
switch v
  case 1
    W=complex(1,0);
    U=W;D=W;
  case 2
    W=[1 0; 0 1];
    U=(1/sqrt(2))*[1 1;1 exp(-1j*pi)];
    D=[1 0;0 exp(-1j*pi*idx)];
  case 4
    k=1+mod(floor(n/4),4);
    switch k
      case 1, un = [1 -1 -1 1].';
      case 2, un = [1 -1 1 -1].';
      case 3, un = [1 1 -1 -1].';
      case 4, un = [1 1 1 1].';
    end
    W = eye(4) - 2*(un*un')./(un'*un);
    switch k % order columns
      case 3
          W = W(:, [3 2 1 4]);
      case 2
```

```
        W = W(:, [1 3 2 4]);
    end
    a=[0*(0:1:3);2*(0:1:3);4*(0:1:3);6*(0:1:3)];
    U=(1/2)*exp(-1j*pi*a/4);
    b=0:1:3;
    D=diag(exp(-1j*2*pi*idx*b/4));
end
```

The following MATLAB function shows the precoding operations used when open-loop spatial multiplexing is selected as the mode of transmission. The function takes as input the modulated symbols organized in layers (*in*), the precoder matrix index (*cbIdx*), and the PDSCH parameter structure (*prmLTEPDSCH*). The output (*out*) is computed in three steps: (i) in the processing loop for each sample of the input, the *PrecoderMatrix* function obtains the three output matrices (*W*, *D*, *U*); (ii) the matrices are multiplied to obtain the transformation matrix (*T*); (iii) finally, the input vector is precoded sample by sample by multiplying it with the transformation matrix.

Algorithm

MATLAB function

```
function out = SpatialMuxPrecoder(in, prmLTEPDSCH)
% Precoder for PDSCH spatial multiplexing
%#codegen
% Assumes the incoming codewords are of the same length
v = prmLTEPDSCH.numLayers;              % Number of layers
% Initialize the output
out = complex(zeros(size(in)));
inLen = size(in, 1);
% Apply the relevant precoding matrix to the symbol over all layers
for n = 1:inLen
    % Compute the precoding matrix
    [W, D, U] = PrecoderMatrix(n, v);
    T=W *D*U;
    temp = T* (in(n, :).');
    out(n, :) = temp.';
end
```

6.7.7.2 MIMO Receiver Operations

The MIMO receiver function in open-loop spatial multiplexing is analogous to those used in closed-loop spatial multiplexing. It takes as input the received signal (*in*), the channel matrix (*chEst*), the PDSCH parameter structure (*prmLTE*), and the noise-variance vector (*nVar*). Depending on the equalization mode specified (*prmLTE.Eqmode*), either of the functions implementing a ZF, MMSE, or SD receiver is then called to generate the output signal (*y*).

Algorithm

MATLAB function

```
function y = MIMOReceiver_OpenLoop(in, chEst, prmLTE, nVar)
%#codegen
v=prmLTE.numTx;
switch prmLTE.Eqmode
   case 1 % ZF receiver
      y = MIMOReceiver_ZF_OpenLoop(in, chEst, v);
   case 2 % MMSE receiver
      y = MIMOReceiver_MMSE_OpenLoop(in, chEst, nVar, v);
   case 3 % Sphere Decoder
      y = MIMOReceiver_SD_OpenLoop(in, chEst, prmLTE, nVar, v);
   otherwise
      error('Function MIMOReceiver: ZF, MMSE, Sphere decoder are only
supported MIMO detectors');
end
```

ZF Receiver

The following MATLAB function shows a MIMO receiver that employs a ZF receiver. The function takes as input the received signal (*in*), the 2D channel matrix (*chEst*), and the number of layers in this subframe (*v*). Based on the ZF equalization method, it generates as output (*y*) the estimated modulated symbols in this subframe.

Algorithm

MATLAB function

```
function y = MIMOReceiver_ZF_OpenLoop(in, chEst, v)
%#codegen
% MIMO Receiver:
%   Based on received channel estimates, process the data elements
%   to equalize the MIMO channel. Uses the ZF detector.
% Get params
numData = size(in, 1);
y = complex(zeros(size(in)));
%% ZF receiver
for n = 1:numData
   [W, D, U] = PrecoderMatrixOpenLoop(n, v);
   iWn = (W *D*U)';
   h = squeeze(chEst(n, :, :)); % numTx x numRx
   h = h.';               % numRx x numTx
   x = h \ (in(n, :).');
   tmp = iWn * x;
   y(n, :) = tmp.';
end
```

MMSE Receiver

The following MATLAB function shows a MIMO receiver that employs a MMSE receiver. The function input and output signatures are very similar to those for the ZF algorithm, but with an additional input parameter corresponding to the noise variance (*nVar*) of the current subframe.

Algorithm

MATLAB function

```
function y = MIMOReceiver_MMSE_OpenLoop(in, chEst, nVar, v)
%#codegen
% MIMO Receiver:
%   Based on received channel estimates, process the data elements
%   to equalize the MIMO channel. Uses the MMSE detector.
% noisFac = numLayers*diag(nVar);
noisFac = diag(nVar);
numData = size(in, 1);
y = complex(zeros(size(in)));
%% MMSE receiver
for n = 1:numData
    [W, D, U] = PrecoderMatrixOpenLoop(n, v);
    iWn = (W *D*U)';          % Orthonormal matrix
    h = chEst(n, :, :);       % numTx x numRx
    h = reshape(h(:), v, v).';   % numRx x numTx
    Q = (h'*h + noisFac)\h';
    x = Q * in(n, :).';
    tmp = iWn * x;
    y(n, :) = tmp.';
end
```

SD Receiver

The following MATLAB function shows a MIMO receiver that employs an Sphere Decoder (SD) receiver. The function input and output signatures are identical to those presented in the MMSE case.

Algorithm

MATLAB function

```
function [y, bittable] = MIMOReceiver_SD_OpenLoop(in, chEst, prmLTE, nVar, v)
%#codegen
% MIMO Receiver:
%   Based on received channel estimates, process the data elements
%   to equalize the MIMO channel. Uses the Sphere detector.
```

```
% Soft-Sphere Decoder
symMap=prmLTE.SymbolMap;
numBits=prmLTE.Qm;
constell=prmLTE.Constellation;
bittable = de2bi(symMap, numBits, 'left-msb');
nVar1=(-1/mean(nVar));
persistent SphereDec
if isempty(SphereDec)
   % Soft-Sphere Decoder
   SphereDec = comm.SphereDecoder('Constellation', constell,...
      'BitTable', bittable, 'DecisionType', 'Soft');
end
% SSD receiver
temp = complex(zeros(size(chEst)));
% Account for precoding
for n = 1:size(chEst,1)
   [W, D, U] =PrecoderMatrixOpenLoop(n, v);
   iWn = (W *D*U).';
   temp(n, :, :) =  iWn * squeeze(chEst(n, :, :)) ;
end
hD = temp;
y = nVar1 * step(SphereDec, in, hD);
```

6.7.8 Downlink Transmission Mode 3

The third downlink transmission mode uses open-loop spatial multiplexing and is intended for transmission in high-mobility scenarios. The following MATLAB function shows a transmitter, receiver, and channel model for this mode featuring single-codeword spatial multiplexing. Using multiple antennas at both the transmitter and the receiver, we showcase both 2×2 and 4×4 MIMO antenna configurations. The key components highlighted in the example include the following:

- Generation of payload data for a single subframe (a transport block)
- DLSCH processing, as described earlier
- PDSCH transmitter processing, including bit-level scrambling, data modulation, layer mapping and precoding for two or four antennas, precoding for spatial multiplexing, resource-element mapping, and OFDM signal generation
- Channel modeling, including a MIMO fading channel followed by an AWGN channel
- PDSCH receiver processing, including an OFDM signal receiver to generate the resource grid, resource-element demapping to separate the CSR signal from the user data, channel estimation, MIMO receiver and layer demapping, soft-decision demodulation, descrambling, and DLSCH decoding.

Algorithm

MATLAB script

```
function [dataIn, dataOut, txSig, rxSig, dataRx, yRec, csr_ref]...
  = commlteMIMO_SM_Mode3_step(nS, snrdB, prmLTEDLSCH, prmLTEPDSCH, prmMdl)
%% TX
persistent hPBer1
if isempty(hPBer1), hPBer1=comm.ErrorRate; end;
% Generate payload
dataIn = genPayload(nS, prmLTEDLSCH.TBLenVec);
% Transport block CRC generation
tbCrcOut1 =CRCgenerator(dataIn);
% Channel coding includes - CB segmentation, turbo coding, rate matching,
% bit selection, CB concatenation - per codeword
[data, Kplus1, C1] = lteTbChannelCoding(tbCrcOut1, nS, prmLTEDLSCH, prmLTEPDSCH);
%Scramble codeword
scramOut = lteScramble(data, nS, 0, prmLTEPDSCH.maxG);
% Modulate
modOut = Modulator(scramOut, prmLTEPDSCH.modType);
% Map modulated symbols to layers
numTx=prmLTEPDSCH.numTx;
LayerMapOut = LayerMapper(modOut, [], prmLTEPDSCH);
% Precoding
PrecodeOut = SpatialMuxPrecoderOpenLoop(LayerMapOut, prmLTEPDSCH);
% Generate Cell-Specific Reference (CSR) signals
csr = CSRgenerator(nS, numTx);
csr_ref=complex(zeros(2*prmLTEPDSCH.Nrb, 4, numTx));
for m=1:numTx
   csr_pre=csr(1:2*prmLTEPDSCH.Nrb,:,:,m);
   csr_ref(:,:,m)=reshape(csr_pre,2*prmLTEPDSCH.Nrb,4);
end
% Resource grid filling
txGrid = REmapper_mTx(PrecodeOut, csr_ref, nS, prmLTEPDSCH);
% OFDM transmitter
txSig = OFDMTx(txGrid, prmLTEPDSCH);
%% Channel
% MIMO Fading channel
[rxFade, chPathG] = MIMOFadingChan(txSig, prmLTEPDSCH, prmMdl);
% Add AWG noise
sigPow = 10*log10(var(rxFade));
nVar = 10.^(0.1.*(sigPow-snrdB));
rxSig = AWGNChannel(rxFade, nVar);
%% RX
% OFDM Rx
```

```
rxGrid = OFDMRx(rxSig, prmLTEPDSCH);
% updated for numLayers -> numTx
[dataRx, csrRx, idx_data] = REdemapper_mTx(rxGrid, nS, prmLTEPDSCH);
% MIMO channel estimation
if prmMdl.chEstOn
    chEst = ChanEstimate_mTx(prmLTEPDSCH, csrRx,  csr_ref, prmMdl.chEstOn);
    hD    = ExtChResponse(chEst, idx_data, prmLTEPDSCH);
else
    idealChEst = IdChEst(prmLTEPDSCH, prmMdl, chPathG);
    hD =  ExtChResponse(idealChEst, idx_data, prmLTEPDSCH);
end
% Frequency-domain equalizer
if (numTx==1)
    % Based on Maximum-Combining Ratio (MCR)
    yRec = Equalizer_simo(dataRx, hD,mean(nVar), prmLTEPDSCH.Eqmode);
else
    % Based on Spatial Multiplexing
    yRec = MIMOReceiver_OpenLoop(dataRx, hD, prmLTEPDSCH, nVar);
end
% Demap received codeword(s)
[cwOut, ~] = LayerDemapper(yRec, prmLTEPDSCH);
if prmLTEPDSCH.Eqmode < 3
    % Demodulate
    demodOut = DemodulatorSoft(cwOut, prmLTEPDSCH.modType, mean(nVar));
else
    demodOut = cwOut;
end
% Descramble received codeword
rxCW = lteDescramble(demodOut, nS, 0, prmLTEPDSCH.maxG);
% Channel decoding includes - CB segmentation, turbo decoding, rate dematching
[decTbData1, ~,~] = lteTbChannelDecoding(nS, rxCW, Kplus1, C1,  prmLTEDLSCH,
prmLTEPDSCH);
% Transport block CRC detection
[dataOut, ~] = CRCdetector(decTbData1);
end
```

6.7.8.1   Structure of the Transceiver Model

The MATLAB script below is the testbench that calls the MIMO transceiver function *commlteMIMO*. First it calls the initialization function (*commlteMIMO_initialize*) to set all relevant parameter structures (*prmLTEDLSCH, prmLTEPDSCH, prmMdl*). Then it uses a while loop to perform subframe processing by calling the MIMO transceiver function *commlteMIMO_SM_Mode3_step*. Finally, it computes the BER and calls the visualization function to illustrate the channel response and modulation constellation before and after equalization.

Algorithm

MATLAB script

```
% Script for MIMO LTE (mode 3)
%
% Single or Two codeword transmission
%
clear all
clear functions
%% Set simulation parameters & initialize parameter structures
commlteMIMO_ params;
[prmLTEPDSCH, prmLTEDLSCH, prmMdl] = commlteMIMO_initialize(txMode, ...
chanBW, contReg, modType, Eqmode,numTx, numRx,cRate,maxIter, fullDecode,
chanMdl, corrLvl, ...
    chEstOn, numCodeWords, snrdB, maxNumErrs, maxNumBits);
clear txMode chanBW contReg modType Eqmode numTx numRx cRate maxIter
fullDecode chanMdl corrLvl chEstOn numCodeWords snrdB maxNumErrs
maxNumBits
%%
disp('Simulating the LTE Mode 3: Multiple Tx & Rx antrennas with Spatial
Multiplexing');
zReport_data_rate(prmLTEPDSCH, prmLTEDLSCH);
hPBer = comm.ErrorRate;
snrdB=prmMdl.snrdB;
maxNumErrs=prmMdl.maxNumErrs;
maxNumBits=prmMdl.maxNumBits;
%% Simulation loop
tic;
nS = 0; % Slot number, one of [0:2:18]
Measures = zeros(3,1); %initialize BER output
while (( Measures(2)< maxNumErrs) && (Measures(3) < maxNumBits))
   [dataIn, dataOut, txSig, rxSig, dataRx, yRec, csr] = ...
      commlteMIMO_SM_Mode3_step(nS, snrdB, prmLTEDLSCH, prmLTEPDSCH,
prmMdl);
   % Calculate bit errors
   Measures = step(hPBer, dataIn, dataOut);
    % Visualize constellations and spectrum
   if (visualsOn && prmLTEPDSCH.Eqmode~=3)
      zVisualize( prmLTEPDSCH, txSig, rxSig, yRec, dataRx, csr, nS);
   end;
   % Update subframe number
   nS = nS + 2; if nS > 19, nS = mod(nS, 20); end;
end
disp(Measures);
toc;
```

6.7.8.2 Verifying Transceiver Performance

By executing the MATLAB script of the MIMO transceiver model (*commlteMIMO*) we can examine various signals to assess the performance of the system. The parameters used in simulation are summarized in the following MATLAB script (*commlteMIMO_params*). This parameter set copies the common MIMO parameters used in Section 6.7.1. The parameters that are different reflect the use of spatial-multiplexing MIMO mode 3 with a single codeword and of the MMSE equalizer for MIMO receiver processing. In this simulation, 1 million bits of user data are processed, the SNR of the AWGN channel is set to 16 dB, and the visualization function is turned on.

Algorithm

MATLAB script

```
% PDSCH
txMode        = 3;   % Transmission mode one of {1, 2, 4}
numTx         = 2;   % Number of transmit antennas
numRx         = 2;   % Number of receive antennas
chanBW        = 4;       % [1,2,3,4,5,6] maps to [1.4, 3, 5, 10, 15, 20]MHz
contReg       = 1;       % {1,2,3} for >=10MHz, {2,3,4} for <10Mhz
modType       = 2;       % [1,2,3] maps to ['QPSK','16QAM','64QAM']
% DLSCH
cRate         = 1/3; % Rate matching target coding rate
maxIter       = 6;    % Maximum number of turbo decoding terations
fullDecode    = 0;    % Whether "full" or "early stopping" turbo decoding is performed
% Channel model
chanMdl       = 'frequency-selective-high-mobility';
% one of {'flat-low-mobility', 'flat-high-mobility','frequency-selective-low-mobility',
% 'frequency-selective-high-mobility', 'EPA 0Hz', 'EPA 5Hz', 'EVA 5Hz', 'EVA 70Hz'}
corrLvl       = 'Medium';
% Simulation parameters
Eqmode        = 2;    % Type of equalizer used [1,2,3] for ['ZF', 'MMSE','Sphere Decoder']
chEstOn       = 1;        % use channel estimation or ideal channel
snrdB         = 16;   % Signal to Noise Ratio in dB
maxNumErrs    = 1e6; % Maximum number of errors found before simulation stops
maxNumBits    = 1e6; % Maximum number of bits processed before simulation stops
visualsOn     = 1;        % Whether to visualize channel response and constellations
numCodeWords  = 1; % Number of codewords in PDSCH
enPMIfback    = 0;    % Enable/Disable Precoder Matrix Indicator (PMI) feedback
cbIdx         = 1;    % Initialize PMI index
```

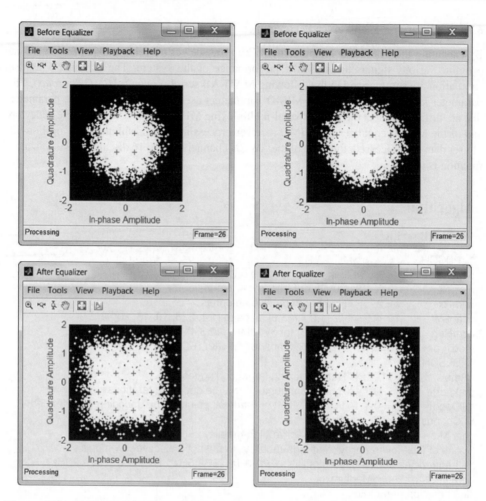

Figure 6.17 LTE model: MIMO spatial-multiplexing constellation diagram of the user data before and after equalization

Figure 6.17 shows the constellation diagrams before (first row) and after (second row) equalization of user data obtained from each of the two receive antennas in a subframe. It shows that the equalizer can compensate for the effect of a fading channel to result in a constellation that more closely resembles that of the 16QAM modulator.

Figure 6.18 illustrates the spectra of user data obtained from each of the two receive antennas in a subframe. It shows the transmitted signal and the received signal before and after equalization. The received signal before equalization (showing the effects of

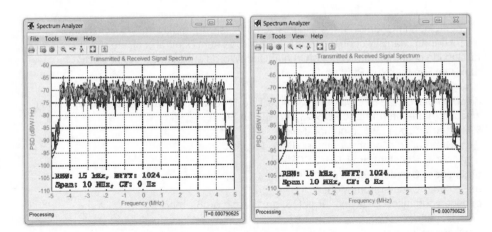

Figure 6.18 LTE MIMO spatial-multiplexing spectra of the transmited signal and of the received signal before and after equalization

frequency-selective fading) is effectively equalized by the open-loop spatial multiplexing used in transmission mode 3 (showing a more frequency-flat nature), which closely resembles the transmitted signal spectrum.

6.7.8.3 BER Measurements

In order to verify the BER performance of the transceiver, we create a testbench called *commlteMIMO_test_timing_ber*, which first initializes the LTE system parameters and then iterates through a range of SNR values and calls the *commlteMIMO_fcn* function in the loop in order to compute the corresponding BER values.

Algorithm

MATLAB script: commlteMIMO_test_timing_ber

```
% Script for MIMO LTE (mode 4)
%
% Single codeword transmission only
%
clear all
clear functions
%% Set simulation parameters & initialize parameter structures
commlteMIMO_params;
maxNumErrs=5e7;
maxNumBits=5e7;
[prmLTEPDSCH, prmLTEDLSCH, prmMdl] = commlteMIMO_initialize(txMode, ...
chanBW, contReg, modType, Eqmode,numTx, numRx,cRate,maxIter, fullDecode,
chanMdl, corrLvl, ...
```

```
    chEstOn, numCodeWords, enPMIfback, cbIdx, snrdB, maxNumErrs, maxNumBits);
clear txMode chanBW contReg modType Eqmode numTx numRx cRate maxIter
fullDecode chanMdl corrLvl chEstOn numCodeWords enPMIfback cbIdx snrdB
maxNumErrs maxNumBits
%%
disp('Simulating the LTE Mode 3: Multiple Tx & Rx antrennas with Spatial Multiplexing');
zReport_data_rate(prmLTEPDSCH, prmLTEDLSCH);
%% Geerate code and setup parallelism
disp('Generating code for commlteMIMO_ fcn.m ...');
arg1=coder.Constant(prmLTEPDSCH);
arg2=coder.Constant( prmLTEDLSCH);
arg3=coder.Constant(prmMdl);
codegen commlteMIMO_ fcn -args {16, arg1, arg2, arg3} -report
disp('Done.');
parallel_setup;
%%
MaxIter=8;
snr_vector=getSnrVector(prmLTEPDSCH.modType, MaxIter);
ber_vector=zeros(size(snr_vector));
maxNumBits=prmMdl.maxNumBits;
tic;
parfor n=1:MaxIter
    fprintf(1,'Iteration %2d out of %2d:  Processing %10d bits. SNR = %3d\n', ...
        n, MaxIter, maxNumBits, snr_vector(n));
    [ber, ~] = commlteMIMO_ fcn_mex(snr_vector(n), prmLTEPDSCH, prmLTEDLSCH,
prmMdl);
    ber_vector(n)=ber;
end;
toc;
semilogy(snr_vector, ber_vector);
title('BER - commlteMIMO SM');xlabel('SNR (dB)');ylabel('ber');grid;
```

Figure 6.19 shows the BER of the transceiver as a function of the SNR values after processing of 50 million bits of user data in each of eight iterations.

6.8   Chapter Summary

In this chapter we studied the multi-antenna MIMO techniques used in the LTE standard. MIMO techniques are integral components of LTE. The nine distinct modes used in downlink transmission, for example, are differentiated based on the choice of MIMO technique they feature. We have focused on MIMO algorithms of the first four transmission modes of the LTE standard and their modeling in MATLAB. These transmission modes exploit two algorithms: (i) transmit diversity (such as SFBC) and (ii) spatial multiplexing, with or without delay-diversity coding. Transmit-diversity techniques improve the link quality and reliability but do not increase the data rate or spectral efficiency of a system. Spatial-multiplexing techniques make possible a substantial boost in data rates.

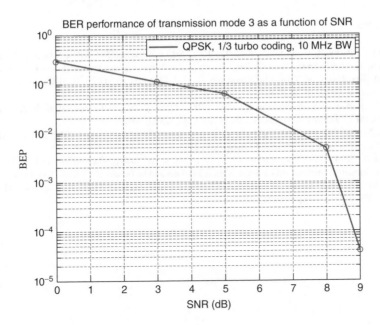

Figure 6.19   BER results: LTE mode 3 spatial-multiplexing single-codeword (2×2) MIMO channel

We first examined the MIMO multipath fading channel models and then presented the functional elements of MIMO transmission schemes that are common between transmit diversity and spatial multiplexing. This involved making updates to the OFDM functional elements presented in the previous chapter, due to the introduction of multiple antennas. Since LTE is a MIMO–OFDM system, we essentially transformed the 2D time–frequency representation of data in a single-antenna scheme into a 3D time–frequency–space representation. Updated common MIMO algorithms included resource-element mapping, channel estimation, and channel-response extraction.

Then we studied those functional elements that are different between transmit-diversity and spatial-multiplexing MIMO techniques. The LTE standard refers to these transmitter-side functional elements as layer-mapping and precoding operations. We examined the receiver-side operations, which invert the transmitter-side operations in order to recover best estimates of the 3D resource grid. We examined three MIMO receivers – ZF, MMSE, and SD algorithms – that provide estimates of the transmitted data on multiple antennas at every subcarrier at a given point in time.

Finally, we integrated all of the functional elements to create in MATLAB a transceiver model for the second, third, and fourth transmission modes of the LTE standard. The second transmission mode is based on transmit diversity, the third uses open-loop spatial multiplexing, and the fourth uses closed-loop spatial multiplexing. Through simulations, we performed both qualitative assessments and BER performance measurements. The results show that the transceiver effectively combats the effects of intersymbol interference caused by multipath fading, and depending on the mode it can achieve high data rates.

References

[1] Dahlman, E., Parkvall, S. and Sköld, J. (2011) *4G LTE/LTE-Advanced for Mobile Broadband*, Elsevier.

[2] Jafarkhani, H. (2005) *Space-Time Coding; Theory and Practice*, Cambridge University Press, Cambridge.

[3] 3GPP Evolved Universal Terrestrial Radio Access (E-UTRA); Base Station (BS) Radio Transmission and Reception, May 2011, TS 36.104.

[4] Scaglione, P., Stoica, S., Barbarossa, G. *et al.* (2002) Optimal designs for space-time linear precoders and decoders. *IEEE Transactions on Signal Processing*, 50, 5, 1051–1064.

[5] Browne, M. and Fitz, M. (2006) Singular value decomposition of correlated MIMO channels. IEEE Global Telecommunications Conference (GLOBECOM) 2006.

[6] Adhikari, S. (2011) Downlink transmission mode selection and switching algorithm for LTE. Proceedings of 3rd International Conference on Communication Systems and Networks (COMSNETS), January 2011.

[7] 3GPP (2011) Evolved Universal Terrestrial Radio Access (E-UTRA); Physical Channels and Modulation V10.0.0. TS 36.211, January 2011.

7

Link Adaptation

So far we have studied the modulation, coding, scrambling, channel-modeling, multicarrier, and multi-antenna transmission schemes used in the LTE (Long Term Evolution) standard. We have examined in detail multiple MIMO (Multiple Input Multiple Output) transmission modes, the best operating condition for each mode, and the peak data rates achievable for the system. We have not yet examined the mechanisms involved in the transition between various transmission modes or the criteria for these changes. In this chapter, we will overview the dynamic nature of the LTE standard and the way in which it chooses various parameters in order to optimize the spectral efficiency in time-varying channel conditions.

Spectral efficiency is an important measure used to evaluate the performance of mobile communications systems. The LTE standard, for example, has specific requirements in terms of average, cell-edge, and overall spectral efficiency relative to 3G (third-generation) standards [1]. Spectral efficiency is defined as the average data rate per bandwidth unit (Hz) per cell. This definition by itself reveals the tradeoffs involved in designing mobile systems. For a given bandwidth allocation, you can increase the spectral efficiency by augmenting the data rate through the use of higher-order modulation or higher-dimension MIMO techniques; in noisy channel conditions, however, such a selection may increase the probability of error and thus have a detrimental effect on the effective throughput.

In order to achieve the desired spectral efficiencies consistently, the 3G and 4G standards, including the LTE, employ techniques that dynamically change system parameters based on channel conditions. These techniques are generally known as channel-aware scheduling or link adaptations.

The basic idea of link adaptation is to adapt certain transmission parameters to varying channel conditions as they are monitored and measured by the system. Typical system parameters that are dynamically adapted include the system bandwidth, MIMO transmission modes, the number of transmission layers, the precoding matrix, Modulation and Coding Schemes (MCSs), and transmission power. With proper selection of these system parameters, we can exploit bandwidth resources more effectively instead of using a fixed parameter set that provides the best performance only in a worst-case channel condition.

Understanding LTE with MATLAB®: From Mathematical Modeling to Simulation and Prototyping, First Edition.
Houman Zarrinkoub.
© 2014 John Wiley & Sons, Ltd. Published 2014 by John Wiley & Sons, Ltd.

In this chapter, we will first review various measurement made in the mobile receiver in order to ascertain the channel conditions as a function of time. These include Channel Quality Indicator (CQI), Precoder Matrix Indicator (PMI), and Rank Indicator (RI) measurements. Then we will discuss adaptation techniques that respond to channel measurements and change various system parameters to maintain a given measure of quality. These include adaptations of MCS, adaptive precoding in closed-loop spatial multiplexing modes, and adaptive MIMO based on rank estimation. Finally, we will provide a short overview of PUCCH (Physical Uplink Control Channel) and PDCCH Physical Downlink Control Channel, which enable communication of the channel measurements and adaptive scheduling between the UE (User Equipment; the mobile terminal) and the eNodeB (enhanced Node Base station).

7.1 System Model

Link adaptation is all about adapting to the channel conditions and changing system parameters based on actual channel quality. The LTE standard enables link adaptations that can help us make use of the spectrum more efficiently. The cost associated with this adaptation is the additional computational complexity involved in implementing link-aware schedulers. Figure 7.1 illustrates the typical operations involved in link adaptation, which are subdivided into downlink and uplink operations.

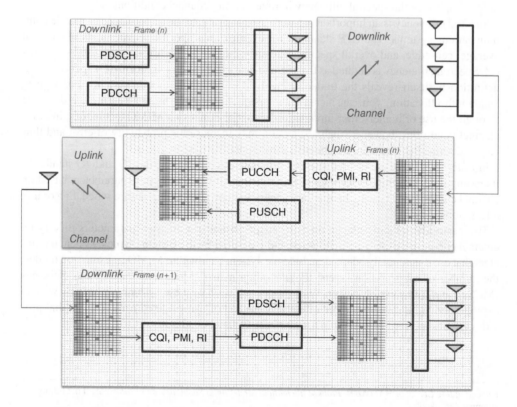

Figure 7.1   Sequence of downlink and uplink operations involved in link adaptations

The series of operations performed in a typical link adaptation scenario can be summarized as follows:

1. At subframe (*n*), the downlink transmitter forms the resource grid from the user data (PDSCH, Physical Downlink Shared Channel) and the Downlink Control Information, DCI (the PDCCH). The DCI contains the scheduling assignments that help the mobile receiver correctly decode the subframe information. The information contained in the PDCCH includes the MCSs, the precoder matrix, rank information, and the MIMO mode used.
2. The mobile receiver can then perform the critical step of channel condition measurement as part of the process of decoding the received resource grid. In this process, it estimates the received channel matrix and performs various channel quality measurements. These measurements include the CQI, the PMI, and the RI.
3. As part of uplink transmission, the mobile (UE) transmitter may embed the channel quality measures within the PUCCH and transmit to the base station (eNodeB) as a closed-loop feedback mechanism.
4. The base station (eNodeB) receiver can then decode the PUCCH information to obtain channel measurements. Having this information available enables the system scheduler to decide whether or not to adapt various system parameters in the next frame as a result of feedback received from downlink channel quality.
5. At the base station (eNodeB) in the downlink transmitter operations for the next subframe (*n* + 1), the scheduling decisions based on channel conditions are encoded into the PDCCH information and transmitted to the mobile. These include the new MCSs, precoder matrix, rank information, and MIMO mode that are now adapted based on the actual channel quality in the last subframe (*n*). This full feedback process is then repeated for each subframe.

7.2   Link Adaptation in LTE

To enable dynamic changes to MCSs and for proper operation of MIMO schemes, the LTE standard provides mechanisms that enable information regarding the channel characteristics to be measured by the mobile unit (UE). This information is then fed back to the base station (eNodeB) to help with scheduling and link adaptation.

At the mobile receiver, three types of channel-state report are generated and transmitted to the base station:

1. The CQI, a measure of downlink radio channel quality that specifies the best modulation constellation and coding rate to match the link quality.
2. The PMI, a measure that indicates the best set of precoding matrices for use in closed-loop single- and multi-user spatial multiplexing modes of the LTE standard.
3. The RI, which signals the number of useful transmission layers that can be used by the transmitter in spatial multiplexing modes.

Next we will discuss each of these reports in detail and provide an overview of various methodologies used in computing these measures.

7.2.1   Channel Quality Estimation

The CQI report gives a measure of the mobile radio channel quality. It provides a recommendation concerning the best MCS for the communication channel. The value of this measure is computed such that the transport block error rate using this recommendation will not exceed 10%. The higher the value of the CQI measure, the higher the modulation order and the higher the coding rate. There are two types of CQI report, based on their granularity: a wideband CQI report assigns a single MCS value for the whole system bandwidth, while a subband CQI report assigns multiple MCS values to different contiguous resource blocks.

There are many formulations for optimal MCS selection in the literature [2–6]. Most of these techniques select the best MCS as a function of the post-detection SINR (Signal-to-Interference and Noise Ratio) measure. This measure is selected such that the Packet Error Rate (PER) experienced in the transmission is less than a given target. This in turn allows the system to avoid frequent retransmissions. The best MCS recommendation can ultimately be selected by quantizing the SINR value using a codebook lookup table [2].

7.2.2   Precoder Matrix Estimation

The PMI report provides a preferred precoding codebook index for use in closed-loop spatial multiplexing of downlink transmission. Like CQI reports, a PMI report can be a single wideband value or multiple subband values. Multiple approaches to PMI selection are discussed in the literature. A summary of typical selection criteria is presented in Reference [7]. These criteria differ based on the metrics that are optimized. Approaches to optimal selection include minimization of singular values and minimization of the Mean Squared Error (MSE) or the capacity.

7.2.3   Rank Estimation

The rank estimation measure (RI) denotes the number of transmission layers or independent data streams for a spatial multiplexing system. Multiple approaches to estimating the rank of the channel matrix are available in the literature, with differing profiles of performance and complexity. Some of these approaches [8] perform the selection based on the post-detection SINR, the same measure used for the MCS selection. Others maximize the mutual information between the transmitted and post-detection signals and therefore directly maximize the capacity [11]. Another, less complex technique exploits the eigenvalues of the channel matrix [7].

7.3   MATLAB® Examples

In this section, we review various MATLAB algorithms used to generate channel-state reports in the receiver. As receiver operations are not specified in the standard, our guiding principles in choosing these algorithms are reasonable computational complexity and suitability for the single-user case. The algorithms featured here provide a starting point and showcase a general framework for the implementation of channel-state measurements and link adaptations of the LTE PHY (Physical Layer) model in MATLAB.

7.3.1 CQI Estimation

The two MATLAB functions in this section implement the channel quality estimation (CQI) measure based on the SINR of the MIMO receiver output and the transmitted signal. The CQI estimation is performed in two steps:

1. **SINR estimation**: The SINR measure is computed as a function of the decoded bits in the receiver and the MIMO receiver output
2. **Spectral efficiency lookup**: The computed SINR values are mapped to a spectral efficiency measure defined as the product of the number of modulated bits per symbol and the coding rate. For each SINR measure, distinct modulation schemes and coding rates are found through a table lookup.

7.3.1.1 SINR Estimation

Let us define G as the optimum equalizer that transforms the received signal $Y(n)$ into the equalized signal $\widehat{X}(n)$ as the best linear estimate of the transmitted signal $X(n)$. The error signal $e(n)$ is then expressed as:

$$e(n) = \widehat{X}(n) - X(n) = GY(n) - X(n) \tag{7.1}$$

For the CQI estimation, we compute a very simplified approximation of the SINR measure, defined as the ratio of the transmitted signal power σ_x^2 to the error signal power σ_e^2.

$$SINR = 10 \log_{10}\left(\frac{\sigma_x^2}{\sigma_e^2}\right) \tag{7.2}$$

The following function (*CQIselection.m*) computes the SINR measure, taking as inputs the decoded bits at the receiver (*bits*), the post-detection MIMO receiver output (*equalized*), the current subframe number (*nS*), and the PDSCH and DLSCH (Downlink Shared Channel Processing) parameter structures (*prmLTEDLSCH, prmLTEPDSCH*). The function output (*sinr*) is the SINR estimate. Note that to compute the SINR we need the best estimate of the transmitted signal $X(n)$, which is denoted by the variable *modOut* in the function. The function computes this signal by operating on the decoded bits at the receiver (input variable *bits*). The receiver output bits are best estimates of the transmitted input bits in every subframe. We have used this signal throughout this book to compute and monitor the bit-error rate of the system. The function applies the first few functions of the transmitter on this signal in order to compute the best estimate of the modulated signal. These operations are CRC (Cyclic Redundancy Check) attachment, channel coding, scrambling, and modulation. Finally, it computes the SINR measure, as defined by Equation 7.2.

Algorithm

MATLAB function

```
function sinr=CQIselection(bits, equalized, nS, prmLTEDLSCH, prmLTEPDSCH)
%#codegen
tbCrcOut1 =CRCgenerator(bits);
% Channel coding includes - CB segmentation, turbo coding, rate matching,
% bit selection, CB concatenation - per codeword
data = lteTbChannelCoding(tbCrcOut1, nS, prmLTEDLSCH, prmLTEPDSCH);
%Scramble codeword
scramOut = lteScramble(data, nS, 0, prmLTEPDSCH.maxG);
% Modulate
modOut = Modulator(scramOut, prmLTEPDSCH.modType);
error=modOut-equalized;
sinr=10*log10(var(modOut)./var(error));
```

7.3.1.2   Spectral Efficiency Lookup

The following function implements a transformation that maps the SINR estimate (input variable *sinr*) to the proposed modulation scheme and the coding rate (output variables *Ms* and *Cr*, respectively). Using a 4 bit (16-interval) scalar quantizer, we first map the SINR values to a CQI index. The dsp.ScalarQuantizerEncoder System object of the DSP System Toolbox is used here to perform the mapping operation. The threshold values correspond to the boundary points of the scalar quantizer. Since the quantizer is defined as unbounded (input values have a range from $-inf$ to $+inf$), only 15 threshold values are needed to subdivide the real axis into 16 regions represented by four CQI bits. The threshold values mapping the SINR values to the spectral efficiency are based on a simple lookup table [9].

Algorithm

MATLAB function

```
function [Ms, Cr]=CQI2indexMCS(sinr)
%#codegen
% Table of SINR threshold values, 15 boundary points for an unbounded quantizer
thresh=[-6.7,-4.7,-2.3,0.2,2.4,4.3,5.9,8.1,10.3,11.7,14.1,16.3,18.7,21,22.7];
% Table of coding rate (16 value)
Map2CodingRate=[0.076, 0.076, 0.117, 0.188, 0.301, 0.438, 0.588, 0.369, 0.479,...
0.602, 0.455, 0.554, 0.650, 0.754, 0.853, 0.926];
% Table of modulation type (1=QPSK, 2=QAM16, 3=QAM64)
Map2Modulator=[1*ones(7,1);2*ones(3,1);3*ones(6,1)];
persistent hQ
if isempty(hQ)
   hQ=dsp.ScalarQuantizerEncoder(...
     'Partitioning', 'Unbounded',...
```

```
      'BoundaryPoints', thresh,...
      'OutputIndexDataType','uint8');
end;
indexCQI=step(hQ, sinr);
index1=indexCQI+1;                    % 1-based indexing
% Map CQI index to modulation type
Ms = Map2Modulator (index1);
% Map CQI index to coding rate
Cr = Map2CodingRate (index1);
if Cr < 1/3, Cr=1/3;end;
```

In order to compute the modulation scheme (Ms) and the coding rate (Cr) outputs, we perform a table lookup operation with the CQI index. For the first seven values of the CQI index (indices 0–6), we map to a QPSK (Quadrature Phase Shift Keying) modulation with a modulation rate of 2 bits per symbol. The next three CQI indices (7, 8, and 9) are mapped to the 16QAM (Quadrature Amplitude Modulation) modulator with a modulation rate of 4 bits per symbol. Finally, the last six CQI indices (10–15) are mapped to 64QAM with a 6-bits-per-symbol modulation rate. Technically, the CQI index 0 signals an out-of-range message and does not participate in modulation mapping. For simplicity, we include this index in our MATLAB function with the QPSK set. Note also that the 16 mapping values for the coding-rate (Cr) mapping of spectral efficiency measures to modulation and coding rates are specified by the LTE standard document [10]. The combined information is provided in Table 7.1.

Table 7.1 Lookup table for mapping SINR estimate to modulation scheme and coding rate

CQI index	Modulation	Coding rate	Spectral efficiency (bps/Hz)	SINR estimate (dB)
1	QPSK	0.0762	0.1523	−6.7
2	QPSK	0.1172	0.2344	−4.7
3	QPSK	0.1885	0.3770	−2.3
4	QPSK	0.3008	0.6016	0.2
5	QPSK	0.4385	0.8770	2.4
6	QPSK	0.5879	1.1758	4.3
7	16QAM	0.3691	1.4766	5.9
8	16QAM	0.4785	1.9141	8.1
9	16QAM	0.6016	2.4063	10.3
10	64QAM	0.4551	2.7305	11.7
11	64QAM	0.5537	3.3223	14.1
12	64QAM	0.6504	3.9023	16.3
13	64QAM	0.7539	4.5234	18.7
14	64QAM	0.8525	5.1152	21.0
15	64QAM	0.9258	5.5547	22.7

7.3.2 PMI Estimation

The MATLAB function in this section implements a PMI codebook index selection. It employs the Minimum Mean Squared Error (MMSE) criterion to calculate as the output (*cbIdx*) the PMI codebook index per subframe. The input arguments of the function include the 3D channel matrix at the receiver (*h*), a Boolean signal indicating whether or not PMI closed-loop feedback is performed (*enPMIfback*), the number of transmit antennas (*numTx*), the number of layers (that is to say, the number of operational transmit antennas with sufficient rank) (*numLayers*), and the noise variance (*nVar*). If the PMI closed-loop feedback is turned off, the function has as output a constant value of 1 for the codebook index. Otherwise, it computes a single codebook index for each subframe by minimizing a distance measure.

Algorithm

MATLAB function

```
function cbIdx = PMICbSelect(h, enPMIfback, numTx, numLayers, nVar)
%#codegen
% Codebook selection using minimum MSE criterion
if (enPMIfback)
   if (numTx == 2)
      cbLen = 2; % Only indices 1 and 2 are used for 2-layer closed-loop Spatial MUX
      MSEcb = zeros(cbLen, 1);
      for cbIdx = 1:cbLen
         Wn = PrecoderMatrix(cbIdx, numTx, numLayers);
         MSEcb(cbIdx) = Sinr_MMSE(h, nVar, Wn);
      end
      [~, cbIdx] = min(MSEcb); % 0-based, note 0 and 3 are not used
   else % for numTx=4
      cbLen = 2^numLayers;
      MSEcb = zeros(cbLen, 1);
      for cbIdx = 1:cbLen
         Wn = PrecoderMatrix(cbIdx-1, numTx, numLayers);
         MSEcb(cbIdx) = Sinr_MMSE(h, nVar, Wn);
      end
      [~, cbIdx] = min(MSEcb);   % 1-based
      cbIdx = cbIdx-1;        % 0-based
   end
else
   cbIdx = 1;
end
end
% Helper function
function out = Sinr_MMSE(chEst, nVar, Wn)
%#codegen
% post-detection SNR computation
% Based on received channel estimates
```

```
% Per layer noise variance
% Precoder matrix
% Uses the MMSE detector.
% Get params
persistent Gmean
if isempty(Gmean), Gmean=dsp.Mean('RunningMean', true);end
noisFac = diag(nVar);
numData = size(chEst, 1);
numLayers = size(Wn,1);
F = inv(Wn);
%% MMSE receiver
for n = 1:numData
    h = chEst(n, :, :);                    % numTx x numRx
    h = reshape(h(:), numLayers, numLayers).';  % numRx x numTx
    Ht= inv((F'*(h'*h)*F) + noisFac);
    % Post-detection SINR
    g=real((1./(diag(Ht).*(nVar.')))-1);
    Gamma=step(Gmean,g);
end
out=mean(Gamma);
reset(Gmean);
end
```

The measure chosen here is the MSE between the post-detection MIMO receiver estimate and the transmitted modulator. This measure is formulated as a quadratic form involving the MIMO channel matrix and the precoder matrix. Each precoder matrix is computed by iterating through all PMI codebook entries; in other words, through a full search. This measure is computed for each codebook index and for each time sample (first dimension) of the 3D channel matrix. The codebook index that minimizes the MSE measure is the selected codebook index output. Note that for a four-antenna transmission, we have to search through 16 codebook indices, and for a two-antenna case, a subset of a codebook represented by a 2 bit index.

7.3.3   RI Estimation

The MATLAB function in this section implements an RI estimation algorithm based on the condition number of the channel matrix. The condition number is defined as the ratio of maximum and minimum eigenvalues of the channel matrix. It is a good indicator of the accuracy of matrix inversion and of the availability of a solution for a system of linear equation. Condition number values near 1 indicate a well-conditioned matrix, whereas very high values indicate an ill-conditioned matrix, for which the system of linear equations associated with the measure cannot be solved. A simple call to the *cond* function in MATLAB provides us access to a numerically reliable condition number for the channel matrix.

Algorithm

MATLAB function

```
function y = RIestimate(Q)
%#codegen
y=cond(Q);   % Condition number of a matrix
```

The rank estimation operation must be performed in the receiver on each 2D channel matrix. This 2D matrix is a sample of the 3D channel matrix computed at each sample time (the first dimension). Therefore, the best place to perform the operation is within the for loop of MIMO receiver operation, where we iterate through the first dimension of the 3D channel matrix to compute the post-detection MIMO receiver output sample by sample. As a result, we need to modify our MIMO receiver functions to include the rank estimation within the body of the function and to provide the estimates as an additional output.

7.3.3.1   RI Computation in MIMO Receiver Functions

The following MIMO receiver functions have been updated to include the rank estimation within their processing loop. In Chapter 6, we developed three different MIMO receivers based on either a Zero-Forcing (ZF), an MMSE, or a Sphere Decoder (SD) algorithm.

The following function computes the RI estimate sample by sample inside the ZF MIMO receiver. The rank estimation output (*ri*) is a column vector composed of all condition numbers computed for MIMO channels, with one value per sample of post-detection receiver output (*y*).

Algorithm

MATLAB function

```
function [y, ri] = MIMOReceiver_ZF(in, chEst, Wn)
%#codegen
% MIMO Receiver:
%    Based on received channel estimates, process the data elements
%    to equalize the MIMO channel. Uses the ZF detector.
% Get params
numData = size(in, 1);
y = complex(zeros(size(in)));
ri=zeros(numData,1);
iWn = inv(Wn);
%% ZF receiver
for n = 1:numData
    h = squeeze(chEst(n, :, :)); % numTx x numRx
    h = h.';                % numRx x numTx
    ri(n) = RIestimate(h);
```

```
   Q = inv(h);
   x = Q * in(n, :).';%#ok
   tmp = iWn * x; %#ok
   y(n, :) = tmp.';
end
```

Similarly, the following function computes the RI estimator output (*ri*) sample by sample inside the MMSE MIMO receiver. The rank estimation output (*ri*) is a column vector composed of all condition numbers computed for MIMO channels, with one value per sample of post-detection receiver output (*y*).

Algorithm

MATLAB function

```
function [y, ri] = MIMOReceiver_MMSE(in, chEst, nVar, Wn)
%#codegen
% MIMO Receiver:
%   Based on received channel estimates, process the data elements
%   to equalize the MIMO channel. Uses the MMSE detector.
% Get params
numLayers = size(Wn,1);
% noisFac = numLayers*diag(nVar);
noisFac = diag(nVar);
numData = size(in, 1);
y = complex(zeros(size(in)));
ri=zeros(numData,1);
iWn = inv(Wn);
%% MMSE receiver
for n = 1:numData
   h = chEst(n, :, :);                       % numTx x numRx
   h = reshape(h(:), numLayers, numLayers).';  % numRx x numTx
   ri(n) = RIestimate(h);
   Q = (h'*h + noisFac)\h';
   x = Q * in(n, :).';
   tmp = iWn * x; %#ok
   y(n, :) = tmp.';
end
```

Finally, the following function computes the RI estimator output (*ri*) sample by sample inside the sphere-decoder MIMO receiver. The rank estimation output (*ri*) is a column vector composed of all condition numbers computed for MIMO channels, with one value per sample of post-detection receiver output (*y*).

Algorithm

MATLAB function

```
function [y, ri, bittable] = MIMOReceiver_SphereDecoder(in, chEst, prmLTE, nVar, Wn)
%#codegen
% MIMO Receiver:
%   Based on received channel estimates, process the data elements
%   to equalize the MIMO channel. Uses the Sphere detector.
% Soft-Sphere Decoder
symMap=prmLTE.SymbolMap;
numBits=prmLTE.Qm;
constell=prmLTE.Constellation;
bittable = de2bi(symMap, numBits, 'left-msb');
iWn=Wn.';
nVar1=(-1/mean(nVar));
ri=zeros(numData,1);
persistent SphereDec
if isempty(SphereDec)
   % Soft-Sphere Decoder
   SphereDec = comm.SphereDecoder('Constellation', constell,...
      'BitTable', bittable, 'DecisionType', 'Soft');
end
% SSD receiver
temp = complex(zeros(size(chEst)));
% Account for precoding
for n = 1:size(chEst,1)
   h= squeeze(chEst(n, :, :));
   temp(n, :, :) = iWn * h;
   ri(n) = RIestimate(h);
end
hD = temp;
y = nVar1 * step(SphereDec, in, hD);
```

The following function uses a threshold approach to update the transmission mode as a function of an average rank estimation measure.

Algorithm

MATLAB function

```
function y=RIselection(ri, threshold)
Ri=mean(ri);
% RI estimation
if Ri > threshold, y = 4; else y=2; end
```

7.4 Link Adaptations between Subframes

In Sections 7.5–7.8, we look at various ways of using the Channel-State Information (CSI: CQI, PMI, and RI estimates) to adapt various transceiver parameters in successive subframes. In this section we highlight what can be regarded as some very simple scheduling scenarios. These algorithms are meant to provide a framework for the implementation of adaptation algorithms in MATLAB. In most realistic implementations, however, scheduling decisions are based on algorithms that take into account a variety of factors, including the CSI, quality of service, and type of data being transmitted.

In the following link adaptation exercises, we let the channel-estimate measures directly affect the scheduled system parameters of the following subframe. In all cases we apply a given adaptation to all resource blocks of the following subframe. This is known as a wideband adaptation. Alternatively, the LTE standard allows different adaptations to be set to different resource blocks in each subframe. This is known as a subband adaptation. In order to reduce the complexity of the algorithm, subband adaptations are not featured in this chapter.

Next, we show four types of adaptation applied to a single-codeword closed-loop spatial multiplexing system (single-codeword model for LTE transmission mode 4). We will first show an adaptive modulation and then an adaptive modulation and coding mechanism by using the CQI measure. Then we will combine adaptive modulation and coding with adaptive precoder selection based on the PMI measure. Finally, we will combine all adaptations by adding adaptive layer mapping using the RI measure.

7.4.1 Structure of the Transceiver Model

The following MATLAB script is the testbench calling the MIMO transceiver function. First it calls the initialization function (*commlteMIMO_initialize*) to set all the relevant parameter structures (*prmLTEDLSCH, prmLTEPDSCH, prmMdl*), then it uses a while loop to perform subframe processing by calling the MIMO transceiver function.

Algorithm

MATLAB script

```
% Script for MIMO LTE (mode 4)
%
% Single codeword transmission
%
clear functions
%% Set simulation parameters & initialize parameter structures
commlteMIMO_ params;
[prmLTEPDSCH, prmLTEDLSCH, prmMdl] = commlteMIMO_initialize(txMode, ...
    chanBW, contReg, modType, Eqmode,numTx, numRx,cRate,maxIter, fullDecode,
chanMdl, Doppler, corrLvl, ...
    chEstOn, numCodeWords, enPMIfback, cbIdx, snrdB, maxNumErrs, maxNumBits);
clear txMode chanBW contReg modType Eqmode numTx numRx cRate maxIter
fullDecode chanMdl Doppler corrLvl chEstOn numCodeWords enPMIfback cbIdx snrdB
maxNumErrs maxNumBits
```

```
%%
disp('Simulating the LTE Mode 4: Multiple Tx & Rx antennas with Spatial Multiplexing');
zReport_data_rate(prmLTEPDSCH, prmLTEDLSCH);
hPBer = comm.ErrorRate;
snrdB=prmMdl.snrdB;
maxNumErrs=prmMdl.maxNumErrs;
maxNumBits=prmMdl.maxNumBits;
%% Simulation loop
nS = 0; % Slot number, one of [0:2:18]
Measures = zeros(3,1); %initialize BER output
while (Measures(3) < maxNumBits)
    % Insert one subframe step processing
    %% including adaptations here
end
BER=Measures(1);
BITS=Measures(3);
```

Note that we have left the body of the while loop as a placeholder. In this script, in place of commented lines that read "Insert one subframe step processing including adaptations here," we will place three different code segments that implement the three different adaptation scenarios.

7.4.2   Updating Transceiver Parameter Structures

The following function implements the adaptation mechanism by updating the three parameter structures (*prmLTEDLSCH*, *prmLTEPDSCH*, *prmMdl*) used in transceiver models. It takes as input the initial parameter structures (given as input arguments *p1*, *p2*, and *p3*) and a variable number of additional input arguments. As output it generates the updated parameter structures.

If only one more input argument is given besides the parameter structures (*p1*, *p2*, and *p3*), we update the modulation scheme by setting the (*modType*) parameter. If two additional input arguments are given, the first updates the modulation scheme and the second updates the coding rate (*cRate*). So far we have only used the CQI measures for adaptive modulation and coding. With three additional input arguments, besides updating *modType* and *cRate* we also adapt the PMI codebook index (*cbIdx*). Finally, by providing four additional input arguments we adapt all parameters, including the final one (*txMode*), which determines whether we are using a spatial-multiplexing mode (*txMode* = 4) or a transmit-diversity mode (*txMode* = 2).

Algorithm

MATLAB function

```
function [p1, p2, p3] = commlteMIMO_update(p1,p2, p3, varargin)
switch nargin
    case 1, modType=varargin{1}; cRate=p2.cRate; cbIdx=p3.cbIdx; txMode=p1.txMode;
```

```
   case 2, modType=varargin{1}; cRate=varargin{2}; cbIdx=p3.cbIdx; txMode=p1.txMode;
   case 3, modType=varargin{1}; cRate=varargin{2}; cbIdx=varargin{3};
txMode=p1.txMode;
   case 4, modType=varargin{1}; cRate=varargin{2}; cbIdx=varargin{3};
txMode=varargin{4};
   otherwise
      error('commlteMIMO_update has 1 to 4 arguments!');
end
% Update PDSCH parameters
tmp = prmsPDSCH(txMode, p1.chanBW, p1.contReg, modType,p1.numTx, p1.numRx, ...
   p1.numCodeWords,p1.Eqmode);
p1=tmp;
[SymbolMap, Constellation]=ModulatorDetail(p1.modType);
p1.SymbolMap=SymbolMap;
p1.Constellation=Constellation;
% Update DLSCH parameters
p2 = prmsDLSCH(cRate, p2.maxIter, p2.fullDecode, p1);
% Update channel model parameters
tmp = prmsMdl(txMode, p1.chanSRate, p3.chanMdl, p3.Doppler, p1.numTx, p1.numRx, ...
   p3.corrLvl, p3.chEstOn, p3.enPMIfback, cbIdx, p3.snrdB, p3.maxNumErrs,
p3.maxNumBits);
p3=tmp;
```

7.5   Adaptive Modulation

In this section we take advantage of the CQI channel-state report to adaptively change
the modulation scheme of the transceiver in successive subframes. We implement a wideband
modulation selection, in which in any given subframe all resource blocks will have the same
modulation scheme and the change occurs between subframes.

To understand the design tradeoffs, we need to compare adaptive modulation with alternative implementations. We feature three algorithms that apply different adaptation scenarios:
(i) baseline (with no adaptation), (ii) adaptation through random changing of the modulation
type, and (iii) adaptation by exploitation of the CQI channel measurements. Next we show the
operations performed in each scenario in the body of the processing while loop.

7.5.1   No Adaptation

The following MATLAB script segment shows the body of the processing while loop. Without
any link adaptation, the while loop includes only five operations: (i) calling the transceiver step
function to process one subframe of data; (ii) reporting the average and instantaneous data rates
together with average values of the coding rate and the modulation rate (number of modulation
bits per symbol); (iii) measuring the BER (Bit Error Rate); (iv) visualizing the post-detection
received signals and the transmitted and received OFDM (Orthogonal Frequency Division
Multiplexing) signals; and (v) updating the subframe number.

Algorithm

MATLAB script segment

```
%% One subframe step processing
[dataIn, dataOut, txSig, rxSig, dataRx, yRec, csr] = ...
    commlteMIMO_SM_step(nS, snrdB, prmLTEDLSCH, prmLTEPDSCH, prmMdl);
 %% Report average data rates
ADR=zReport_data_rate_average(prmLTEPDSCH, prmLTEDLSCH);
%% Calculate  bit errors
Measures = step(hPBer, dataIn, dataOut);
%% Visualize results
if (visualsOn && prmLTEPDSCH.Eqmode~=3)
   zVisualize( prmLTEPDSCH, txSig, rxSig, yRec, dataRx, csr, nS);
end;
% Update subframe number
nS = nS + 2; if nS > 19, nS = mod(nS, 20); end;
%% No adaptations here
```

7.5.2   Changing the Modulation Scheme at Random

The following MATLAB script segment shows the body of the processing while loop. In addition to operations performed in the case of no adaptations, the while loop includes the following two operations: (i) it assigns to the modulation-type parameter (*modType*) a random integer value of 1, 2, or 3, corresponding to QPSK, 16QAM, and 64QAM, respectively; and (ii) it calls the *commlteMIMO_update* function, which updates and recomputes all *LTEPDSCH* and *LTEDLSCH* parameters based on the new modulation type.

Algorithm

MATLAB script segment

```
%% One subframe step processing
[dataIn, dataOut, txSig, rxSig, dataRx, yRec, csr] = ...
    commlteMIMO_SM_step(nS, snrdB, prmLTEDLSCH, prmLTEPDSCH, prmMdl);
 %% Report average data rates
ADR=zReport_data_rate_average(prmLTEPDSCH, prmLTEDLSCH);
%% Calculate  bit errors
Measures = step(hPBer, dataIn, dataOut);
%% Visualize results
if (visualsOn && prmLTEPDSCH.Eqmode~=3)
   zVisualize( prmLTEPDSCH, txSig, rxSig, yRec, dataRx, csr, nS);
end;
% Change of modulation scheme randomly
modType=randi([1 3],1,1);
```

```
[prmLTEPDSCH, prmLTEDLSCH] = commlteMIMO_update( prmLTEPDSCH, prmLT-
EDLSCH, modType);
% Update subframe number
nS = nS + 2; if nS > 19, nS = mod(nS, 20); end;
%% No adaptations here
```

7.5.3   CQI-Based Adaptation

The following MATLAB script segment shows the body of the processing while loop. In addition to operations performed in the case of no adaptations, the while loop contains the following four adaptation operations: (i) CQI measure reporting, which computes the *SINR* measure by calling the *CQIselection* function; (ii) new modulation-type selection, carried out by calling the *CQI2indexMCS* function that maps the *SINR* measure into a modulation type; (ii) calling the *commlteMIMO_update* function that updates and recomputes *LTEPDSCH*, *LTEDLSCH*, and *prmMdl* parameters based on the new modulation type; and (iv) visualizing the variations in the channel quality with time by calling the *zVisSinr* function that plots the present and the 24 past values of the SINR measure.

Algorithm

MATLAB script segment

```
%% One subframe step processing
[dataIn, dataOut, txSig, rxSig, dataRx, yRec, csr] = ...
    commlteMIMO_SM_step(nS, snrdB, prmLTEDLSCH, prmLTEPDSCH, prmMdl);
%% Report average data rates
ADR=zReport_data_rate_average(prmLTEPDSCH, prmLTEDLSCH);
%% CQI feedback
sinr=CQIselection(dataOut, yRec,  nS, prmLTEDLSCH, prmLTEPDSCH);
indexMCS=CQI2indexMCS(sinr);
%% Calculate  bit errors
Measures = step(hPBer, dataIn, dataOut);
%% Visualize results
if (visualsOn && prmLTEPDSCH.Eqmode˜=3)
    zVisualize( prmLTEPDSCH, txSig, rxSig, yRec, dataRx, csr, nS);
    zVisSinr(sinr);
end;
% Update subframe number
nS = nS + 2; if nS > 19, nS = mod(nS, 20); end;
% Adaptive change of modulation
modType=indexMCS;
[prmLTEPDSCH, prmLTEDLSCH, prmMdl] = commlteMIMO_update
( prmLTEPDSCH, prmLTEDLSCH, prmMdl, modType);
```

7.5.4 Verifying Transceiver Performance

The parameters used in simulation are summarized in the following MATLAB script, called *commlteMIMO_params*. During the simulation, all of these parameters remain constant except for the modulation scheme specified by the variable *modType*.

Algorithm

MATLAB script

```
% PDSCH
txMode        = 4;   % Transmission mode one of {1, 2, 4}
numTx         = 2;   % Number of transmit antennas
numRx         = 2;   % Number of receive antennas
chanBW        = 6;        % [1,2,3,4,5,6] maps to [1.4, 3, 5, 10, 15, 20]MHz
contReg       = 1;        % {1,2,3} for >=10MHz, {2,3,4} for <10Mhz
modType       = 3;        % [1,2,3] maps to ['QPSK','16QAM','64QAM']
% DLSCH
cRate         = 1/3; % Rate matching target coding rate
maxIter       = 6;   % Maximum number of turbo decoding iterations
fullDecode    = 0;   % Whether "full" or "early stopping" turbo decoding is performed
% Channel
chanMdl       =  'frequency-selective'; % Channel model
Doppler       = 70;                     % Average Doppler shift
% one of {'flat-low-mobility', 'flat-high-mobility','frequency-selective-low-mobility',
% 'frequency-selective-high-mobility', 'EPA 0Hz', 'EPA 5Hz', 'EVA 5Hz', 'EVA 70Hz'}
corrLvl       = 'Medium';
% Simulation parameters
Eqmode        = 2;   % Type of equalizer used [1,2,3] for ['ZF', 'MMSE','Sphere Decoder']
chEstOn       = 1;        % use channel estimation or ideal channel
snrdB         = 20;  % Signal to Noise Ratio in dB
maxNumErrs    = 2e7; % Maximum number of errors found before simulation stops
maxNumBits    = 2e7; % Maximum number of bits processed before simulation stops
visualsOn     = 0;       % Whether to visualize channel response and constellations
numCodeWords  = 1; % Number of codewords in PDSCH
enPMIfback    = 0;   % Enable/Disable Precoder Matrix Indicator (PMI) feedback
cbIdx         = 1;   % Initialize PMI index
```

The function *zReport_data_rate_average* reports the average and instantaneous values of data rates, the coding rate, and the modulation rate. The average values are computed as running means by the *dsp.Mean* System object of the DSP System Toolbox. The instantaneous data rate is computed as the sum of input bits in all 10 subframes of a frame multiplied by a constant factor of 100 frames per second.

Algorithm

MATLAB function

```
function t = zReport_data_rate_average(p2, p1)
persistent Rmean Rmod Rcod
if isempty(Rmean), Rmean=dsp.Mean('RunningMean', true);end
if isempty(Rmod), Rmod=dsp.Mean('RunningMean', true);end
if isempty(Rcod), Rcod=dsp.Mean('RunningMean', true);end
y=(1/10.0e-3)*(p1.TBLenVec(1)+p1.TBLenVec(2)+8*p1.TBLenVec(3));
z=y/1e6;
t=step(Rmean,z);
mm=step(Rmod,2*p2.modType);
cc=step(Rcod,p1.cRate);
Mod={'QPSK','16QAM','64QAM'};
fprintf(1,'Modulation              = %s\n',Mod{p2.modType});
fprintf(1,'Instantaneous Data rate     = %.2f Mbps\n',z);
fprintf(1,'Average Data rate        = %.2f Mbps\n',t);
fprintf(1,'Instantaneous Modulation rate = %4.2f\n',2*p2.modType);
fprintf(1,'Average Modulation rate     = %4.2f\n',mm);
fprintf(1,'Instantaneous Coding rate    = %.4f\n',p1.cRate);
fprintf(1,'Average Coding rate        = %.4f \n\n',cc);
end
```

By executing the MATLAB script of the MIMO transceiver model with adaptive modulation, we can look at various signals in order to assess the performance of the system. As Figure 7.2 illustrates, changes in the modulation scheme affect the instantaneous data rate and thus the average data rate.

7.5.5 Adaptation Results

For each of the adaptation scenarios, we compute the BERs by processing 20 million bits. Table 7.2 summarizes the results. As expected, adaptive modulation that responds to channel quality performs best. In scenarios without adaptations, as we use higher modulation rates, such as 64QAM, we obtain high data rates at the cost of higher BERs. When using lower modulation rates, such as QPSK, we obtain lower BERs but lower data rates. When selecting the modulation scheme randomly, without any correlation to the channel quality, both the average data rate and the BER are average values of the cases without any adaptations.

However, as we select a modulation scheme based on channel quality, we obtain the best compromise in conditions of both low and high channel qualities. In subframes with higher channel quality, we choose higher modulation rates. Although we are using modulation schemes with smaller minimum constellation distances, as the channel is deemed clean, the probability of error in these subframes is low, so we enjoy the highest rate without too much

```
Modulation                        = 64QAM
Instantanous Data rate            = 30.58 Mbps
Average Data rate                 = 30.58 Mbps
Instantanous Modulation rate      = 6.00
Average Modulation rate           = 6.00
Instantanous Coding rate          = 0.3333
Average Coding rate               = 0.3333


Modulation                        = QPSK
Instantanous Data rate            = 10.30 Mbps
Average Data rate                 = 20.44 Mbps
Instantanous Modulation rate      = 2.00
Average Modulation rate           = 4.00
Instantanous Coding rate          = 0.3333
Average Coding rate               = 0.3333


Modulation                        = 16QAM
Instantanous Data rate            = 19.85 Mbps
Average Data rate                 = 20.24 Mbps
Instantanous Modulation rate      = 4.00
Average Modulation rate           = 4.00
Instantanous Coding rate          = 0.3333
Average Coding rate               = 0.3333
```

Figure 7.2   Link adaptation: data rates, modulation, and coding modes subframe by subframe

Table 7.2   Adaptive modulation: BER, data rates, and modulation rates in different scenarios

Type of modulation	Average data rate (Mbps)	Modulation rate	Coding rate	Bit error rate
QPSK – no adaptation	20.61	2	0.3333	$1.2e^{-06}$
16QAM – no adaptation	39.23	4	0.3333	$1.4e^{-06}$
64QAM – no adaptation	61.66	6	0.3333	0.0033
Random selection	40.75	2 or 4 or 6	0.3333	0.0014
Adaptive modulation	52.61	2 or 4 or 6	0.3333	0.0009

cost in bit errors. In subframes with lower channel quality, we revert to lower modulation rates. These rates are associated with higher distances between constellation points and as a result the probability of error is low. These subframes result in a reduction in the overall rates but maintain the quality within an acceptable range. As a result, the average BER with adaptive modulation (0.0009) is lower than the random selection (0.0016) and the average data rate with adaptive modulation (52.61 Mbps) is higher than that with random selection (40.75 Mbps). We observe that with adaptation based on channel quality we obtain the best tradeoff in terms of highest rate and reasonable error rate.

7.6 Adaptive Modulation and Coding Rate

In this section, we use the CQI channel-state report to adaptively change both the modulation scheme and the coding rate of the transceiver in successive subframes. We will compare the channel quality-based (CQI-based) adaptive approach with two algorithms that apply different adaptation scenarios: (i) baseline (without adaptation) and (ii) adaptation through random changing of the modulation type and coding rate.

In the scenario where no adaptation is performed, we examine each of the three LTE modulation schemes with a coding rate equal to the average coding rate used in the CQI-based adaptive scenario. In the random-adaptation scenario, in each subframe we randomly select one of the LTE modulation schemes and choose a random value for the coding rate within the same range of values used in the CQI-based adaptive scenario.

7.6.1 No Adaptation

The following MATLAB script segment shows the body of the processing while loop. Since no adaptation is performed here, the MATLAB code is identical to that presented in Section 7.5.

Algorithm

MATLAB script segment

```
%% One subframe step processing
[dataIn, dataOut, txSig, rxSig, dataRx, yRec, csr] = ...
    commlteMIMO_SM_step(nS, snrdB, prmLTEDLSCH, prmLTEPDSCH, prmMdl);
%% Report average data rates
ADR=zReport_data_rate_average(prmLTEPDSCH, prmLTEDLSCH);
%% Calculate  bit errors
Measures = step(hPBer, dataIn, dataOut);
%% Visualize results
if (visualsOn && prmLTEPDSCH.Eqmode~=3)
    zVisualize( prmLTEPDSCH, txSig, rxSig, yRec, dataRx, csr, nS);
end;
% Update subframe number
nS = nS + 2; if nS > 19, nS = mod(nS, 20); end;
%% No adaptations here
```

7.6.2 Changing Modulation Scheme at Random

The following MATLAB code segment shows the body of the processing while loop in the case where the modulation and coding rate are randomly selected. In addition to operations performed in the case of no adaptations, the while loop includes the following three operations: (i) it assigns to the modulation-type parameter (*modType*) a random integer value of 1, 2, or 3, corresponding to QPSK, 16QAM, and 64QAM, respectively; (ii) it assigns to the coding-rate parameter (*cRate*) a normal random value in the range between 1/3 and 0.95; and (iii) it calls the *commlteMIMO_update* function that updates and recomputes all parameters based on the new modulation type and the coding rate.

Algorithm

MATLAB script segment

```
%% One subframe step processing
[dataIn, dataOut, txSig, rxSig, dataRx, yRec, csr] = ...
    commlteMIMO_SM_step(nS, snrdB, prmLTEDLSCH, prmLTEPDSCH, prmMdl);
 %% Report average data rates
ADR=zReport_data_rate_average(prmLTEPDSCH, prmLTEDLSCH);
%% Calculate  bit errors
Measures = step(hPBer, dataIn, dataOut);
%% Visualize results
if (visualsOn && prmLTEPDSCH.Eqmode˜=3)
   zVisualize( prmLTEPDSCH, txSig, rxSig, yRec, dataRx, csr, nS);
end;
% Change of modulation and coding rates randomly
Average_cRate=0.4932;
modType=randi([1 3],1,1);
cRate = Average_cRate + (1/6)*randn;
if cRate > 0.95, cRate=0.95;end; if cRate < 1/3, cRate=1/3;end;
[prmLTEPDSCH, prmLTEDLSCH, prmMdl] = commlteMIMO_update( ...
    prmLTEPDSCH, prmLTEDLSCH, prmMdl, modType, cRate);
% Update subframe number
nS = nS + 2; if nS > 19, nS = mod(nS, 20); end;
```

7.6.3   CQI-Based Adaptation

The following MATLAB script segment shows the body of the processing while loop that implements the CQI-based adaptive modulation and coding. In addition to operations characterizing the no-adaptation scenario, the following three tasks are performed: (i) by calling the functions *CQIselection* and *CQI2indexMCS* in sequence, the CQI report is computed, estimating the best modulation scheme (*modType*) and coding rate (*cRate*) for the following subframe; (ii) the *commlteMIMO_update* function that updates and recomputes the *LTEPDSCH*, *LTEDLSCH*, and *prmMdl* parameters based on the new modulation type and coding rate is called; and (iii) the variations in channel quality with time are visualized by calling the *zVisSinr* function, which plots a vector comprising the present and the 24 past values of the SINR measure.

Algorithm

MATLAB script segment

```
%% One subframe step processing
[dataIn, dataOut, txSig, rxSig, dataRx, yRec, csr] = ...
    commlteMIMO_SM_step(nS, snrdB, prmLTEDLSCH, prmLTEPDSCH, prmMdl);
%% Report average data rates
ADR=zReport_data_rate_average(prmLTEPDSCH, prmLTEDLSCH);
%% CQI feedback
sinr=CQIselection(dataOut, yRec, nS, prmLTEDLSCH, prmLTEPDSCH);
[modType, cRate]=CQI2indexMCS(sinr);
%% Calculate  bit errors
Measures = step(hPBer, dataIn, dataOut);
%% Visualize results
if (visualsOn && prmLTEPDSCH.Eqmode˜=3)
    zVisualize( prmLTEPDSCH, txSig, rxSig, yRec, dataRx, csr, nS);
    zVisSinr(sinr);
end;
% Adaptive change of modulation and coding rate
[prmLTEPDSCH, prmLTEDLSCH, prmMdl] = commlteMIMO_update(...
    prmLTEPDSCH, prmLTEDLSCH, prmMdl, modType, cRate);
% Update subframe number
nS = nS + 2; if nS > 19, nS = mod(nS, 20); end;
```

7.6.4   Verifying Transceiver Performance

The parameters used in the simulation are the same as used in the adaptive-modulation case and are summarized in the script *commlteMIMO_params*. During the simulation, all of these parameters remain constant except the modulation scheme (*modType*) and the coding rate (*cRate*). By executing the MATLAB script of the MIMO transceiver model with adaptive modulation and coding rate, we can assess the performance of the system. As the simulation is running, the message shown in Figure 7.3 appears on the MATLAB screen. As we can see, changes in both the modulation scheme and the coding rate affect the instantaneous data rate and thus the average data rate of the transceiver.

7.6.5   Adaptation Results

For each of the three adaptation scenarios, we compute the BER by processing 20 million bits. For the first four experiments (no adaptations in the case of QPSK, 16QAM, and 64QAM modulation and a random selection of modulation in each subframe), we choose a constant coding rate of 0.4932. This coding rate is the average of rates in the adaptive case and is chosen in order to make fair comparisons. Table 7.3 summarizes the results.

```
Modulation                        = 64QAM
Instantanous Data rate            = 61.66 Mbps
Average Data rate                 = 61.66 Mbps
Instantanous Modulation rate      = 6.00
Average Modulation rate           = 6.00
Instantanous Coding rate          = 0.3333
Average Coding rate               = 0.3333


Modulation                        = 16QAM
Instantanous Data rate            = 39.23 Mbps
Average Data rate                 = 50.45 Mbps
Instantanous Modulation rate      = 4.00
Average Modulation rate           = 5.00
Instantanous Coding rate          = 0.3685
Average Coding rate               = 0.3509


Modulation                        = QPSK
Instantanous Data rate            = 31.70 Mbps
Average Data rate                 = 44.20 Mbps
Instantanous Modulation rate      = 2.00
Average Modulation rate           = 4.00
Instantanous Coding rate          = 0.6225
Average Coding rate               = 0.4415
```

Figure 7.3 Adaptive modulation and coding: data rates, modulation, and coding modes subframe by subframe

Table 7.3 Adaptive modulation and coding: BER, data rates, modulation, and coding rates in different scenarios

Type of modulation	Average data rate (Mbps)	Modulation	Coding rate	Bit error rate
QPSK – no adaptation	28.34	2	0.4932	$2.8e^{-06}$
16QAM – no adaptation	57.34	4	0.4932	$7.9e^{-04}$
64QAM – no adaptation	87.01	6	0.4932	$3.6e^{-02}$
Random selection	56.81	2 or 4 or 6	0.5037	$2.5e^{-02}$
Adaptive modulation and coding	64.73	2 or 4 or 6	0.333–0.94	$4.7e^{-03}$

The results in this case are very similar to those in the case where only adaptive modulation is used. With fixed modulation and coding rates, we obtain higher rates and higher-order modulations at the cost of a much lower achievable BER. Changing the modulation based on random selection provides the average results of the three fixed modulation cases. Adaptive modulation and coding based on channel quality provides the best compromise. The average data rate in the CQI-based adaptive approach (64.73 Mbps) is higher than in the case of random selection (56.81 Mbps). The BER of adaptive coding (0.0047) is lower than that in the case of random selection (0.0250).

7.7   Adaptive Precoding

In this section we use the PMI channel-state report to adaptively change the precoding matrix index in successive subframes. This adaptation is only available in the closed-loop spatial multiplexing mode of transmission. We use a parameter called *enPMIfback*, which enables or disables the PMI mechanism. When this parameter is turned on, the PMI index is selected within the receiver and fed back to the transmitter for use at the next time step. Otherwise, the fixed user-specified codebook index is used for the duration of the simulation. The feedback granularity is modeled once for the whole subframe (wideband) and applied to the next transmission subframe.

Algorithm

MATLAB function

```
function [dataIn, dataOut, txSig, rxSig, dataRx, yRec, csr_ref, cbIdx]...
   = commlteMIMO_SM_PMI_step(nS, snrdB, prmLTEDLSCH, prmLTEPDSCH, prmMdl)
%% TX
persistent hPBer1
if isempty(hPBer1), hPBer1=comm.ErrorRate; end;
%  Generate payload
dataIn = genPayload(nS,  prmLTEDLSCH.TBLenVec);
% Transport block CRC generation
tbCrcOut1 =CRCgenerator(dataIn);
% Channel coding includes - CB segmentation, turbo coding, rate matching,
% bit selection, CB concatenation - per codeword
[data, Kplus1, C1] = lteTbChannelCoding(tbCrcOut1, nS, prmLTEDLSCH, prmLTEPDSCH);
%Scramble codeword
scramOut = lteScramble(data, nS, 0, prmLTEPDSCH.maxG);
% Modulate
modOut = Modulator(scramOut, prmLTEPDSCH.modType);
% Map modulated symbols  to layers
numTx=prmLTEPDSCH.numTx;
LayerMapOut = LayerMapper(modOut, [], prmLTEPDSCH);
usedCbIdx = prmMdl.cbIdx;
% Precoding
[PrecodeOut, Wn] = SpatialMuxPrecoder(LayerMapOut, prmLTEPDSCH, usedCbIdx);
% Generate Cell-Specific Reference (CSR) signals
csr = CSRgenerator(nS, numTx);
csr_ref=complex(zeros(2*prmLTEPDSCH.Nrb, 4, numTx));
for m=1:numTx
   csr_pre=csr(1:2*prmLTEPDSCH.Nrb,:,:,m);
   csr_ref(:,:,m)=reshape(csr_pre,2*prmLTEPDSCH.Nrb,4);
```

```
end
% Resource grid filling
txGrid = REmapper_mTx(PrecodeOut, csr_ref, nS, prmLTEPDSCH);
% OFDM transmitter
txSig = OFDMTx(txGrid, prmLTEPDSCH);
%% Channel
% MIMO Fading channel
[rxFade, chPathG] = MIMOFadingChan(txSig, prmLTEPDSCH, prmMdl);
% Add AWG noise
sigPow = 10*log10(var(rxFade));
nVar = 10.^(0.1.*(sigPow-snrdB));
rxSig = AWGNChannel(rxFade, nVar);
%% RX
% OFDM Rx
rxGrid = OFDMRx(rxSig, prmLTEPDSCH);
% updated for numLayers -> numTx
[dataRx, csrRx, idx_data] = REdemapper_mTx(rxGrid, nS, prmLTEPDSCH);
% MIMO channel estimation
if prmMdl.chEstOn
    chEst = ChanEstimate_mTx(prmLTEPDSCH, csrRx, csr_ref, prmMdl.chEstOn);
    hD    = ExtChResponse(chEst, idx_data, prmLTEPDSCH);
else
    idealChEst = IdChEst(prmLTEPDSCH, prmMdl, chPathG);
    hD = ExtChResponse(idealChEst, idx_data, prmLTEPDSCH);
end
% PMI codebook selection
if (prmMdl.enPMIfback)
cbIdx = PMICbSelect( hD, prmMdl.enPMIfback, prmLTEPDSCH.numTx, ...
    prmLTEPDSCH.numLayers, nVar );
else
    cbIdx=prmMdl.cbIdx;
end
% Frequency-domain equalizer
if (numTx==1)
    % Based on Maximum-Combining Ratio (MCR)
    yRec = Equalizer_simo(dataRx, hD, nVar, prmLTEPDSCH.Eqmode);
else
    % Based on Spatial Multiplexing
    yRec = MIMOReceiver(dataRx, hD, prmLTEPDSCH, nVar, Wn);
end
% Demap received codeword(s)
[cwOut, ~] = LayerDemapper(yRec, prmLTEPDSCH);
if prmLTEPDSCH.Eqmode < 3
    % Demodulate
    demodOut = DemodulatorSoft(cwOut, prmLTEPDSCH.modType, max(nVar));
else
```

```
    demodOut = cwOut;
end
% Descramble received codeword
rxCW = lteDescramble(demodOut, nS, 0, prmLTEPDSCH.maxG);
% Channel decoding includes - CB segmentation, turbo decoding, rate dematching
[decTbData1, ˜,˜] = lteTbChannelDecoding(nS, rxCW, Kplus1, C1,  prmLTEDLSCH,
prmLTEPDSCH);
% Transport block CRC detection
[dataOut, ˜] = CRCdetector(decTbData1);
end
```

7.7.1   PMI-Based Adaptation

The following MATLAB script segment shows the body of the processing while loop that implements the PMI codebook index selection without adaptive modulation and coding. The script uses the last output argument (*cbIdx*) of the *commlteMIMO_SM_PMI_step* function, which is the updated PMI codebook index in the current subframe. By providing this index as the third input argument to the *commlteMIMO_update* function, we set the PMI index for the following subframe.

Algorithm

MATLAB script segment

```
%% One subframe step processing
    [dataIn, dataOut, txSig, rxSig, dataRx, yRec, csr_ref, cbIdx]...
    = commlteMIMO_SM_PMI_step(nS, snrdB, prmLTEDLSCH, prmLTEPDSCH, prmMdl);
    %% Report average data rates
    ADR=zReport_data_rate_average(prmLTEPDSCH, prmLTEDLSCH);
    fprintf(1,'PMI codebook index = %2d\n', cbIdx);
    %% Calculate  bit errors
    Measures = step(hPBer, dataIn, dataOut);
    %% Visualize results
    if (visualsOn && prmLTEPDSCH.Eqmode˜=3)
        zVisualize( prmLTEPDSCH, txSig, rxSig, yRec, dataRx, csr, nS);
    end;
    %% Update subframe number
    nS = nS + 2; if nS > 19, nS = mod(nS, 20); end;
    % Adaptive PMI
    modType=prmLTEPDSCH.modType;
    cRate=prmLTEDLSCH.cRate;
    [prmLTEPDSCH, prmLTEDLSCH, prmMdl] = commlteMIMO_update(...
        prmLTEPDSCH, prmLTEDLSCH, prmMdl, modType, cRate, cbIdx);
    %% including adaptations here
```

```
PMI codebook index =  1
Modulation                        = 16QAM
Instantanous Data rate            = 19.85 Mbps
Average Data rate                 = 19.85 Mbps
Instantanous Modulation rate = 4.00
Average Modulation rate           = 4.00
Instantanous Coding rate          = 0.3333
Average Coding rate               = 0.3333


PMI codebook index =  2
Modulation                        = 16QAM
Instantanous Data rate            = 19.85 Mbps
Average Data rate                 = 19.85 Mbps
Instantanous Modulation rate = 4.00
Average Modulation rate           = 4.00
Instantanous Coding rate          = 0.3333
Average Coding rate               = 0.3333
```

Figure 7.4   Adaptive PMI: change of PMI codebook index subframe by subframe

7.7.2   Verifying Transceiver Performance

The parameters used in the simulation are the same as those used in the adaptive PMI, as summarized in the script *commlteMIMO_params*. During the simulation, all of these parameters remain constant except the PMI codebook index (*cbIdx*). By executing the MATLAB script of the MIMO transceiver model, we can assess the performance of the system. As the simulation is running, the message in Figure 7.4 appears on the MATLAB screen, showing variations in the PMI codebook index.

Table 7.4   Adaptive precoding: BER, data rates, modulation, and coding rates in different scenarios

Type of modulation	Average data rate (Mbps)	Modulation rate	Coding rate	Bit error rate
No modulation and coding adaptation	35.16	4	1/3	0.1278
No modulation and coding adaptation + adaptive PMI	35.16	4	1/3	0.01191

7.7.3   Adaptation Results

For each of the adaptation scenarios, we compute the BERs by processing 20 million bits. Table 7.4 illustrates the results. These PMI index variations do not affect the modulation scheme, the coding rate, the instantaneous data rate, or the average data rate of the transceiver. As we expect the relative effect on the BER reflects the benefits of adaptive precoding.

7.8   Adaptive MIMO

In this section we use the RI channel-state report to adaptively toggle the transmission mode between transmit diversity and spatial multiplexing. If the estimated rank is equal to the number of transmit antennas, we perform spatial multiplexing. For the sake of simplicity, if the rank is less than the number of transmit antennas then we revert to transmit-diversity mode. We will use the same number of antennas but we forego the increased data rate associated with spatial multiplexing in favor of the greater link reliability associated with transmit diversity.

By applying a threshold to the condition number, we propose a wideband rank value for the whole subframe. The function takes as input a 2D channel matrix at the receiver (h). This is either a 2×2 or a 4×4 matrix, depending on whether two- or four-antenna transmission is used.

Algorithm

MATLAB function

```
function [dataIn, dataOut, txSig, rxSig, dataRx, yRec, csr_ref, cbIdx, ri]...
    = commlteMIMO_SM_PMI_RI_step(nS, snrdB, prmLTEDLSCH, prmLTEPDSCH,
prmMdl)
%% TX
persistent hPBer1
if isempty(hPBer1), hPBer1=comm.ErrorRate; end;
%  Generate payload
dataIn = genPayload(nS,  prmLTEDLSCH.TBLenVec);
% Transport block CRC generation
tbCrcOut1 =CRCgenerator(dataIn);
% Channel coding includes - CB segmentation, turbo coding, rate matching,
% bit selection, CB concatenation - per codeword
[data, Kplus1, C1] = lteTbChannelCoding(tbCrcOut1, nS, prmLTEDLSCH, prmLTEPDSCH);
%Scramble codeword
scramOut = lteScramble(data, nS, 0, prmLTEPDSCH.maxG);
% Modulate
modOut = Modulator(scramOut, prmLTEPDSCH.modType);
% Map modulated symbols  to layers
if (prmLTEPDSCH.txMode ==4)
LayerMapOut = LayerMapper(modOut, [], prmLTEPDSCH);
usedCbIdx = prmMdl.cbIdx;
% Precoding
```

```
[PrecodeOut, Wn] = SpatialMuxPrecoder(LayerMapOut, prmLTEPDSCH, usedCbIdx);
else
% TD with SFBC
PrecodeOut = TDEncode(modOut(:,1),prmLTEPDSCH.numTx);
end
% Generate Cell-Specific Reference (CSR) signals
numTx=prmLTEPDSCH.numTx;
csr = CSRgenerator(nS, numTx);
csr_ref=complex(zeros(2*prmLTEPDSCH.Nrb, 4, numTx));
for m=1:numTx
    csr_pre=csr(1:2*prmLTEPDSCH.Nrb,:,:,m);
    csr_ref(:,:,m)=reshape(csr_pre,2*prmLTEPDSCH.Nrb,4);
end
% Resource grid filling
txGrid = REmapper_mTx(PrecodeOut, csr_ref, nS, prmLTEPDSCH);
% OFDM transmitter
txSig = OFDMTx(txGrid, prmLTEPDSCH);
%% Channel
% MIMO Fading channel
[rxFade, chPathG] = MIMOFadingChan(txSig, prmLTEPDSCH, prmMdl);
% Add AWG noise
sigPow = 10*log10(var(rxFade));
nVar = 10.^(0.1.*(sigPow-snrdB));
rxSig = AWGNChannel(rxFade, nVar);
%% RX
% OFDM Rx
rxGrid = OFDMRx(rxSig, prmLTEPDSCH);
% updated for numLayers -> numTx
[dataRx, csrRx, idx_data] = REdemapper_mTx(rxGrid, nS, prmLTEPDSCH);
% MIMO channel estimation
if prmMdl.chEstOn
    chEst = ChanEstimate_mTx(prmLTEPDSCH, csrRx,  csr_ref, prmMdl.chEstOn);
    hD   = ExtChResponse(chEst, idx_data, prmLTEPDSCH);
else
    idealChEst = IdChEst(prmLTEPDSCH, prmMdl, chPathG);
    hD = ExtChResponse(idealChEst, idx_data, prmLTEPDSCH);
end
% Frequency-domain equalizer
if (numTx==1)
    % Based on Maximum-Combining Ratio (MCR)
    yRec = Equalizer_simo(dataRx, hD, nVar, prmLTEPDSCH.Eqmode);
else
    if (prmLTEPDSCH.txMode ==4)
        % Based on Spatial Multiplexing
        [yRec, ri] = MIMOReceiver_ri(dataRx, hD, prmLTEPDSCH, nVar, Wn);
    else
% Based on Transmit Diversity  with SFBC combiner
    [yRec, ri] = TDCombine_ri(dataRx, hD, prmLTEPDSCH.numTx, prmLTEPDSCH.numRx);
    end
```

```
end
% PMI codebook selection
if (prmMdl.enPMIfback)
cbIdx = PMICbSelect( hD, prmMdl.enPMIfback, prmLTEPDSCH.numTx, ...
  prmLTEPDSCH.numLayers, snrdB);
else
  cbIdx = prmMdl.cbIdx;
end
% Demap received codeword(s)
[cwOut, ~] = LayerDemapper(yRec, prmLTEPDSCH);
if prmLTEPDSCH.Eqmode < 3
  % Demodulate
  demodOut = DemodulatorSoft(cwOut, prmLTEPDSCH.modType, max(nVar));
else
  demodOut = cwOut;
end
% Descramble received codeword
rxCW = lteDescramble(demodOut, nS, 0, prmLTEPDSCH.maxG);
% Channel decoding includes - CB segmentation, turbo decoding, rate dematching
[decTbData1, ~,~] = lteTbChannelDecoding(nS, rxCW, Kplus1, C1, prmLTEDLSCH,
prmLTEPDSCH);
% Transport block CRC detection
[dataOut, ~] = CRCdetector(decTbData1);
end
```

7.8.1   RI-Based Adaptation

The following MATLAB script segment shows the body of the processing while loop that implements the RI-based adaptation without any previously presented adaptation. The script uses the updated rank-estimation index in the current subframe, as represented by the last output argument (*ri*) of the *commlteMIMO_SM_PMI_RI_step* function. By providing this index as the fourth input argument to the *commlteMIMO_update* function, we set the RI index for the following subframe.

Algorithm

MATLAB script segment

```
%% One subframe step
  [dataIn, dataOut, txSig, rxSig, dataRx, yRec, csr, cbIdx, ri] ...
  = commlteMIMO_SM_PMI_RI_step(nS, snrdB, prmLTEDLSCH, prmLTEPDSCH,
prmMdl);

  Ri=RIselection(ri, threshold);
```

```
ADR_a=zReport_data_rate_average(prmLTEPDSCH, prmLTEDLSCH);
fprintf(1,'PMI codebook index = %2d\nTransmission mode  = %2d\n', cbIdx, Ri);

%% Calculate  bit errors
Measures = step(hPBer, dataIn, dataOut);
%% Visualize results
if (visualsOn && prmLTEPDSCH.Eqmode~=3)
    zVisualize( prmLTEPDSCH, txSig, rxSig, yRec, dataRx, csr, nS);
end;
% Update subframe number
nS = nS + 2; if nS > 19, nS = mod(nS, 20); end;
% Adaptive RI
modType=prmLTEPDSCH.modType;
cRate=prmLTEDLSCH.cRate;
cbIdx=prmMdl.cbIdx;
[prmLTEPDSCH, prmLTEDLSCH, prmMdl] = commlteMIMO_update(...
    prmLTEPDSCH, prmLTEDLSCH, prmMdl, modType, cRate,  cbIdx, Ri);
```

7.8.2   Verifying Transceiver Performance

The parameters used in the simulation are the same as those used in the adaptive modulation case, as summarized in the script *commlteMIMO_params*. During the simulation, all of these parameters remain constant except the rank-estimation index (*ri*). As the simulation is running, the message shown in Figure 7.5 appears on the MATLAB screen. As we can see, transmission-mode changes based on RI estimation affect the instantaneous data rate and thus the average data rate of the transceiver.

7.8.3   Adaptation Results

The results indicate that, as expected, we obtain higher data rates when using spatial multiplexing than when using transmit diversity (Table 7.5). By using the rank-estimation method, we obtain an average rate that is closer to that obtained with spatial multiplexing. This reflects the greater proportion of full-rank subframes (86.5%) as compared to lower-rank subframes (13.5%).

7.9   Downlink Control Information

As we saw earlier, link adaptation involves the following three components: (i) channel-state estimation at the receiver; (ii) scheduling operations to create scheduling assignments; and (iii) transmission of the scheduling assignments in the following subframe. In downlink, the scheduling assignments are captured as DCI. There are multiple DCI formats for various LTE-standard transmission modes. A DCI format may contain many types of information, including multi-antenna characteristics such as the precoding matrix indicator index, modulation scheme, and transport block size and HARQ (Hybrid Automatic Repeat Request) process details such as the process number, the redundancy version, and a new data indicator.

```
PMI codebook index =   1
Transmission mode  =   4
Modulation                            = 16QAM
Instantanous Data rate                = 19.85 Mbps
Average Data rate                     = 19.14 Mbps
Instantanous Modulation rate = 4.00
Average Modulation rate               = 4.00
Instantanous Coding rate              = 0.3333
Average Coding rate                   = 0.3333


PMI codebook index =   1
Transmission mode  =   2
Modulation                            = 16QAM
Instantanous Data rate                = 9.91 Mbps
Average Data rate                     = 19.03 Mbps
Instantanous Modulation rate = 4.00
Average Modulation rate               = 4.00
Instantanous Coding rate              = 0.3333
Average Coding rate                   = 0.3333
```

Figure 7.5   Adaptive MIMO: change of transmission mode subframe by subframe

Table 7.5   Adaptive rank estimation and MIMO: BER, data rates, modulation, and coding rates in different scenarios

Type of modulation	Average data rate (Mbps)	Modulation rate	Coding rate	Bit error rate
Fixed mode: transmit diversity	15.26	4	2/3	$3.4e^{-07}$
Fixed mode: spatial multiplexing	19.85	4	1/3	$1.3e^{-03}$
Adaptive mode based on RI feedback	19.23	4	1/3	$7.1e^{-04}$

In this book we focus on user-plane information and the single-user processing case. As such, we will not discuss in detail the DCI formats or their implementation in MATLAB. However, we will present two topics with reference to the control information: (i) how CQI information is mapped to the MCS information and (ii) how the PDCCH is encoded before it is transmitted within the first few OFDM symbols of the downlink resource grid. Finally, we perform a quick review of the reliability of the PDCCH transmission by comparing its BER with that of the PDSCH, which we have examined in detail in this book.

7.9.1 MCS

The scheduler provides the downlink transmitter with the modulation type and coding rate of the data in the current subframe. This information is encoded as a 5 bit MCS index and incorporated in all different DCI formats. The MCS index jointly codes the modulation scheme and the transport block size (*tbSize*). The coding rate (R) is defined as the ratio of the number of input bits (K) to the number of output bits (N): K/N. Since a 24 bit CRC is used prior to DLSCH processing, the number of input bits at the encoder is equal to *tbSise* + 24. The number of DLSCH coder output bits (N) is equal to the size of each PDSCH codeword (*numPDSCHbits*); that is, $N = numPDSCHbits$. Since the number of bits in each PDDSCH codeword is completely specified by the number of resource blocks, knowledge of the transport block size (*tbSize*) gives the coding rate as $R = \frac{tbSise+24}{numPDSCHbits}$.

The following MATLAB function shows how the CQI information (modulation scheme and target coding rate) can be mapped to the transport block size, the index of the transport-block-size lookup table, and the actual coding rate.

Algorithm

MATLAB function

```
function [tbSize, TBSindex, ActualRate] = getTBsizeMCS(modType, TCR, Nrb,
numLayers, numPDSCHBits)
% Get the transport block size for a specified configuration.
% Inputs:
%   modType    : 1 (QPSK), 2 (16QAM), 3 (64QAM)
%   TCR        : Target Code Rate
%   Nrb        : number of resource blocks
%   numLayers  : number of layers
%   numPDSCHBits: number of PDSCH bits (G)
% Output:
%   tbSize : transport block length
% Example: R.10 of A.3.3.2.1 in 36.101
%   tbLen = getTBsizeRMC(1, 1/3, 50, 1, 12384)
% Reference:
%   1) Section 7.1.7 of 36.213, for TB sizes.
%      Uses preloaded Tables 7.1.7.1-1, 7.1.7.2.1-1, 7.1.7.2.2 and 7.1.7.2.5.
%   2) Section A.3.1 of 36.101 for TB size selection criteria.
switch modType
  case 1 % QPSK
     numTBSizes = 10;
     stIdx = 0;
  case 2 % 16QAM
     numTBSizes = 7;
     stIdx = 9;
  case 3 % 64QAM
     numTBSizes = 12;
     stIdx = 15;
```

```
end
numBitsPerLayer=numPDSCHBits/numLayers;
% Load saved entries for Tables 7.1.7.2.1-1, 7.1.7.2.2 and 7.1.7.2.5.
load TBSTable.mat
tbVec = baseTBSTab(stIdx+(1:numTBSizes), Nrb);       % for 1-based indexing
ProposedRates=(tbVec+24)./numBitsPerLayer;
ProposedRates(ProposedRates<1/3)=10;
[~, c] = min(abs(TCR-ProposedRates));
tbSize = tbVec(c);
if (numLayers==2) % Section 7.1.7.2.2
   if (Nrb <= 55)
      tbSize = baseTBSTab(TBSindex, 2*Nrb);
   else
      index=(layer2TBSTab(:,1)==tbSize);
      tbSize = layer2TBSTab(index, 2);
   end
elseif (numLayers==4) % Section 7.1.7.2.5
   if (Nrb <= 27)
      tbSize = baseTBSTab(TBSindex, 4*Nrb);
   else
      index=(layer4TBSTab(:,1)==tbSize);
      tbSize = layer4TBSTab(index, 2);
   end
end
ActualRate=(tbSize+24)./numPDSCHBits;
TBSindex= stIdx+c;
```

The following function shows how the transport-block-size index and the modulation type
are mapped to a unique MCS index that can be encoded using a 5 bit scalar quantizer.

Algorithm

MATLAB function

```
function MCSindex=map2MCSindex(TBSindex, modType)
%#codegen
% Assume 1-based indexing
if ((TBSindex < 1) || (TBSindex >27)),
error('map2MCSindex function: Wrong TBSindex.');end
switch TBSindex
   case 10
      switch modType
         case 1
            MCSindex=10;
         case 2
```

```
            MCSindex=11;
        otherwise
            error('Wrong combination of TBSindex and modulation type');
    end
  case 16
    switch modType
      case 2
          MCSindex=17;
      case 3
          MCSindex=18;
      otherwise
          error('Wrong combination of TBSindex and modulation type');
    end
  otherwise
    if TBSindex <10
      MCSindex=TBSindex;
    elseif ((TBSindex >10) && (TBSindex <16))
      MCSindex=TBSindex+1;
    else
      MCSindex=TBSindex+2;
    end
end
```

7.9.2   Rate of Adaptation

The scheduling decision affecting the downlink can be updated once every subframe. But the DCI containing the MCS, for example, does not need to adapt every 1 ms. The granularity and rate of adaptation are left to the discretion of the scheduling algorithm, which takes into account various factors, including the totality of the link quality for all users, the base-station interference profile, quality-of-service requirements, service, and service priorities [9].

7.9.3   DCI Processing

In this section we will examine the performance of the transceiver applied to the DCI. DCI is encoded and transmitted in the first few OFDM symbols of each subframe. Reliable decoding of the DCI is a critical requirement for proper recovery of the user data placed within the ensuing OFDM symbols of the subframe. Since DCI is usually transmitted in small packets, a convolutional code is used to encode them. The LTE standard specifies a combination of tail-biting convolutional encoding and transmit diversityfor reliable link performance by the DCI.

To stay true to our user-plane processing focus, we will not discuss in detail all the components of the DCI (which including the Physical Hybrid ARQ Indicator Channel (PHICH) and the Physical Control-Format Indicator Channel (PCFICH). We will also forego a detailed description of the placement of the information within the resource grid and the OFDM transmission. Instead, we will focus on the signal processing chain prior to transmission of the signal through a combination flat-fading and AWGN (Additive White Gaussian Noise) channel in order to evaluate the performance of the control information.

7.9.3.1   Transceiver Function

The following MATLAB function (*commlteMIMO_DCI_step*) summarizes the DCI process-ing chain. In the transmitter, we first apply a DCI-specific CRC generation. The DCI data are processed by a tail-biting convolutional encoder with the same trellis structure as the one used in the user plane for turbo coding. Then we apply to the encoded bits the same rate-matching, scrambling, and modulation operations that were performed on the user-plane bits. Finally, transmit diversity is performed on the modulated signal. In this function we forego the resource-grid-mapping and OFDM-signal-generation operations. We apply channel mod-eling directly to the outputs of the transmit diversity encoder. At the receiver, we perform the inverse operation to that of the transmitter, including ideal-channel estimation, transmit-diversity combining, soft-decision demodulation, descrambling, rate dematching, and Viterbi decoding. Finally, by performing the DCI-specific CRC detection we find the best output esti-mate of the DCI bits. The function *commlteMIMO_DCI_step* represents a simplified version of the transceiver operations applied to the DCI.

Algorithm

MATLAB function

```
function [dataIn, dataOut, modOut, rxSig]...
   = commlteMIMO_DCI_step(nS, snrdB, prmLTEDLSCH, prmLTEPDSCH, prmMdl)
%% TX
numBitsDCI = 205;
%  Generate payload
dataIn    = genPayload(nS, numBitsDCI);
% Transport block CRC generation
CrcOut1   = CRCgeneratorDCI(dataIn);
% Channel coding includes - tail biting convolutional coding, rate matching
data      = TailbitingConvEnc(CrcOut1, prmLTEDLSCH.cRate);
%Scramble codeword
scramOut  = lteScramble(data, nS, 0, prmLTEPDSCH.maxG);
% Modulate
modOut    = Modulator(scramOut, prmLTEPDSCH.modType);
% Transmit diversity encoder
PrecodeOut = TDEncode(modOut, prmLTEPDSCH.numTx);
%% Channel
% MIMO fading channel
[fadeOut, pathGain]   = MIMOFadingChan(PrecodeOut, prmLTEPDSCH, prmMdl);
nVar             = real(var(fadeOut(:)))/(10.^(0.1*snrdB));
pathG            = squeeze(pathGain);
% AWGN
recOut           = AWGNChannel(fadeOut, nVar);
%% RX
% Transmit diversity combiner
rxSig = TDCombine(recOut, pathG, prmLTEPDSCH.numTx, prmLTEPDSCH.numRx);
% Demodulate
```

```
demodOut = DemodulatorSoft(rxSig, prmLTEPDSCH.modType, nVar);
% Descramble both received codewords
rxCW1 = IteDescramble(demodOut, nS, 0, prmLTEPDSCH.maxG);
% Channel decoding includes - tail biting Viterbi decoding, rate dematching
L=numel(CrcOut1);
decData1=TailbitingViterbiSoft( rxCW1, L);
% Transport block CRC detection
[dataOut, ˜] = CRCdetectorDCI(decData1);
end
```

The transceiver function features two functions: *TailbitingConvEnc* and *TailbitingViterbiSoft*. These implement tail-biting convolutional coding and its inverse Viterbi decoding with System objects from the Communications System Toolbox. In the encoder, we create two convolutional coders: one that maintains the state of the encoder and one that streams each frame of data using the state provided by the encoder. Finally, we compute the output using the rate-matching operations introduced in the previous chapter.

Algorithm

MATLAB function

```
function y=TailbitingConvEnc(u, codeRate)
%#codegen
trellis=poly2trellis(7, [133 171 165]);
L=numel(u);
C=6;
persistent ConvEncoder1 ConvEncoder2
if isempty(ConvEncoder1)
    ConvEncoder1=comm.ConvolutionalEncoder('TrellisStructure', trellis,
'FinalStateOutputPort', true, ...
        'TerminationMethod','Truncated');
    ConvEncoder2 = comm.ConvolutionalEncoder('TerminationMethod','Truncated',
'InitialStateInputPort', true,...
        'TrellisStructure', trellis);
end
u2        = u((end-C+1):end);                    % Tail-biting convolutional coding
[˜, state] = step(ConvEncoder1, u2);
u3        = step(ConvEncoder2, u,state);
y         = fcn_RateMatcher(u3, L, codeRate);  % Rate matching
```

In the decoder, we first perform rate dematching and then concatenate the received likelihood measures and perform soft-decision Viterbi decoding based on the state value that is embedded within the concatenated input signal. We compute the output by extracting the subset of the Viterbi-decoding output samples.

Algorithm

MATLAB function

```
function y=TailbitingViterbiSoft(u, L)
%#codegen
trellis=poly2trellis(7, [133 171 165]);
Index=[L+1:(3*L/2) (L/2+1):L];
persistent Viterbi
if isempty(Viterbi)
   Viterbi=comm.ViterbiDecoder(...
     'TrellisStructure', trellis,
'InputFormat','Unquantized','TerminationMethod','Truncated','OutputDataType','logical');
end
uD          = fcn_RateDematcher(u, L);          % Rate de-matching
uE          = [uD;uD];                          % Tail-biting
uF          = step(Viterbi, uE);                % Viterbi decoding
y           = uF(Index);
```

7.9.3.2   Testbench for the DCI Transceiver

The following function (*commlteMIMO_DCI*) represents the testbench for evaluating performance. The function takes as its inputs the *Eb/N0* value, the specified maximum number of errors observed, and the maximum number of bits processed. As its outputs, it produces the BER measure and the actual number of process bits. First it calls the initialization function (*commlteMIMO_initialize*) in order to set all the relevant parameter structures (*prmLT-EDLSCH, prmLTEPDSCH, prmMdl*). Then the testbench uses a while loop to perform subframe processing by calling the DCI-processing-chain function.

Algorithm

MATLAB function

```
function [ber, bits]=commlteMIMO_DCI(EbNo, maxNumErrs, maxNumBits)
%
clear functions
%% Set simulation parameters & initialize parameter structures
commlteMIMO_ params_DCI;
codeRate=cRate;
k=2*modType;
snrdB = EbNo + 10*log10(codeRate) + 10*log10(k);
[prmLTEPDSCH, prmLTEDLSCH, prmMdl] = commlteMIMO_initialize(txMode, ...
   chanBW, contReg, modType, Eqmode,numTx, numRx,cRate,maxIter, fullDecode, chan-
Mdl, Doppler, corrLvl, ...
   chEstOn, numCodeWords, enPMIfback, cbIdx, snrdB, maxNumErrs, maxNumBits);
```

```
clear txMode chanBW contReg modType Eqmode numTx numRx cRate maxIter
fullDecode chanMdl Doppler corrLvl chEstOn numCodeWords enPMIfback cbIdx
%%
hPBer = comm.ErrorRate;
%% Simulation loop
nS = 0; % Slot number, one of [0:2:18]
Measures = zeros(3,1); %initialize BER output
while (( Measures(2)< maxNumErrs) && (Measures(3) < maxNumBits))
    [dataIn, dataOut, modOut, rxSig] = ...
        commIteMIMO_DCI_step(nS, snrdB, prmLTEDLSCH, prmLTEPDSCH, prmMdl);
    % Calculate  bit errors
    Measures = step(hPBer, dataIn, dataOut);
    % Visualize results
    if visualsOn, zVisConstell( prmLTEPDSCH, modOut, rxSig, nS); end;
end
ber=Measures(1);
bits=Measures(3);
reset(hPBer);
```

7.9.3.3   BER Measurements

What characterizes the DCI transceiver is the combination of a low-constellation modulator (QPSK), a tail-biting convolutional encoder, and transmit-diversity coding. As illustrated by Figure 7.6, this combination provides a high BER performance for the processing of DCI information as designated by the LTE standard to process the DCI. This is critical, as excessive errors in the control information will have catastrophic effects on the quality of the recovered user data.

Figure 7.6 illustrates the result of processing or testbenching with $Eb/N0$ values of between 0 and 4 dB and the maximum number of errors set to 1000 and the maximum number of bits to $1e^7$. We process the transmitted data using a flat-fading channel in which the CSI is completely known. This implements an ideal channel estimator. Note that the BER measure is quite low even at low SNR (Signal-to-Noise Ratio) values. For example, at an $Eb/N0$ of 0 dB we have a BER of $1e^{-4}$, and at 2 dB we have a BER of around $1e^{-6}$.

These results are compatible with the high performance of the PDSCH processing discussed in Chapter 4. Even without transmit diversity, the combination of LTE turbo coding, scrambling, and modulation processed by the channel model presented here allows us to obtain BER performances that are compatible with DCI processing.

7.10   Chapter Summary

In this chapter we studied some of the LTE-standard specifications related to link adaptations. The operations related to link adaptation are performed in the receiver and involve the selection and feedback of various system parameters for use in the transmitter in the following subframes. They are applied to system parameters specifying: (i) the modulation and coding schemes, (ii) the number of transmission layers used in spatial multiplexing, and (iii) the choice

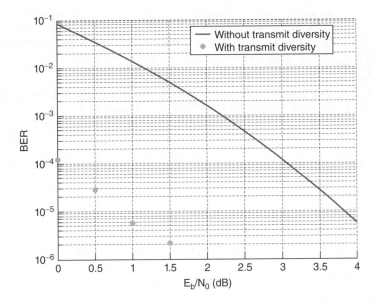

Figure 7.6 BER performance of the DCI: combination of AWGN and a flat-fading channel model with ideal channel-response estimation

of precoding matrices used in closed-loop spatial-multiplexing modes. The selection criteria for link adaptation are usually maximizing user data rates and increased spectral efficiency.

We reviewed the channel-quality measurements needed to perform the adaptations. These include CQI, PMI, and RI measurements. We then introduced algorithms that implement adaptations based on the channel-quality measures, including algorithms for adaptive modulation and coding, adaptive precoding for determination of the best precoder matrix, and adaptive MIMO for resolving occasional rank deficiencies in spatial multiplexing modes of transmission. Finally, we reviewed the signal-processing chain involved in the transmission of DCI. We showed that by exploiting a combination of transmit diversity and tail-biting convolutional coding, the LTE standard provides a reliable mechanism for the transmission of control information.

References

[1] 3GPP (2009) Requirements for Evolved UTRA (E-UTRA) and Evolved UTRAN (E-UTRAN) v9.0.0. TR 25.913, December 2009.

[2] Ding, L., Tong, F., Chen, Z., and Liu, Z. (2011) A novel MCS selection criterion for VOIP in LTE. International Conference on Wireless Communications, Networking and Mobile Computing - WiCom, pp. 1–4.

[3] Pande, A., Ramamurthi, V., and Mohapatra, P. (2011) Quality-oriented video delivery over LTE using adaptive modulation and coding. IEEE Global Telecommunications Conference (GLOBECOM), pp. 1–5.

[4] Tan, P., Wu, Y. and Sun, S. (2008) Link adaptation based on adaptive modulation and coding for multiple-antenna OFDM system. *IEEE Journal on Selected Areas in Communications*, **26**, 8, 1599–1606.

[5] Kim, J., Lee, K., Sung, C. and Lee, I. (2009) A simple SNR representation method for AMC schemes of MIMO systems with ML detector. *IEEE Transactions on Communications*, **57**, 2971–2976.

[6] Flahati, S., Svensson, A., Ekman, T. and Sternad, M. (2004) Adaptive modulation systems for predicted wireless channels. *IEEE Transactions on Communications*, **52**, 307–316.

[7] Ohlmer, E. and Fettweis, G. (2009) Link adaptation in linearly precoded closed-loop MIMO-OFDM systems with linear receivers. IEEE International Conference on Communications (ICC), June 2009.

[8] Love, D. and Heath, R. (2005) Limited feedback unitary precoding for spatial multiplexing systems. *IEEE Transactions on Information Theory*, **51**, 8, 2967–2976.

[9] Ghosh, A. and Ratasuk, R. (2011) *Essentials of LTE and LTE-A*, Cambridge University Press.

[10] 3GPP (2013) Physical Layer Procedures v11.3.0. TR 25.213, June 2013.

[11] Jiang, M., Prasad, N., Yue, G., and Rangarajan, S. (2011) Efficient link adaptation for precoded multi-rank transmission and turbo SIC receivers. IEEE ICC, 2011.

8

System-Level Specification

So far we have presented the four main enabling technologies of the LTE (Long Term Evolution) standard individually: OFDM (Orthogonal Frequency Division Multiplexing) multicarrier transmission, MIMO (Multiple Input Multiple Output) multi-antenna techniques, channel coding based on turbo coders, and link adaptations. OFDM in downlink and its single-carrier counterpart SC-FDM (Single-Carrier Frequency Division Multiplexing) in uplink provide the backbone of the LTE transmission strategy. As examined in Chapter 7, adaptive modulation and coding provide superior spectral efficiency. Arguably the defining feature that sets LTE apart from the previous standards is the incorporation of various types of MIMO technique. At any given time, the operating mode of the LTE downlink transceiver system is uniquely defined by one of the nine MIMO multi-antenna techniques specified in the LTE standard. As a result, when evaluating the performance of an LTE system we should pay special attention to the type of MIMO technique used and the corresponding channel operating conditions.

In this chapter we will put together a simulation model for the PHY (Physical Layer) of the LTE standard. So far, our pedagogic step-by-step approach has demanded that we focus on a single transmission mode at a time. In this chapter, we develop a simulation model that incorporates multiple transmission modes in both the transmitter and the receiver. We also evaluate various aspects related to the performance of the system.

We will highlight both the common set of processing performed in multiple modes and the specific features used in any particular transmission mode. To stay true to our focus on examining user-plane processing and single-carrier operations, our simulation model will include the first four transmission modes of the LTE standard. Then we will evaluate the performance of the system under changing operating conditions; this includes examining system quality and throughput when using different types of MIMO mode, channel model, channel estimation technique, and MIMO receiver algorithm. Finally, we will create a Simulink model for our LTE PHY system. We will show how easy it is to incorporate the MATLAB® algorithms developed throughout this book as distinct blocks within the Simulink model. Expressing the design as a Simulink model has the added benefit of having Simulink as the simulation testbench. Instead of developing MATLAB scripts to perform various operations in the loop and

Understanding LTE with MATLAB®: From Mathematical Modeling to Simulation and Prototyping, First Edition. Houman Zarrinkoub.

call the algorithms and visualization functions iteratively, we will use the Simulink model to perform these tasks. In the final section of this chapter, we will undertake a qualitative assessment of the performance of our LTE model. We will use the LTE Simulink model to stream audio signals as the inputs to the downlink transmitter and will listen to and assess the quality of the recovered voice at the receiver after a certain amount of processing delay.

8.1   System Model

In this section, we compose a system model for the PHY of the LTE standard by integrating various enabling technologies. The system model is composed of a transmitter, a channel model, and a receiver. The processing chain in the transmitter is specified in detail by the standard. The standard also specifies various channel models for performance evaluations. The receiver operations, however, are not specified, which provides the opportunity for various system designers to distinguish their implementation with distinct performance profiles.

Figure 8.1 illustrates the overall structure of the LTE downlink transceiver. The payload bits provided by the transport channel are processed by the transmitter. The transmitter output is composed of a sequence of the symbols transmitted on various available transmit antennas. The channel operates on the transmitted symbols of multiple transmit antennas to produce the set of received symbols arriving at each of the receive antennas. The receiver operates on the received symbols. By implementing various strategies aimed at inverting the operations of the transmitter, the receiver generates its best estimate of the transmitted payload bits.

8.1.1   Transmitter Model

In the transmitter, the signal processing chain is applied to the payload bits provided by the transport channel. The processing depends on the transmission mode, defined by the downlink scheduler. The transmission mode manifests itself as the choice of MIMO technique used at any given subframe. In this book, we have focused on the first four of the nine transmission modes used in the LTE standard. Figure 8.2 illustrates the processing chain in the downlink transmitter.

In each subframe, the scheduler selects one of the four transmission modes: the first implements a single-antenna transmission and can be referred to as a SIMO (Single Input Multiple Output) mode. The second employs transmit diversity, where redundant information is transmitted on multiple antennas to boost the overall link reliability. The third and fourth both employ spatial-multiplexing MIMO techniques to boost data rates. The third mode uses open-loop spatial multiplexing and is intended for transmission in high-mobility scenarios, while the fourth uses closed-loop spatial multiplexing intended mostly for low-mobility scenarios and is responsible for the highest data rates achievable by the LTE standard.

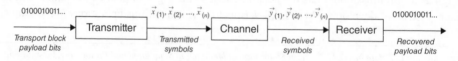

Figure 8.1   LTE downlink PHY transceiver model

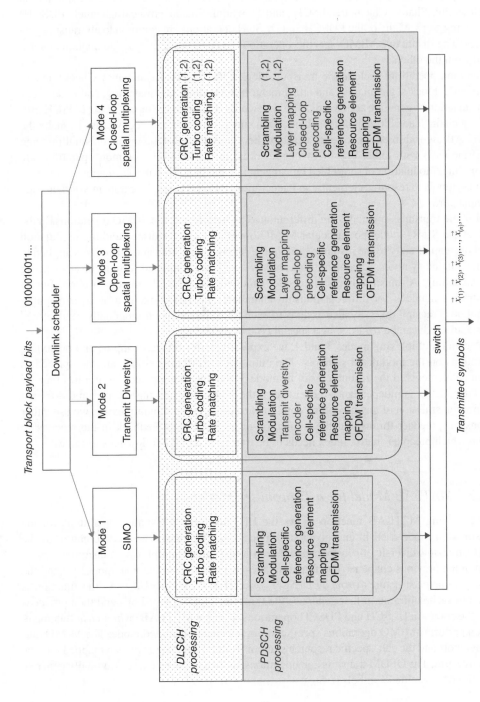

Figure 8.2 LTE downlink system model: transmission modes 1–4, transmitter operations

In each transmission mode, we go through a series of operations that are a combination of Downlink Shared Channel (DLSCH) and Downlink Shared Physical Channel (PDSCH) processing steps. Many of the PDSCH and DLSCH operations are common to all transmission modes. The distinctive MIMO-specific operations in each mode are the elements that set it apart from the others.

Common operations include the transport block CRC (Cyclic Redundancy Check) attachment, code-block segmentation and attachment, turbo coding, rate matching, and code-block concatenation to generate codewords. The codewords constitute the inputs of the PDSCH processing chain. The LTE downlink specification supports one or two codewords. To make the MATLAB algorithm easier to read, we show here only the closed-loop spatial-multiplexing case (mode 4) with either one or two codewords. In the PDSCH, common operations are scrambling, modulation of scrambled bits to generate modulation symbols, mapping of modulation symbols to resource elements, and generation of OFDM signal for transmission on each antenna port.

Different transmission modes are differentiated by their respective MIMO operations. In the case of the SIMO mode, no particular MIMO operations are performed and the modulation symbols map directly to the resource grid. In the second mode, transmit diversity is applied to the modulated symbols. This operation can be viewed as a combined form of layer mapping and precoding. A transmit-diversity encoder subdivides the modulated stream into different substreams intended for different transmit antennas. This is analogous to a layer-mapping operation. Transmit diversity also assigns to each transmit antenna orthogonally transformed substreams that can be regarded as a special type of precoding.

In spatial multiplexing modes 3 and 4, we explicitly perform layer mapping and precoding operations on the modulated symbols. The outputs of MIMO operations are then placed as resource elements in the resource grid. In mode 3, the open-loop precoding implies that the precoder matrix is independently generated at the transmitter and the receiver and does not require transmission of any precoding data. In closed-loop spatial multiplexing, we use either a constant precoder throughout the simulation or a closed-loop feedback in the receiver to signal which precoder matrix index needs to be transmitted in the following subframe.

8.1.2   MATLAB Model for a Transmitter Model

The following MATLAB function shows the LTE downlink transmitter operations for the transmission mode used in any given subframe. It can be viewed as a combination of the SIMO, transmit-diversity, open-loop, and closed-loop spatial-multiplexing transmitters developed in the previous chapters. The function takes as input the subframe number (nS) and the three parameter structures (*prmLTEDLSCH*, *prmLTEPDSCH*, *prmMdl*). As the function is called in each subframe, it first generates the transport-block payload bits and then proceeds with the common DLSCH and PDSCH operations. Using a MATLAB *switch-case* statement, it then performs MIMO operations specific to the selected transmitted mode. The MIMO output symbols and the cell-specific resource symbols (pilots) generated are then mapped into the resource grid. The OFDM transmission operations applied to the resource grid finally generate the output transmitted symbols (*txSig*).

Algorithm

MATLAB function

```
function [txSig, csr_ref] = commlteMIMO_Tx(nS, dataIn, prmLTEDLSCH, prmLTEPDSCH,
prmMdl)
%#codegen
%  Generate payload
dataIn1 = genPayload(nS,  prmLTEDLSCH.TBLenVec);
% Transport block CRC generation
tbCrcOut1 =CRCgenerator(dataIn1);
% Channel coding includes - CB segmentation, turbo coding, rate matching,
% bit selection, CB concatenation - per codeword
[data1, Kplus1, C1] = lteTbChannelCoding(tbCrcOut1, nS, prmLTEDLSCH,
prmLTEPDSCH);
%Scramble codeword
scramOut = lteScramble(data1, nS, 0, prmLTEPDSCH.maxG);
% Modulate
modOut = Modulator(scramOut, prmLTEPDSCH.modType);
%%%%%%%%%%%%%%%%%%%%%%%%%%%%%%
% MIMO transmitter based on the mode
%%%%%%%%%%%%%%%%%%%%%%%%%%%%%%%
numTx=prmLTEPDSCH.numTx;
dataIn= dataIn1;
Kplus=Kplus1;
C=C1;
Wn=complex(ones(numTx,numTx));
switch prmLTEPDSCH.txMode
   case 1 % Mode 1: Single-antenna (SIMO mode)
      PrecodeOut =modOut;

   case 2 % Mode 2: Transmit diversity
      % TD with SFBC
      PrecodeOut = TDEncode(modOut(:,1),prmLTEPDSCH.numTx);

   case 3 % Mode 3: Open-loop Spatial multiplexing
      LayerMapOut = LayerMapper(modOut, [], prmLTEPDSCH);
      % Precoding
      PrecodeOut = SpatialMuxPrecoderOpenLoop(LayerMapOut, prmLTEPDSCH);

   case 4 % Mode 4: Closed-loop Spatial multiplexing
      if  prmLTEPDSCH.numCodeWords==1
         % Layer mapping
         LayerMapOut = LayerMapper(modOut, [], prmLTEPDSCH);
      else
         dataIn2 = genPayload(nS,  prmLTEDLSCH.TBLenVec);
         tbCrcOut2 =CRCgenerator(dataIn2);
```

```
      [data2, Kplus2, C2] = lteTbChannelCoding(tbCrcOut2, nS, prmLTEDLSCH,
prmLTEPDSCH);
      scramOut2 = lteScramble(data2, nS, 0, prmLTEPDSCH.maxG);
      modOut2 = Modulator(scramOut2, prmLTEPDSCH.modType);
      % Layer mapping
      LayerMapOut = LayerMapper(modOut, modOut2, prmLTEPDSCH);
      dataIn= [dataIn1;dataIn2];
      Kplus=[Kplus1;Kplus2];
      C=[C1; C2];
   end
   % Precoding
   usedCbIdx = prmMdl.cbIdx;
   [PrecodeOut, Wn] = SpatialMuxPrecoder(LayerMapOut, prmLTEPDSCH, usedCbIdx);
end
% Generate Cell-Specific Reference (CSR) signals
numTx=prmLTEPDSCH.numTx;
csr = CSRgenerator(nS, numTx);
csr_ref=complex(zeros(2*prmLTEPDSCH.Nrb, 4, numTx));
for m=1:numTx
   csr_pre=csr(1:2*prmLTEPDSCH.Nrb,:,:,m);
   csr_ref(:,:,m)=reshape(csr_pre,2*prmLTEPDSCH.Nrb,4);
end
% Resource grid filling
txGrid = REmapper_mTx(PrecodeOut, csr_ref, nS, prmLTEPDSCH);
% OFDM transmitter
txSig = OFDMTx(txGrid, prmLTEPDSCH);
```

8.1.3   Channel Model

Channel modeling is performed by combining a MIMO fading channel with an AWGN (Additive White Gaussian Noise) channel. MIMO channels specify the relationships between signals transmitted over multiple transmit antennas and signals received at multiple receive antennas. Typical parameters of MIMO channels include the antenna configurations, multipath delay profiles, maximum Doppler shifts, and spatial correlation levels within the antennas in both the transmitter side and the receiver side. The AWGN channel is usually specified using the SNR (Signal-to-Noise Ratio) value or the noise variance. Figure 8.3 illustrates the operations performed within a fading-channel model, showing an example in which a 4×4 MIMO channel connects a single path of the transmitted signals from four transmit antennas to four receive antennas. The uncorrelated white Gaussian noise is then added to each MIMO received signal to produce the channel-modeling output signal.

8.1.4   MATLAB Model for a Channel Model

The following MATLAB function shows how channel modeling is performed by combining a multipath MIMO fading channel with an AWGN channel. First, by calling the *MIMOFadingChan* function, we generate the faded version of the transmitted signal (*rxFade*) and the

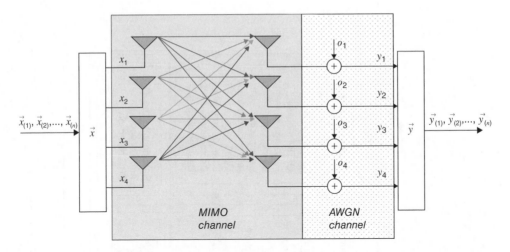

Figure 8.3   LTE downlink system model: channel model per path

corresponding channel matrix (*chPathG*). The MIMO fading channel computes the faded signal as a linear combination of multiple transmit antennas. As a result, the output signal (*rxFade*) may not have an average power (signal variance) of one. To compute the noise variance needed to execute the *AWGNChannel* function, we need to first compute the signal variance (*sigPow*) and then derive the noise variance as the difference between the signal power and the SNR value in dB.

Algorithm

MATLAB function

```
function [rxSig, chPathG, nVar] = commlteMIMO_Ch(txSig, prmLTEPDSCH, prmMdl )
%#codegen
snrdB = prmMdl.snrdB;
% MIMO Fading channel
[rxFade, chPathG] = MIMOFadingChan(txSig, prmLTEPDSCH, prmMdl);
% Add AWG noise
sigPow = 10*log10(var(rxFade));
nVar = 10.^(0.1.*(sigPow-snrdB));
rxSig =  AWGNChannel(rxFade, nVar);
```

8.1.5   Receiver Model

In the receiver, the signal processing chain is applied to the received symbols following channel modeling. At the receiver, essentially the inverse operations to those of the transmitter are performed in order to obtain a best estimate of the transmitted payload bits. Figure 8.4 illustrates the processing chain in the downlink receiver.

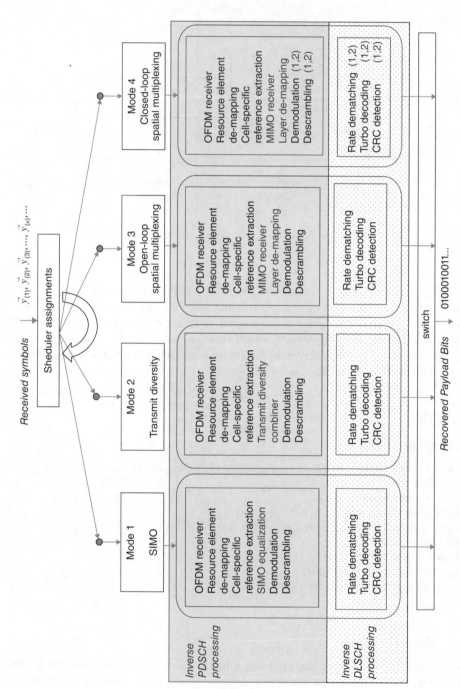

Figure 8.4   LTE downlink system model: transmission modes 1–4, receiver operations

At the receiver, the first few operations are independent of the transmission mode. These include the OFDM receiver, resource element demapping, and Cell-Specific Reference (CSR) signal extraction. Following these common operations, we reconstruct the resource grid at the receiver and extract the user data and the CSR signal from it. Based on the received CSR signals, we then estimate the channel response matrices in each subframe. The channel estimation can be based on multiple algorithms: the ideal estimation algorithm exploits a complete knowledge of channel-state information, while other algorithms are based on various forms of interpolation and averaging.

At this point, we perform the MIMO detection operations based on the scheduled transmission mode in order to recover the best estimates of the modulated symbols. In SIMO mode, receiver detection is same as frequency-domain equalization. In transmit-diversity mode, the transmit-diversity combiner operation is performed. In spatial-multiplexing modes, the MIMO receiver operations are performed in order to solve the MIMO equation for each received symbol given the estimated channel matrix and then different substreams are explicitly mapped back to a single modulated stream using the layer-demapping operation.

The difference between the open-loop (mode 3) and the closed-loop (mode 4) MIMO receivers relates to the precoder matrix. The open-loop algorithm uses different precoder matrix values for different predetection received symbols, whereas the closed-loop algorithm uses a common precoder matrix for all received symbols. After we obtain the best estimates of the modulated symbols, we perform demodulation, descrambling, channel-decoding, and CRC-detection operations in order to obtain best estimates of the payload bits at the receiver. Where a sphere decoder is used as part of the MIMO receiver algorithm, we bypass the demodulation as it is included within the sphere-decoding algorithm. The operations following MIMO detection are repeated according to the number of transmitted codewords.

8.1.6 MATLAB Model for a Receiver Model

The following MATLAB function shows the LTE downlink receiver operations for a given transmission mode used in any subframe. The function takes as input the subframe number (nS), the OFDM signal processed by the channel ($rxSig$), an estimate of the noise variance per received channel ($nVar$), the channel-path gain matrices ($chPathG$), the transmitted cell-specific reference signals (csr_ref), and the three parameter structures ($prmLTEDLSCH$, $prmLTEPDSCH$, $prmMdl$). It generates as its output a best estimate of the transport-block payload bits ($dataOut$). As described in the last section, the function first performs common OFDM receiver and demapping operations in order to recover the resource grid and estimate the channel response and then, using a MATLAB *switch-case* statement, performs MIMO receiver operations specific to the selected transmitted mode (represented by the *prmLTEPDSCH.txMode* variable). Finally, by performing common demodulation, descrambling, channel decoding, and CRC-detection operations, the function computes its output signal.

Algorithm

MATLAB function

```
function dataOut = commlteMIMO_Rx(nS, rxSig, chPathG, nVar, csr_ref, prmLTEDLSCH,
prmLTEPDSCH, prmMdl)
%#codegen
% OFDM Rx
rxGrid = OFDMRx(rxSig, prmLTEPDSCH);
% updated for numLayers -> numTx
[dataRx, csrRx, idx_data] = REdemapper_mTx(rxGrid, nS, prmLTEPDSCH);
% MIMO channel estimation
if prmMdl.chEstOn
    chEst = ChanEstimate_mTx(prmLTEPDSCH, csrRx,  csr_ref, prmMdl.chEstOn);
    hD   = ExtChResponse(chEst, idx_data, prmLTEPDSCH);
else
    idealChEst = IdChEst(prmLTEPDSCH, prmMdl, chPathG);
    hD  = ExtChResponse(idealChEst, idx_data, prmLTEPDSCH);
end
%%%%%%%%%%%%%%%%%%%%%%%%%%%%%%%%%%%%%
% MIMO Receiver based on the mode
%%%%%%%%%%%%%%%%%%%%%%%%%%%%%%%%%%%%%
dataOut=false(size(dataIn));
switch prmLTEPDSCH.txMode
    case 1
        % Based on Maximum-Combining Ratio (MCR)
        yRec = Equalizer_simo(dataRx, hD, max(nVar), prmLTEPDSCH.Eqmode);
        cwOut = yRec;

    case 2
        % Based on Transmit Diversity  with SFBC combiner
        yRec = TDCombine(dataRx, hD, prmLTEPDSCH.numTx, prmLTEPDSCH.numRx);
        cwOut = yRec;

    case 3
        yRec = MIMOReceiver_OpenLoop(dataRx, hD, prmLTEPDSCH, nVar);
        % Demap received codeword(s)
        [cwOut, ~] = LayerDemapper(yRec, prmLTEPDSCH);

    case 4
        % Based on Spatial Multiplexing
        yRec = MIMOReceiver(dataRx, hD, prmLTEPDSCH, nVar, Wn);
        % Demap received codeword(s)
        [cwOut1, cwOut2] = LayerDemapper(yRec, prmLTEPDSCH);
        if  prmLTEPDSCH.numCodeWords==1
            cwOut = cwOut1;
        else
            cwOut = [cwOut1, cwOut2];
        end
```

```
end
% Codeword processing
Len=numel(dataOut)/prmLTEPDSCH.numCodeWords;
index=1:Len;
for n = 1:prmLTEPDSCH.numCodeWords
    % Demodulation
    if prmLTEPDSCH.Eqmode == 3
        % not necessary in case of Sphere Decoding
        demodOut = cwOut(:,n);
    else
        % Demodulate
        demodOut = DemodulatorSoft(cwOut(:,n), prmLTEPDSCH.modType, max(nVar));
    end
    % Descramble received codeword
    rxCW =  Descramble(demodOut, nS, 0, prmLTEPDSCH.maxG);
    % Channel decoding includes CB segmentation, turbo decoding, rate dematching
    [decTbData, ~,~] = TbChannelDecoding(nS, rxCW, Kplus(n), C(n),  prmLT-
EDLSCH, prmLTEPDSCH);
    % Transport block CRC detection
    [dataOut(index), ~] = CRCdetector(decTbData);
    index = index +Len;
end
```

8.2   System Model in MATLAB

In this section, we showcase the MATLAB testbench (*commlteSystem*) that represents the system model for the PHY of the LTE standard. First it calls the initialization function (*commlteSystem_initialize*) to set all the relevant parameter structures (*prmLTEDLSCH*, *prmLTEPDSCH*, *prmMdl*). Then it uses a while loop to perform subframe processing by calling the MIMO transceiver function composed of the transmitter (*commlteSystem_Tx*), the channel model (*commlteSystem_Channel*), and the receiver (*commlteSystem_Rx*). Finally, it updates the Bit Error Rate (BER) and calls the visualization function to illustrate the channel response and modulation constellation before and after equalization. By comparing the transmitted and received bits, we can then compute various measures of performance based on the simulation parameters.

Algorithm

MATLAB function

```
% Script for LTE (mode 1 to 4, downlink transmission)
%
% Single or double codeword transmission for mode 4
%
clear functions
```

```
%%
commlteSystem_params;
[prmLTEPDSCH, prmLTEDLSCH, prmMdl] = commlteSystem_initialize(txMode, ...
   chanBW, contReg, modType, Eqmode,numTx, numRx,cRate,maxIter, fullDecode,
chanMdl, Doppler, corrLvl, ...
   chEstOn, numCodeWords, enPMIfback, cbIdx);
clear txMode chanBW contReg modType Eqmode numTx numRx cRate maxIter
fullDecode chanMdl Doppler corrLvl chEstOn numCodeWords enPMIfback cbIdx
%%
disp('Simulating the LTE Downlink - Modes 1 to 4');
zReport_data_rate_average(prmLTEPDSCH, prmLTEDLSCH);
hPBer = comm.ErrorRate;
%% Simulation loop
tic;
SubFrame =0;
nS = 0; % Slot number, one of [0:2:18]
Measures = zeros(3,1); %initialize BER output
while (Measures(3) < maxNumBits) && (Measures(2) < maxNumErrs)
   %% Transmitter
   [txSig, csr, dataIn] = commlteSystem_Tx(nS, prmLTEDLSCH, prmLTEPDSCH, prmMdl);
    %% Channel model
   [rxSig, chPathG, ~] = commlteSystem_Channel(txSig, snrdB, prmLTEPDSCH,  prmMdl );
    %% Receiver
    nVar=(10.^(0.1.*(-snrdB)))*ones(1,size(rxSig,2));
   [dataOut, dataRx, yRec] = commlteSystem_Rx(nS, csr, rxSig, chPathG, nVar, ...
       prmLTEDLSCH, prmLTEPDSCH, prmMdl);
   %% Calculate  bit errors
   Measures = step(hPBer, dataIn, dataOut);
   %% Visualize results
   if (visualsOn && prmLTEPDSCH.Eqmode~=3)
      zVisualize( prmLTEPDSCH, txSig, rxSig, yRec, dataRx, csr, nS);
   end;
   fprintf(1,'Subframe no. %4d ; BER = %g \r', SubFrame, Measures(1));
   %% Update subframe number
   nS = nS + 2; if nS > 19, nS = mod(nS, 20); end;
   SubFrame =SubFrame +1;
end
toc;
```

8.3   Quantitative Assessments

In this section we look at performance from various different perspectives. By executing the system model in MATLAB with different simulation parameters we can assess the performance of the LTE standard. First we look at performance as a function of transmission mode. Then, given a particular transmission mode, we observe the effect of varying channel models. Next, we validate the proper implementation of MIMO–OFDM equalizers by looking at the BER as a function of the link SNR. Then we verify how the link delay spread and the Cyclic

Prefix (CP) of the OFDM transmitter relate to the overall performance. Finally, we observe the effects of receiver operations including the channel estimation and MIMO receiver algorithms on the overall performance.

8.3.1   Effects of Transmission Modes

In this experiment, we examine the BER performance as a function of the transmission mode. We iterate through nine test cases, where each test case is characterized by a transmission mode and a valid antenna configuration. For example, in the SIMO case (mode 1) we examine three valid antenna configurations (1×1, 1×2, and 1×4). For transmit diversity (mode 2) and spatial multiplexing (modes 3 and 4), we examine only antenna configurations of 2×2 and 4×4. The experiment assigns a common parameter set to the transmitter and receiver and is performed twice, once for a low-distortion channel model and once for a channel with a lot of distortion. The following MATLAB scripts show how easy it is to sweep through these configurations and perform the experiment.

Algorithm

MATLAB scripts

```
clear all                                        clear all
TestCases=[...                                   TestCases=[...
   1,1,1;                                            1,1,1;
   1,1,2;                                            1,1,2;
   1,1,4;                                            1,1,4;
   2,2,2;                                            2,2,2;
   2,4,4;                                            2,4,4;
   3,2,2;                                            3,2,2;
   3,4,4;                                            3,4,4;
   4,2,2;                                            4,2,2;
   4,4,4];                                           4,4,4];
NumCases=size(TestCases,1);                      NumCases=size(TestCases,1);
Ber_vec_Experiment1=zeros(NumCases,1);           Ber_vec_Experiment2=zeros(NumCases,1);
for Experiment = 1:NumCases                      for Experiment = 1:NumCases
   txMode      = TestCases(Experiment,1);           txMode      = TestCases(Experiment,1);
   % Transmisson mode one of {1, 2, 3, 4}           % Transmisson mode one of {1, 2, 3, 4}
   numTx       = TestCases(Experiment,2);           numTx       = TestCases(Experiment,2);
   % Number of transmit antennas                    % Number of transmit antennas
   numRx       = TestCases(Experiment,3);           numRx       = TestCases(Experiment,3);
   % Number of receive antennas                     % Number of receive antennas
```

```
   copyfile('commlteSystem_params               copyfile('commlteSystem_params_clean.m',
   _distorted.m','commlteSystem_params.m');        'commlteSystem_params.m');
   commlteSystemModel;                             commlteSystemModel;
```

```
   Ber_vec_Experiment1(Experiment)=             Ber_vec_Experiment2(Experiment)=
   Measures(1);                                    Measures(1);
end                                              end
```

The common transmitter and receiver parameters certainly may vary but in this experiment we have chosen the following parameters: a 16QAM modulation scheme, turbo coding with a 1/2 rate, a bandwidth of 10 MHz, two PDCCH (Physical Downlink Control Channel) symbols per subframe, and a single codeword. Receiver parameters include the following: turbo decoder with four maximum decoding iterations with early termination decoding, no feedback for the precoder matrix, channel estimation based on interpolation, and the MMSE (Minimum Mean Squared Error) MIMO receiver.

The distorted channel uses flat fading with a 70 Hz maximum Doppler shift, antennas with high spatial correlations, and an SNR value of 5 dB. Table 8.1 shows BER and data-rate measures as a function of transmission mode and antenna configuration in noisy channel conditions. The clean channel condition is characterized by a frequency-selective channel model with antennas of low spatial correlation, a maximum Doppler shift of zero, and an SNR value of 15 dB. Table 8.2 shows BER and data-rate measures as a function of transmission mode and antenna configuration in low-distortion channel conditions.

Table 8.1   BER performance and data rate as a function of transmission mode: high-distortion channel profile

Performance results	Antenna configuration	Data rate (Mbps)	BER
Mode 1	1×1	13.88	0.2123
	1×2	13.88	0.0098
	1×4	13.88	0.0004
Mode 2	2×2	12.96	0.0075
	4×4	12.81	0.0013
Mode 3	2×2	25.46	0.3392
	4×4	50.46	0.4067
Mode 4	2×2	25.46	0.2621
	4×4	50.46	0.4167

Table 8.2   BER performance and data rate as a function of transmission mode: low-distortion channel profile

Performance results	Antenna configuration	Data rate (Mbps)	BER
Mode 1	1×1	13.88	0.0032
	1×2	13.88	$4.2e^{-05}$
	1×4	13.88	0.0
Mode 2	2×2	12.96	0.0
	4×4	12.81	0.0
Mode 3	2×2	25.46	0.1341
	4×4	50.46	0.2748
Mode 4	2×2	25.46	0.0967
	4×4	50.46	0.1948

Based on the results, we can make the following observations:

- The performance in each mode is consistently better in a clean channel than in a noisy channel.
- In the SIMO case, performance improves as a result of diversity when multiple receive antennas are employed. The BER profile matches what is expected from Maximum Ratio Combining (MRC) [1].
- Transmit diversity improves performance and is comparable to receive diversity. The theoretical bound for TD performance presented in [1] matches our results.
- In Spatial Multiple (SM) modes 3 and 4, performance seems rather low under both channel conditions. Since SM modes are responsible for highest data rates, the question is what parameter set results in acceptable BER performance in spatial multiplexing modes. The next section attempts to answer this question.

8.3.2   BER as a Function of SNR

In this section we will develop criteria to validate the performance results of our LTE PHY simulator. Specifically, we want to find out whether the SM results satisfy the minimum requirements of the standard. The LTE standard is a MIMO–OFDM system. It combats the effects of multipath fading and the resulting intersymbol interference by using frequency-domain equalization. As we examine the transmitter system model, we realize that the combination of MIMO and OFDM techniques operates on the modulated streams following coding and scrambling operations. In the receiver, by inverting the OFDM and MIMO operations first, we recover the received modulated substreams and the MIMO receiver detection operation recovers the best estimate of the modulated stream at the receiver.

This means that by looking at the constellation of received signals (Figure 8.5) before MIMO detection we can readily see the effects of multipath fading in terms of rotations and attenuations in each of the modulated symbols. If the MIMO and OFDM operations are implemented correctly and do what they are designed to do, the MIMO detector inverts and compensates for the effects of fading channel. A good MIMO detector computes a channel-aware equalizer by first estimating the channel response and then providing the equalization as a counter-measure to all the rotations and attenuations incurred in each symbol. After applying an effective equalizer, the constellation diagram of the recovered symbols following MIMO detection resembles that of the transmitted symbols with additive Gaussian noise around them. As a result, although the channel involves multipath fading following equalization, the effective channel can be approximated by an AWGN channel.

In Chapter 4, we showed that the turbo-coded and modulated symbols transmitted on an AWGN channel are characterized by BER curves that show sharp improvements following a cutoff SNR value. Since after successful MIMO detection the effective channel is an AWGN channel in the LTE system, the system will have the same pattern of BER curve as a function of the SNR.

Figure 8.6 shows BER curves from a system simulation model executed for a range of SNR values. The model operates in transmission mode 4, using a frequency-selective channel with all the parameters of the lower-distortion channel model specified in the last section. Note how they reflect the same structure and pattern, as if the AWGN channel were the only channel model present. This means that the frequency-domain equalization furnished by the

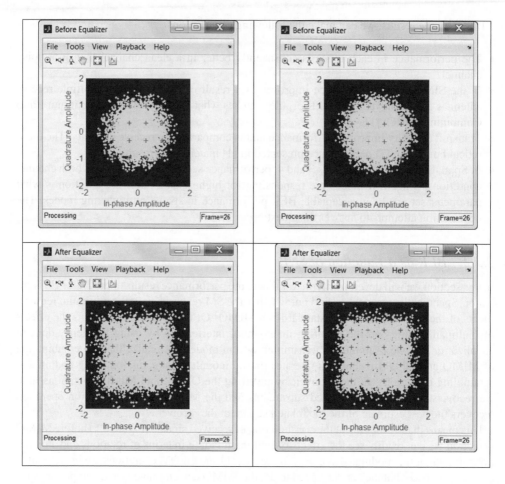

Figure 8.5   Constellation of received signals: before and after MIMO detection

MIMO detection effectively compensates for the effects of multipath fading. As we can see in Figure 8.6, the results are quite prominent for the QPSK and 16QAM modulation schemes. In the case of 64QAM modulation, if we use an approximate channel-estimation technique we will have a less prominent drop in the BER curve.

8.3.3   *Effects of Channel-Estimation Techniques*

In this section we examine BER performance as a function of channel-estimation methods. The experiment provides similar results in different transmission modes. For example, we run the system model in transmission mode 4 with a frequency-selective channel and with all parameters of the lower-distortion channel model previously specified. The experiment

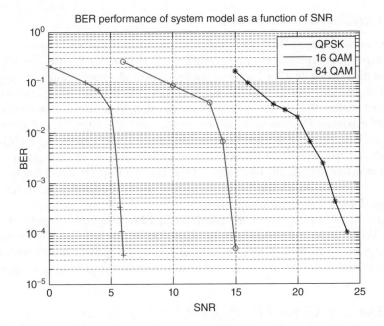

Figure 8.6 BER as a function of an SNR–LTE simulation model with frequency-selective fading channel: QPSK (left), 16QAM (middle), 64QAM (right)

is performed by changing one of four channel-estimation techniques, simply by changing the *prmMdl.chEst* parameter in the *commlteSystem_params* MATLAB script. The results are summarized in Table 8.3. As expected, the ideal channel estimator provides the best BER performance. The estimation based on interpolation allows for the most variation within slots and subframes and thus has a higher performance in an MMSE error-minimization context. When averaging in time across slots or subframes, we smooth out the spectrum. These averaging methods do not therefore improve the BER performance results but may provide the continuity and smoothness needed for better perceptual results. We will verify this effect in the next section, where we process actual voice data through our simulation model and focus on perceived voice quality.

Table 8.3 Sample of BER results as a function of channel-estimation technique

Channel-estimation method	BER
Ideal estimation	0.0001
Interpolation-based	0.0056
Averaging over each slot	0.0076
Averaging over a subframe	0.0086

8.3.4   Effects of Channel Models

In this section we examine BER performance as a function of different channel models. The experiment provides similar results in different transmission modes. We use transmission mode 2 (transmit diversity) with a 2×2 antenna configuration and iterate through multiple channel models. This includes user-defined flat and frequency-selective fading models and all 3GPP LTE channel models. We also sweep through various mobility measures expressed as either 0, 5, or 70 Hz maximum Doppler shifts and iterate through different profiles of antenna spatial correlations. Keeping the link SNR to a constant value of 13 dB and using a 16QAM modulation scheme, the experiment iterates through channel models by changing the *chanMdl* parameter structure, specifically by changing the maximum Doppler shift (*chanMdl.Doppler*) and the spatial correlation (*chanMdl.corrLvl*) parameters in the *commlteSystem_params* MATLAB script. Table 8.4 summarizes the results.

As we increase the extent of the noise profile (using EPA, Extended Pedestrian A, and EVA, Extended Vehicular A), the performance becomes consistently reduced. Higher mobility has an adverse effect on the performance. Also, by increasing the spatial correlations, we increase the chance of rank deficiency in the channel matrix, with detrimental effects on quality.

8.3.5   Effects of Channel Delay Spread and Cyclic Prefix

As discussed earlier, the CP length is an important parameter in the OFDM transmission methodology. If path delays in a channel model correspond to values greater than the CP length, the OFDM transmission cannot maintain orthogonality between subcarriers in the receiver. This will have a detrimental effect on the BER quality of the transceiver.

Table 8.4   BER performance as a function of channel model

Performance results	Maximum Doppler shift (Hz)	Spatial correlation	BER
Flat fading	0	Low	0.0
	5	Medium	$1.3821e^{-02}$
	70	High	$1.1538e^{-02}$
Frequency-selective fading (user-defined)	0	Low	0.0
	5	Medium	$8.0994e^{-06}$
	70	High	$3.4419e^{-03}$
EPA	5	Low	0.0
	5	Medium	$1.5399e^{-03}$
	5	High	$6.6134e^{-03}$
EVA (5 Hz)	5	Low	0.0
	5	Medium	$4.6661e^{-07}$
	5	High	$2.0997e^{-06}$
EVA (70 Hz)	70	Low	0.0
	70	Medium	$1.1854e^{-07}$
	70	High	$7.0629e^{-04}$

Table 8.5   Mapping of CP lengths in samples to the channel bandwidth

Channel bandwidth	Number of resource blocks	Smallest CP length (samples)	Channel sample rate (MHz)	Maximum supported delay (µs)
1.4	6	9	1.92	4.6875
3	15	18	3.84	4.6875
5	25	36	7.68	4.6875
10	50	72	15.36	4.6875
15	75	108	23.04	4.6875
20	100	144	30.72	4.6875

The LTE standard allows the scheduler to provide both normal and extended CP lengths. In propagation environments with high delay spreads, we can switch to transmissions that employ extended CP lengths, which can help improve performance.

As shown in Table 8.5, all transmission bandwidths with normal CP lengths have a constant maximum delay spread of about 4.6875 µs. We verify in this section that when employing user-defined frequency-selective channel models, setting the path delays to different values relative to the CP lengths can have a significant impact on BER performance. We implement this experiment by altering the way in which the path delays of the user-specified frequency-selective channel models are specified. In the following MATLAB code, we determine the path delays as equally-spaced samples between 0 and the maximum delay value.

Algorithm

MATLAB code segment in the prmsMdl function

```
% Channel parameters
prmMdl.PathDelays = floor(linespace(0,DelaySpread,5))*(1/chanSRate);
```

In the first case, we use as delay spread a value within the CP length range. The following MATLAB code segment initializes the delay spread as just less than the value of the CP length.

Algorithm

MATLAB code segment in initialization function – Low delay spread length

```
% Channel parameters
chanSRate  = prmLTEPDSCH.chanSRate;
DelaySpread = prmLTEPDSCH.cpLenR - 2;
 prmMdl = prmsMdl(chanSRate,  DelaySpread, chanMdl, Doppler, numTx, numRx, ...
   corrLvl, chEstOn, enPMlfback, cbIdx);
```

In the second case, we use as delay spread a value outside the CP length range. The following MATLAB code segment initializes the delay spread as twice the value of the CP length.

Algorithm

MATLAB code segment in initialization function – High delay spread length

```
% Channel parameters
chanSRate  = prmLTEPDSCH.chanSRate;
DelaySpread = 2* prmLTEPDSCH.cpLenR;
 prmMdl = prmsMdl(chanSRate,  DelaySpread, chanMdl, Doppler, numTx, numRx, ...
   corrLvl, chEstOn, enPMIfback, cbIdx);
```

The experiment provides similar results in different transmission modes. We have used transmission mode 1 with a 1×2 antenna configuration. We applied a constant link SNR value of 15 dB and used a 16QAM modulation scheme. The results are summarized in Table 8.6. We observe that having delay spread values that occasionally exceed the maximum CP length can result in severe performance degradations.

8.3.6   Effects of MIMO Receiver Algorithms

In this section, we examine BER performance as a function of different MIMO receiver algorithms. Simply by changing the equalization mode (represented by the *prmLTEPDSCH. Eqmode* parameter) to a value of 1, 2, or 3, we can examine a Zero Forcing (ZF), MMSE, and sphere-decoder algorithm, respectively. Table 8.7 summarizes the results.

Although the ZF algorithm provides the simplest implementation, by ignoring the noise power at the receiver, it results in the lowest BER performance. The BER performance of the MMSE algorithm is better than its ZF performance. It is formulated to essentially invert the channel matrix while taking into account the power of the noise. However, the best

Table 8.6   Effect of delay spread range on BER performance

Delay spread value	BER
Low	0.00019
High	0.02440

Table 8.7   Effect of a MIMO detection algorithm on BER performance

MIMO detection method	BER
ZF algorithm	0.0001
MMSE algorithm	0.0056
Soft-sphere decoding	0.0076

performance is furnished by the sphere decoder, which uses maximum-likelihood decoding to optimize for the modulation symbols based on their symbol mapping. A sphere decoder is an algorithm of relatively high computational complexity and the time it takes to process a sphere-decoder receiver can be substantially greater than for an MMSE receiver. As such, the choice between a MIMO receiver based on MMSE and on a sphere decoder represents a classical tradeoff between complexity and performance.

8.4 Throughput Analysis

LTE-standard documents provide not only the transmitter specifications but also channel conditions for testing and the minimum performance criteria needed to qualify for standard compliance. For example, the standard document TS 36.101 provides all minimum performance requirements for downlink transmission. An excerpt from this document is illustrated in Figure 8.7. As an example, a single throughput requirement for the SIMO transmission mode is captured as a set of test cases in which various parameters are given as inputs and the expected throughput is given as output. Input specifications include the bandwidth, reference channel, propagation (channel mode), antenna spatial correlation matrix, and reference SNR values. Throughput is defined as the average data rate for which successful transmission occurs. Maximum throughput corresponds to the case where no input block with errors is received. The relative throughput is the fraction of successful transmission with respect to maximum throughput. For example, test case 1, corresponding to QPSK (Quadrature Phase Shift Keying) modulation, expects a 70% relative throughput for an SNR value of -1 dB when the EVA channel model with a Doppler shift of 5 Hz is used with low spatial antenna correlations and a transmission mode of 1 is used with a 1×2 antenna configuration and a bandwidth of 10 MHz.

We have modified our receiver and the system MATLAB functions for computation of the throughput. In the receiver, as the last processing step we compute CRC detection. When any error is found in CRC detection, the block of output is deemed to have been received in error. By excluding all erroneous blocks from the total number of blocks processed we can find the

Test number	Bandwidth (MHz)	Reference channels	OCNG pattern	Propagation condition	Correlation matrix & Antenna configuration	Fraction of Max. Throughput (%)	SNR (dB)	UE category
1	10	R.2 FDD	OP.1 FDD	EVA5	1x2 Low	70	−1.0	1–8
1A	2x10	R.2 FDD	OP.1 FDD (Note 1)	EVA5	1x2 Low	70	−1.1	3–8
2	10	R.2 FDD	OP.1 FDD	ETU70	1x2 Low	70	−0.4	1–8
3	10	R.2 FDD	OP.1 FDD	ETU300	1x2 Low	70	0.0	1–8
4	10	R.2 FDD	OP.1 FDD	HST	1x2 Low	70	−2.4	1–8
5	1.4	R.4 FDD	OP.1 FDD	EVA5	1x2 Low	70	0.0	1–8
...	...	...	...	...	...	...	...	...

Figure 8.7 Test cases for LTE downlink compliance: TS 36.101 excerpt on minimum requirement testing [2]. Courtesy of 3GPP documentation

relative throughput as the ratio of correctly received blocks to total received blocks. The following MATLAB function uses this definition to compute and display the relative throughput.

Algorithm

MATLAB function

```
function Throughput=getThroughput( ber, CbFlag, SubFrame)
persistent ErrorBlk
if isempty(ErrorBlk)
    ErrorBlk=0;
end
ErrorBlk = ErrorBlk + CbFlag;
Throughput=1-(ErrorBlk/SubFrame);
fprintf(1,'Subframe %4d ; BER = %6.4f ; ErrorFrame = %4d ; Throughput = %4.2f \r', ...
    SubFrame, ber, ErrorBlk, Throughput );
end
```

8.5   System Model in Simulink

So far we have presented MATLAB algorithms and testbenches to simulate the PHY of the LTE standard. In this section we show how expressing the same system model as a Simulink model facilitates our design process. Simulink models naturally provide a simulation testbench, which enables us to focus on algorithms and their updates rather than having to maintain the testbench. Let us take a look at our MATLAB system model in order to distinguish its algorithmic portions (i.e., code related to system processing) from its testbench components (i.e., code related to maintaining a simulation framework).

Algorithm

MATLAB function

```
clear functions
commlteSystem_params;
[prmLTEPDSCH, prmLTEDLSCH, prmMdl] = commlteSystem_initialize(txMode, ...
    chanBW, contReg, modType, Eqmode,numTx, numRx,cRate,maxIter, fullDecode,
chanMdl, Doppler, corrLvl, ...
    chEstOn, numCodeWords, enPMIfback, cbIdx);
clear txMode chanBW contReg modType Eqmode numTx numRx cRate maxIter
fullDecode chanMdl Doppler corrLvl chEstOn numCodeWords enPMIfback cbIdx
%%
disp('Simulating the LTE Downlink - Modes 1 to 4');
zReport_data_rate_average(prmLTEPDSCH, prmLTEDLSCH);
hPBer = comm.ErrorRate;
```

```
%% Simulation loop
tic;
SubFrame =0;
nS = 0; % Slot number, one of [0:2:18]
Measures = zeros(3,1); %initialize BER output
while (Measures(3) < maxNumBits) && (Measures(2) < maxNumErrs)

    %% Transmitter
    [txSig, csr, dataIn] = commlteSystem_Tx(nS, prmLTEDLSCH, prmLTEPDSCH, prmMdl);
     %% Channel model
    [rxSig, chPathG, ~] = commlteSystem_Channel(txSig, snrdB, prmLTEPDSCH, prmMdl );
    %% Receiver
    nVar=(10.^(0.1.*(-snrdB)))*ones(1,size(rxSig,2));
    [dataOut, dataRx, yRec] = commlteSystem_Rx(nS, csr, rxSig, chPathG, nVar, ...
        prmLTEDLSCH, prmLTEPDSCH, prmMdl);
-------------------------------------------------------------

    %% Calculate  bit errors
    Measures = step(hPBer, dataIn, dataOut);
    %% Visualize results
    if (visualsOn && prmLTEPDSCH.Eqmode~=3)
      zVisualize( prmLTEPDSCH, txSig, rxSig, yRec, dataRx, csr, nS);
    end;
    fprintf(1,'Subframe no. %4d ; BER = %g \r', SubFrame, Measures(1));
    %% Update subframe number
    nS = nS + 2; if nS > 19, nS = mod(nS, 20); end;
    SubFrame =SubFrame +1;
end
-------------------------------------------------------------
toc;
```

We can identify three main portions in the MATLAB system model:

1. **Initialization**: Operations that set various system parameters and are performed once before the processing loop starts
2. **Scheduling**: A while loop that schedules iterative subframe processing and the extra operations needed to update the conditions for while-loop execution
3. **In-loop processing**: Subframe processing on input bits in order to perform transmitter, channel-modeling, and receiver operations, as well as extra code to compare the input and output bits and visualize various signals.

When we express the same model in Simulink, we essentially focus on modeling the in-loop processing. The scheduling is handled by Simulink. With its time-based simulation engine, Simulink iterates through samples or frames of data until the specified simulation time or a stopping condition has been reached. Since MATLAB and Simulink share data in MATLAB workspace, we can either perform the initialization commands manually before simulating the Simulink model or set the Simulink model up to perform the initialization code when the model opens or at any time before the simulation starts.

In the following sections, we go through a step-by-step process of expressing the transceiver in Simulink. First, we create a Simulink model and integrate the MATLAB algorithms developed so far as distinct blocks within it. Then we set the initialization routines up automatically and make parameterizations easier by creating a parameter dialog.

8.5.1   Building a Simulink Model

We can build the Simulink model expressing the LTE transceiver system by using blocks from the Simulink library. The Simulink library can be accessed by clicking on the Simulink Library icon in the MATLAB environment, as illustrated in Figure 8.8. Within the Simulink library browser, various collections of blocks from Simulink and other MathWorks products can be found. We will be using mostly blocks from Simulink, the DSP System Toolbox, and the Communications System Toolbox. As an example, Figure 8.9 illustrates a library of Simulink blocks called the User-Defined Functions.

To start building a new Simulink model, we can use the Menu bar of the Simulink library browser and make the following selections: File → New → Model. An empty Simulink model will appear, as shown in Figure 8.10.

8.5.2   Integrating MATLAB Algorithms in Simulink

Next, we populate this system with blocks step by step in order to represent the LTE transceiver in Simulink using previously developed MATLAB algorithms.

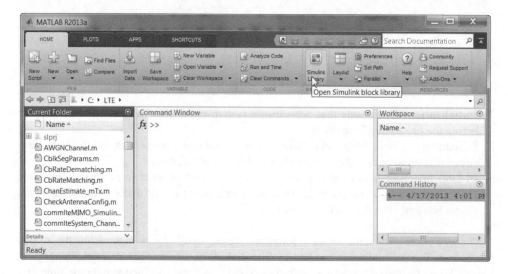

Figure 8.8   Accessing Simulink library from within the MATLAB environment

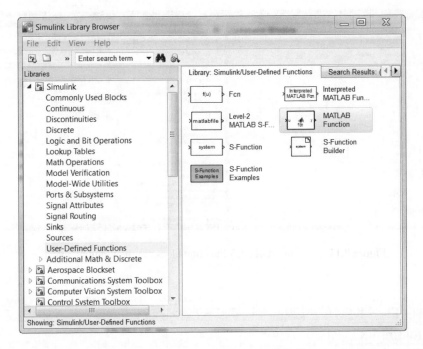

Figure 8.9   Simulink block libraries, organized as a main Simulink library and other product libraries

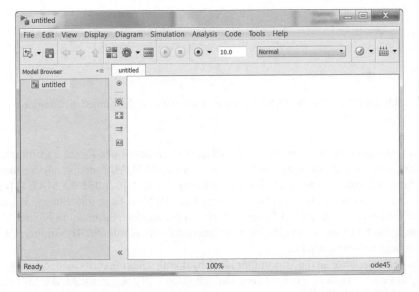

Figure 8.10   Building a new Simulink model for an LTE transceiver: start with an empty model

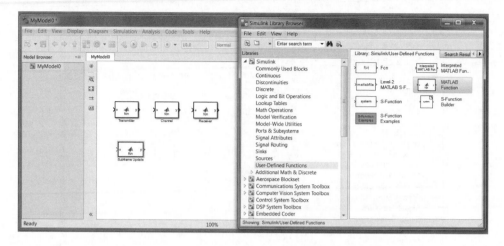

Figure 8.11   Adding MATLAB Function blocks to the Simulink model

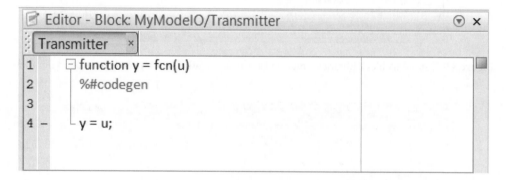

Figure 8.12   The MATLAB Function block, with a default function definition

Since we are planning to reuse the MATLAB algorithms developed for the transmitter, channel model, and receiver, we need blocks that can turn a MATLAB function into a component
of the Simulink model. The block that can perform this task is called the MATLAB Function block and can be found under the Simulink User-Defined Functions library. We need to
add four copies of the MATLAB Function block to our model. As shown in Figure 8.11, we
can change the block names to identify their functionalities as follows: Transmitter, Channel,
Receiver, and Subframe Update.

As the next step we open the Transmitter block by double clicking on its icon. As we open any
MATLAB Function block, a default function definition will open in the MATLAB Editor, as
illustrated in Figure 8.12. Now we can modify the default function definition to implement our
function. As we define the function, the input arguments of the function become the input ports
of the block and the output arguments of the function become the output ports of the block.

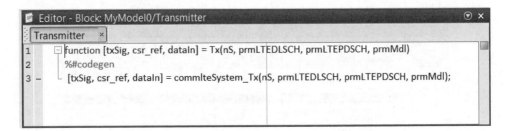

Figure 8.13   MATLAB Function block: updating function definition to implement the transmitter

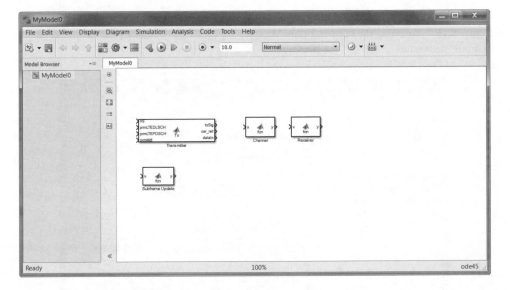

Figure 8.14   Simulink model with an updated Transmitter block definition without any parameters

At this point we have two choices. We can either copy the body of our transmitter function (*commlteSystem_Tx*) and paste it under the function definition line or we can make a function call to our transmitter function, as illustrated by Figure 8.13, in order to relate the input variables of the function definition to its output variables.

After saving this function, we can go back to the parent model by clicking on the Go to Diagram icon. As shown in Figure 8.14, we will see that the Transmitter block is transformed to reflect the new function definition. At this stage, by default all function arguments are mapped to corresponding input and output ports.

We would like to turn all the relevant parameter structures (*prmLTEDLSCH*, *prmL-TEPDSCH*, *prmMdl*) into parameters of the Simulink model. These parameter structures will have constant values during the simulation and will be accessed by the Simulink model as three variables in the MATLAB workspace. We open the Transmitter block and in the MATLAB Editor click on the Edit Data icon above the Editor, as illustrated in Figure 8.15.

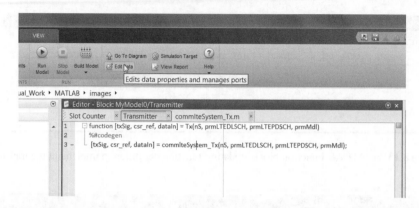

Figure 8.15   Accessing the Ports and Data Manager to express transmitter LTE structure arguments as model parameters

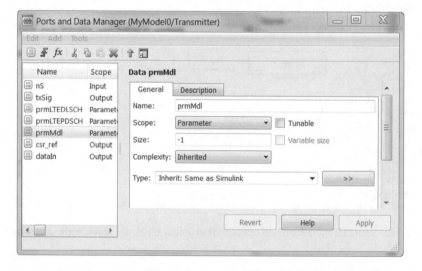

Figure 8.16   Setting LTE parameter structures as nontunable parameters

A dialog called Ports and Data Manager appears, which enables us to edit data properties and manage ports and parameters. This dialog is shown in Figure 8.16. Next, we click on each of the port names (*prmLTEDLSCH*, *prmLTEPDSCH*, *prmMdl*), change the Scope property into a Parameter, uncheck the Tunable checkbox to mark the parameter as a constant, and click on the Apply button.

We then repeat these operations for the Channel and the Receiver MATLAB Function blocks. Figure 8.17 illustrates simple MATLAB function calls made inside the Channel and the Receiver MATLAB Function blocks, which allow us to directly integrate algorithms developed previously in MATLAB as blocks in Simulink.

At this stage we can connect the output signal of the Transmitter block (*txSig*) to the input port of the Channel block. We also connect the three output signals of the Channel block (*rxSig*, *chPathG*, *nVar*) to the input ports of the Receiver block, as illustrated in Figure 8.18.

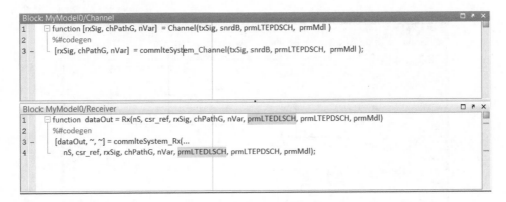

Figure 8.17 Calling Channel Model and Receiver functions inside corresponding MATLAB Function blocks

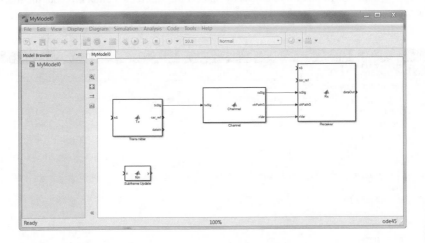

Figure 8.18 Connecting Transmitter, Channel Model, and Receiver blocks

Now we have to update the subframe number (*nS*). The subframe number is the common input to both the Transmitter and the Receiver blocks. To avoid excessive wiring in the model, we use the GoTo and the From blocks from the Simulink Signal Routing library. The output of the Subframe Update MATLAB Function block is the slot number of the current frame. We connect this output signal to a GoTo block and, after clicking on the block, assign this signal an identifier, otherwise known as the tag of the block. Now, any From block in the model bearing the same tag can route the signal to multiple blocks in the model. As illustrated in Figure 8.19, we have used two *From* blocks to connect the subframe number to both the Transmitter and the Receiver blocks.

When we double click the Subframe Update MATLAB Function block we find the MATLAB function illustrated in Figure 8.20. This is the same code used previously in our MATLAB testbenches to update the subframe number. We have used a persistent variable here to implement the subframe update operation as a counter that resets its value in every 10 ms frame. Now

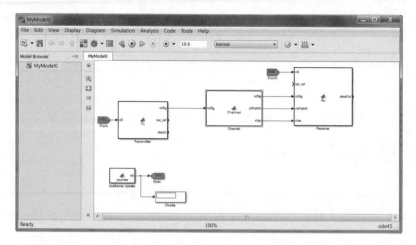

Figure 8.19 Using *GoTo* and *From* blocks to connect the subframe number to both the transmitter and the receiver

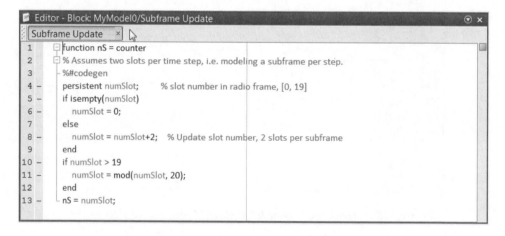

Figure 8.20 Subframe Update MATLAB function block implemented as a counter

we use additional GoTo and From blocks to finalize block connectivity in the model. As illustrated in Figure 8.21, the pair of GoTo and the From blocks is identified by a selected tag (*csr*) and a purple background color. This ensures that the same cell-specific reference signal (pilot) computed in the transmitter is also used in the receiver in the same subframe. We also use two other GoTo blocks to collect the transmitted input bit stream (*dataIn*) and the receiver output bit stream (*dataOut*). We use the tags *Input* and *Output* for the signals *dataIn* and *dataOut*, respectively, in the GoTo blocks, and use the same background color. To compute the BER of the system, we use the Error Rate Calculation block from the *CommSinks* Simulink library of the Communications System Toolbox. Using two From blocks with the tags *Input* and *Output*, we route the transmitted input stream and receiver output stream to the Error Rate Calculation block, as illustrated in Figure 8.22. The Error Rate Calculation block compares the decoded

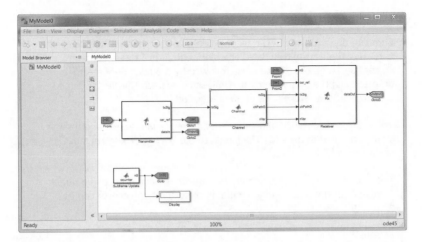

Figure 8.21   Improved block port connectivities using GoTo and From block pairs

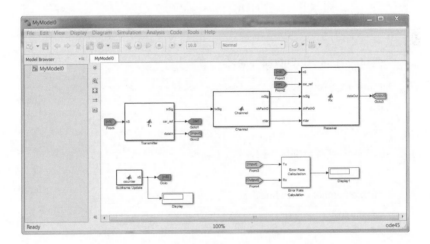

Figure 8.22   Completing the in-loop processing specification in Simulink with BER calculation

bits with the original source bits per subframe and dynamically updates the BER measure throughout the simulation. The output of this block is a three-element vector containing the BER, the number of error bits observed, and the number of bits processed.

The model checks the *Stop Simulation* parameter of the Error Rate Calculation block in order to control the duration of the simulation (Figure 8.23). The simulation stops upon detection of the target value for whichever of the following two parameters comes first: the maximum number of errors (specified by the *maxNumErrs* parameter) or the maximum number of bits (specified by the *maxNumBits* parameter).

At this point we have completed the in-loop processing specification in our Simulink model. The next steps include initializing the model with LTE parameter structures and running the Simulink model. Simulink model parameters can be initialized in multiple ways, but we will

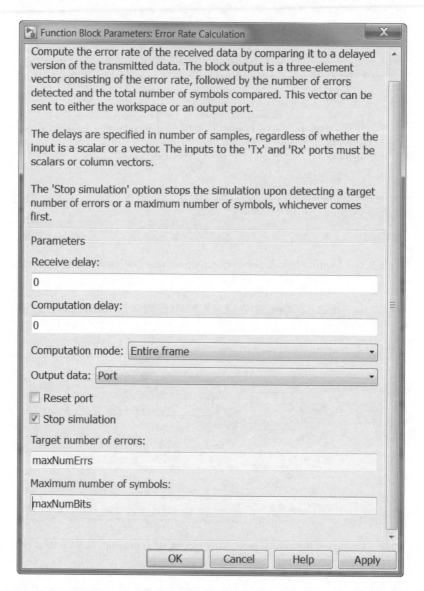

Figure 8.23   The Error Rate Calculation block controlling the duration of the simulation

look at two here: (i) setting model properties as the model is opened and (ii) using a mask
subsystem to provide a parameter dialog.

8.5.3   Parameter Initialization

Some operations in a system model get executed only once before the simulation loop starts.
Such operations represent system initialization. All the operations specified so far in our
Simulink model are part of the processing loop and are repeated in simulation iterations. One

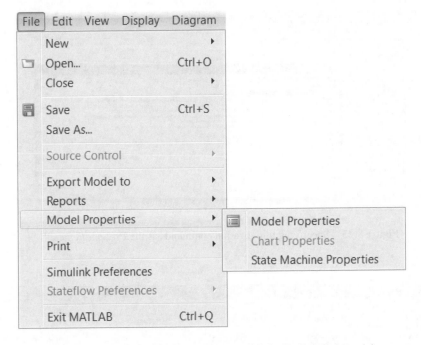

Figure 8.24   Accessing the Model Properties of a Simulink model

way of specifying initialization operations in Simulink is to use the Model Properties and in particular the Callback functions.

As illustrated in Figure 8.24, we can access the Model Properties by using the Menu bar of our model and making the following selections: File → Model Properties → Model Properties.

As we open the Model Properties dialog, we find multiple Model Callbacks under the Callbacks tab. Associated with each type of Model Callback, we find an edit box where MATLAB functions or commands can be executed. The type of callback relates to each stage of simulation. For example, in Figure 8.25 we have selected the *PreLoadFcn* callback. This means initialization functions, identical to the ones executed before the while loop in our MATLAB model, will execute as we load (or open) the Simulink model.

At this point, the Simulink model is specified both for initialization operations and for in-loop processing. Initialization routines are performed when we open the model and the simulation starts when we click on the Run button. The final step is to specify the Simulink solver and, if need be, set the sample time for the simulation. We can access Simulink solvers by using the Menu bar of our model and making the following selections: Simulation → Model Configuration Parameters → Solvers. Since we are performing a digital baseband simulation of a communications system, the simulation uses discrete-time sampling. Therefore, as illustrated in Figure 8.26, under the Solver Options properties, we should use Fixed-Step as the Type and Discrete (No Continuous State) as the Solver. These options are typical for most Simulink models simulating DSP or Communications System Models with no analog or mixed-signal components.

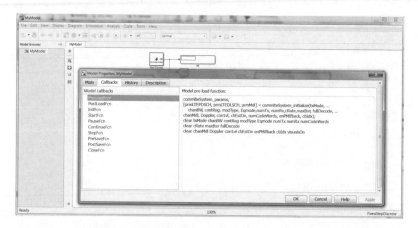

Figure 8.25 Specifying initialization commands as the *PreLoadFcn* callback

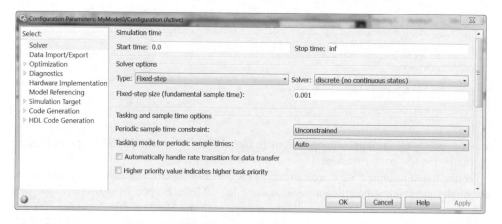

Figure 8.26 Setting Solver Options, Sample Time, and Stop Time for the simulation model

In addition, since we stop the simulation based on parameters of the Error Rate Calculation block, we specify any value as the Stop Time in this dialog. Here we select a maximum value of infinity (specified by the *inf* parameter). Furthermore, since the unit of processing is a subframe with a duration of 1 ms, in the edit box for the property Fixed-Step Size (Fundamental Sample Time) we set a value of 0.001 seconds.

Now the specification phase of our Simulink model is complete and we can run the model to test whether (i) the initialization routines performed as *CallBacks* are implemented correctly and (ii) the transceiver works as intended.

We save the model and then close it. After opening the model again, all three LTE parameter structures (*prmLTEDLSCH*, *prmLTEPDSCH*, *prmMdl*) and three additional parameters (*snrdB*, *maxNumBits*, *maxNumErrs*) are automatically generated in the MATLAB workspace. This verifies the proper operations of our *CallBack* routines, as illustrated in Figure 8.27.

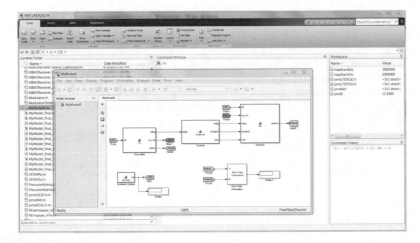

Figure 8.27 Opening the model and automatically generating simulation parameters using *CallBacks*

8.5.4   *Running the Simulation*

Through simulation, we can test the proper operation of the transceiver. We execute the simulation by clicking the Run button on the Model Editor displaying the model. Running the simulation causes the Simulink engine to convert the model to an executable form, in a process known as model compilation. As its first compilation step, the Simulink engine checks for consistency in the model by evaluating the model's block parameter expressions, determining attributes of all signals and verifying that each block can accept the signals connected to its inputs.

Since our model is composed of MATLAB Function blocks, at this stage the Simulink engine first converts the MATLAB code inside the MATLAB Function blocks to C code and then compiles the generated C code to create an executable for each MATLAB Function block of the model. If any MATLAB Function blocks contain MATLAB codes that are not supported by code generation, we have to modify them. Finally, in the linking phase data memory is allocated for each signal and all compiled executable are linked together and made ready for execution.

Simulation errors may occur, either at the model compilation time or at run time. In case of simulation errors, Simulink halts the simulation, opens the component that caused the error, and displays pertinent information regarding the error in the Simulation Diagnostics Viewer. An example is illustrated in Figure 8.28, which shows how the Simulink engine accurately checks the consistency of the port connection between blocks in the model; the Simulink engine has inferred that the subframe number (nS) signal affects the sizes of transmitted and received bit (*dataIn* and *dataOut*) signals. So in each subframe, the size of these two signals may change. However, by default, if we do not specify the sizes of any signal in any MATLAB Function block, the signal is assumed to be of a constant size. This "variable-size" nature of the two signals is the subject of the error message shown in Figure 8.28.

To fix this compilation problem, we need to mark both the *dataIn* and *dataOut* signals as variable-size signals and specify their maximum size. We need to double click on both the

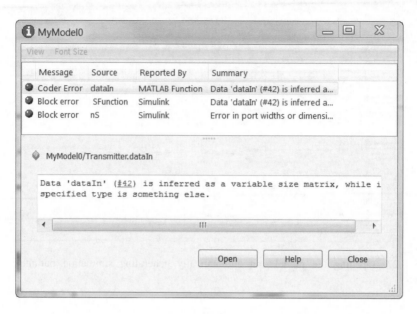

Figure 8.28   Simulation error displayed by Simulink regarding the size of the transmitted signal

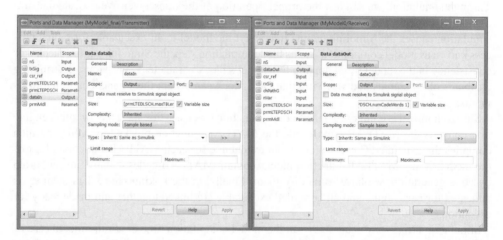

Figure 8.29   Configuring Transmitter and Receiver output signals as variable size and setting their maximum size

Transmitter and Receiver blocks, access their Port and Data Manager dialogs, as described before, select the *dataIn* and *dataOut* output signals, and change their Size properties by clicking on the Variable Size checkbox and specifying the maximum size. These steps are illustrated in Figure 8.29.

This type of variable-size signal handling is typical in Simulink models that express communications systems. This is in part due to the fact that during a simulation the sizes of various signals can change from one frame to another.

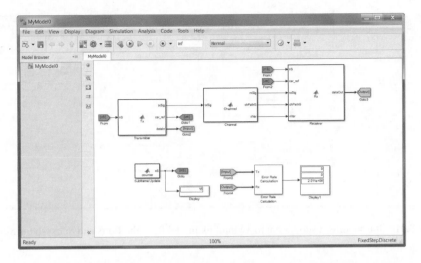

Figure 8.30 Running the Simulink system model and measuring BER performance

Now we can successfully run the simulation by clicking on the Run button. The simulation will proceed until a specified number of bits are processed. The simulation stops as the Error Rate Calculation block detects the stopping criteria. The BER results are shown in Figure 8.30.

The performance results reflect the system parameters specified during initialization in the MATLAB script (*commlteSystem_params*). If we want to run the simulation for a different transmission mode or a different set of operating conditions, we must first modify the system parameters by changing the MATLAB script. Then we need to return to our Simulink model and rerun the simulation. This process of iterating between MATLAB and Simulink for parameter specification can become tedious if we run multiple simulations. To make parameter specification easier, in the next section we develop a parameter dialog in Simulink.

8.5.5 Introducing a Parameter Dialog

Parameter dialogs facilitate updates to the system parameters and enhance the way in which we specify simulation parameters directly and graphically in Simulink. Essentially, we need to introduce a new subsystem in our Simulink model that contains all the model parameters and allows us to easily update them. This special kind of subsystem is known as a masked subsystem.

A subsystem is a collection of one or more blocks in a Simulink model. Every subsystem can be masked; that is, it can have a dialog in which the subsystem parameters are specified. However, in this section we introduce a subsystem whose purpose is to contain and update the model parameters. As illustrated in Figure 8.31, we start by adding a Subsystem block from the Ports and Subsystems Simulink library into our model.

As we double click the Subsystem block icon to open it, we notice that it contains a trivial connection from an input port to an output port. Note that in this section, we are not actually building a subsystem but rather using the Subsystem block to create a masked subsystem in order to hold our system parameters. As a result, our first action is to remove the input

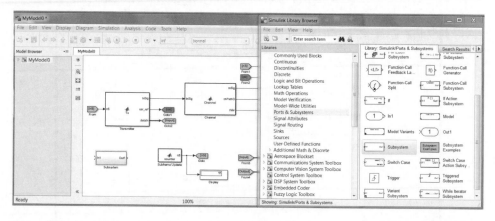

Figure 8.31 Introducing a subsystem to a Simulink model from the Ports and Subsystems Simulink library

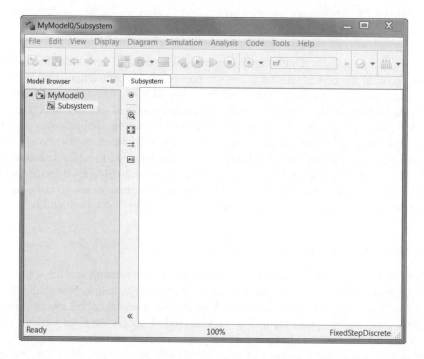

Figure 8.32 Making an empty subsystem to focus on masking and introducing parameters

and output ports and the connector from the Subsystem block. The subsystem will now be empty, as illustrated in Figure 8.32. To distinguish this Subsystem from the other blocks and subsystems of our model, we change its background color and rename it Model Parameters, as shown in Figure 8.33.

The next step is to turn our empty subsystem into a masked subsystem. As illustrated by Figure 8.34, this is easily done by first clicking on the subsystem to select it and then using the

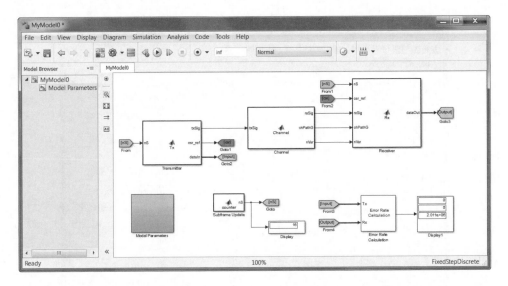

Figure 8.33   Changing our subsystem background color and changing its name to Model Parameters

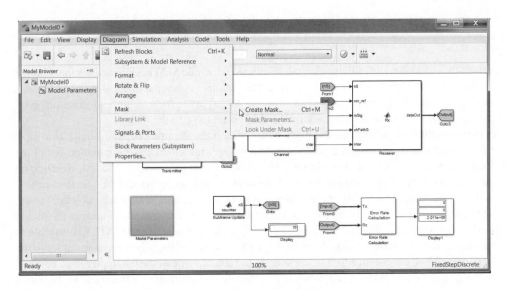

Figure 8.34   Turning the Model Parameters subsystem into a masked subsystem

Menu bar of the model to make the following selections: Diagram → Mask → Create Mask. When we turn a subsystem into a masked subsystem, double clicking on the Subsystem icon no longer opens the subsystem to display its content. Rather, a mask editor opens, enabling us to add and specify various parameters and to initialize them. Figure 8.35 shows an empty mask editor for our Model Parameters subsystem.

A mask editor contains four tabs: the Icon and Ports tab allows us to display text and images on the subsystem icon, the Parameters tab allows us to introduce parameters and specify what

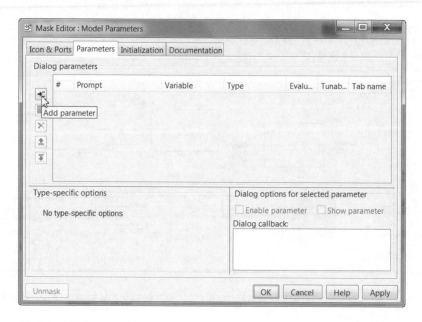

Figure 8.35   Opening the Model Parameters mask editor in order to introduce parameters

values they can take, the Initialization tab allows us to include MATLAB functions that operate on the subsystem parameters and help us create various system parameters in the MATLAB workspace, and the Documentation tab allows us to display text containing information about the subsystem when we double click on the subsystem to open the dialog.

In this section, we customize the mask editor within the Parameters and Initialization tabs. Fist, we introduce our operating parameters one by one under the Dialog Parameters list of the Parameters tab, as illustrated in Figure 8.36. To add a new parameter to the list, we click on the first icon in the upper left-hand side of the Dialog Parameters list, called Add Parameters. A new list item will be added to the Dialog Parameters list, containing fields such as *Prompt*, *Variable*, *Type*, and so on. In the Prompt field, we specify the text that will appear in the dialog next to the parameter. In the Variable field, we specify the name of the MATLAB variable to be used in the initialization stage later. The Type field specifies the way we assign values to the parameter. If we choose an edit box, we specify the actual value of the parameter in an edit box in the dialog. If we choose a check box, the parameter is specified as a Boolean choice with a value of *true* if the condition expressed in the Prompt field holds and a value of *false* otherwise.

If we choose a popup, we specify the parameter as a choice among a finite number of choices. These choices are enumerated in the Type-Specific Options list in the lower left-hand corner of the Parameters tab. For example, for the transmission mode (*txMode*) parameter, as illustrated in Figure 8.36, we choose a popup and then list all four choices for the transmission mode (SIMO, TD, open-loop SM, closed-loop SM) in the Type-Specific Options list associated with the *txMode* parameter.

At this point we can see how the parameter dialog looks after specifying the first parameter. To do so we save and close the mask editor, go back to the model, and double click on the Model Parameters subsystem icon. As illustrated in Figure 8.37, a Block Parameters

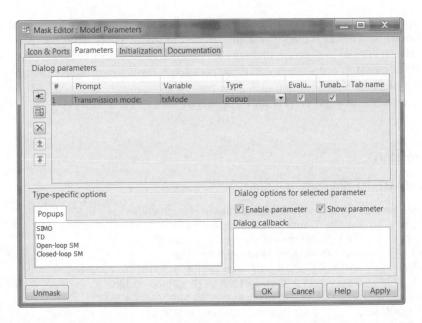

Figure 8.36   Adding parameters one by one to the Dialog Parameters list in the Parameters tab

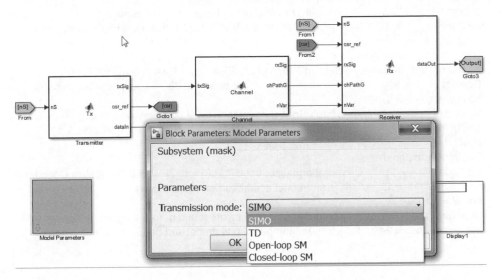

Figure 8.37   Inspecting the parameter dialog and how it reflects parameter lists developed under the mask editor

dialog will appear, bearing the name of the subsystem. So far, only one icon (Transmission Mode) has appeared under the Parameters list, matching exactly what we typed in the parameter prompt field. Note that in front of the prompt, we find a popup menu with the four choices for the transmission mode, matching exactly what we typed in the *Type*-Specific Options list associated with the parameter.

We can now repeat the process of adding parameters to the parameter list in the mask editor. Note that we need to add as many parameters as there are in the MATLAB script (*commlteSystem_params*) in order to generate all three LTE parameter structures (*prmLTEDLSCH*, *prmLTEPDSCH*, *prmMdl*) and all three additional parameters (*snrdB*, *maxNumBits*, *maxNumErrs*) for our system simulation in Simulink. One particular instance of the MATLAB parameter script (*commlteSystem_params*) is shown as a reference:

Algorithm

commlteMIMO_Simulink_init function

```
%% Set simulation parametrs & initialize parameter structures
txMode        = 4;   % Transmisson mode one of {1, 2, 3, 4}
numTx         = 2;   % Number of transmit antennas
numRx         = 2;   % Number of receive antennas
chanBW        = 4;   % [1,2,3,4,5,6] maps to [1.4, 3, 5, 10, 15, 20]MHz
contReg       = 2;   % {1,2,3} for >=10MHz, {2,3,4} for <10Mhz
modType       = 2;   % [1,2,3] maps to ['QPSK','16QAM','64QAM']
numCodeWords  = 1;   % Number of codewords in PDSCH
% DLSCH
cRate         = 1/2; % Rate matching target coding rate
maxIter       = 6;   % Maximum number of turbo decoding terations
fullDecode    = 0;   % Whether "full" or "early stopping" turbo decoding is performed
% Channel
chanMdl       = 'EPA 0Hz';
% one of {'flat','frequency-selective', 'EPA 0Hz', 'EPA 5Hz', 'EVA 5Hz', 'EVA 70Hz'}
Doppler       = 0;              % a value between 0 to 300 = Maximum Doppler shift
corrLvl       = 'Low';
%  one of {'Low', 'Medium', 'High'} Spatial correlation level between antennas
enPMIfback    = 0;   % Enable/Disable Precoder Matrix Indicator (PMI) feedback
cbIdx         = 1;   % Initialize PMI index
% Simulation parametrs
Eqmode        = 2;   % Type of equalizer used [1,2,3] for ['ZF', 'MMSE','Sphere Decoder']
chEstOn       = 1;      % use channel estimation or ideal channel
snrdB = 12.1;
maxNumErrs    = 2e6; % Maximum number of errors found before simulation stops
maxNumBits    = 2e6; % Maximum number of bits processed before simulation stops
```

Figure 8.38 shows how we can add and populate parameters in the mask editor to bear the same variable names as found in the MATLAB parameter script *commlteSystem_param*.

After saving and closing the mask editor, we can inspect the parameter dialog once we have specified all the parameters. As illustrated in Figure 8.39, specifying parameters using a masked subsystem in Simulink is the most convenient approach. We no longer need to edit and save the MATLAB parameter script in the MATLAB editor, come back to the Simulink

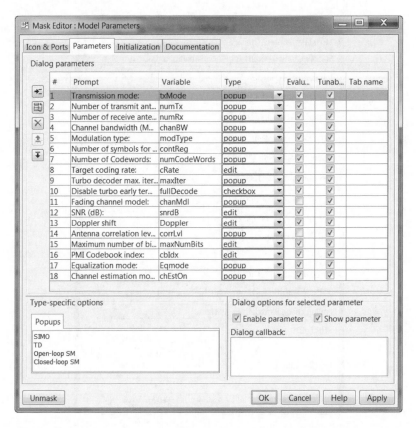

Figure 8.38 Adding to the parameter list and matching variable names to ones specified in the MAT-LAB script

model, and rerun the simulation. All parameters can now be specified in the Simulink model using an intuitive parameter dialog.

It is beneficial to the process of parameter initialization, as we will see shortly, to match variable names in the mask editor to those in the MATLAB parameter script. All that is needed now is to run the initialization commands in the Initialization tab of the mask editor that generates the LTE parameter structures based on the parameter dialog values. As illustrated in Figure 8.40, the parameter dialog variables are listed in the left-hand side of the Initialization tab. On the right-hand side we find an Initialization Commands edit box. In this edit box we can type various MATLAB commands or call a MATLAB function. Here, we are calling a MATLAB function (*commlteMIMO_Simulink_init*) to generate model parameters from the dialog variables and put them in the MATLAB workspace.

Since the variable names we chose in the mask editor are identical to those in the MATLAB parameter script, the masked subsystem initialization function *commlteMIMO_Simulink_init* is almost identical to the MATLAB initialization function we used in our MATLAB system model (*commlteSystem_initialize*).

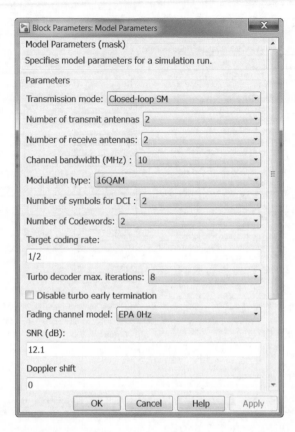

Figure 8.39 A convenient parameter dialog for setting LTE system model parameters in the Simulink model

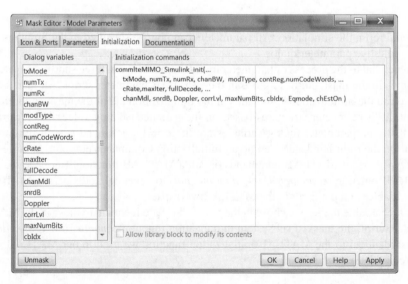

Figure 8.40 Initialization commands in the mask editor generating LTE parameter structures from the parameter dialog

Algorithm

Masked subsystem initialization function (*commlteMIMO_Simulink_init*)

```
function commlteMIMO_Simulink_init(txMode, Tx, Rx, chanBW, modType, contReg,...
    numCodeWords, cRate,maxIter, fullDecode, ...
    chanMdl, snrdB, Doppler, corrLvl, maxNumBits, cbIdx,  Eqmode, chEstOn )
% Create the parameter structures
vector=[1,2,4];
numTx=vector(Tx);
numRx=vector(Rx);
% PDSCH parameters
CheckAntennaConfig(numTx, numRx, txMode, numCodeWords);
prmLTEPDSCH = prmsPDSCH(txMode, chanBW, contReg, mod-
Type, numTx, numRx, numCodeWords,Eqmode);
[SymbolMap, Constellation]=ModulatorDetail(prmLTEPDSCH.modType);
prmLTEPDSCH.SymbolMap=SymbolMap;
prmLTEPDSCH.Constellation=Constellation;
if numTx==1
    prmLTEPDSCH.csrSize=[2*prmLTEPDSCH.Nrb, 4];
else
    prmLTEPDSCH.csrSize=[2*prmLTEPDSCH.Nrb, 4, numTx];
end
% DLSCH parameters
prmLTEDLSCH = prmsDLSCH(cRate,maxIter, fullDecode, prmLTEPDSCH);
% Channel parameters
chanSRate   = prmLTEPDSCH.chanSRate;
DelaySpread = prmLTEPDSCH.cpLenR;
 prmMdl = prmsMdl( chanSRate,  DelaySpread, chanMdl, Doppler, numTx, numRx, ...
    corrLvl, chEstOn-1, 0, cbIdx);
%% Assign parameter structure variables to base workspace
assignin('base', 'prmLTEPDSCH', prmLTEPDSCH);
assignin('base', 'prmLTEDLSCH', prmLTEDLSCH);
assignin('base', 'prmMdl', prmMdl);
assignin('base', 'snrdB', snrdB);
assignin('base', 'maxNumBits', maxNumBits);
assignin('base', 'maxNumErrs', maxNumBits);
```

Note that the main difference between this initialization function and the one used in our MATLAB model is the addition of six lines at the end of the function. These extra lines take the variables defined locally in the scope of the function and write them to the MATLAB workspace.

8.6   Qualitative Assessment

As the last topic in this chapter, we perform a qualitative assessment of our LTE system model. Instead of processing randomly generated payload bits, we can process the bit stream of a voice signal. In a sense, this experiment simulates a phone conversation over the simulated LTE PHY model.

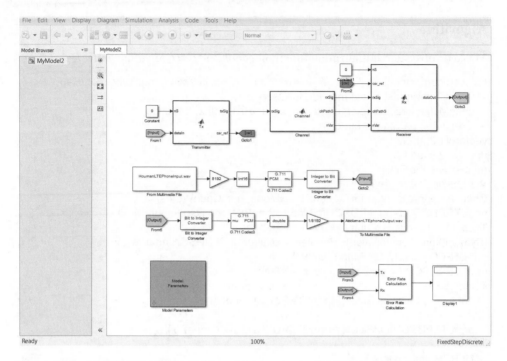

Figure 8.41 Modified Simulink model combining speech encoding and decoding with an LTE transceiver model to measure voice quality

8.6.1 Voice-Signal Transmission

The first step of qualitative assessment is to introduce speech coding. In this step, we encode the voice signal and pass the encoded bit stream as input to the LTE transceiver model. We are using one of the simplest voice-coding algorithms, based on either A-law or μ-law Pulse Code Modulation (PCM) coding. At the receiver, we apply the corresponding A-law or μ-law decoder to the recovered bit stream of the LTE model in order to obtain the output voice signal and listen to the speech. The quality of the recovered voice signal reflects all the degradations introduced by the channel and the receiver.

To simulate the LTE phone call, we use the Simulink model we developed in the last section. The only modifications needed (Figure 8.41) are related to the encoding and decoding of the voice signal:

1. Remove the payload bit generator function from the transmitter subsystem.
2. Introduce the encoded speech as an input to the transmitter subsystem.
3. Speech coding: generate encoded speech bits subframe by subframe by introducing blocks from the DSP System Toolbox.
4. Speech decoding: decode the output bits of the LTE system model and recover the output speech signal.

The speech coding sequence is implemented as follows:

1. Stream the speech from an audio file (any audio file format supported by MATLAB) subframe by subframe using the From Multimedia File block of the DSP System Toolbox.
2. Scale the normalized audio sample outputs of the From Multimedia File block with the range of the μ-law coder (a value of 8192) and cast the result as an integer (int16) MATLAB data type.
3. Process the integer input through the G.711 PCM coder block from the DSP System Toolbox.
4. Unpack the bytes of compressed data into individual bits using the Integer-to-Bit conversion block from the Communications System Toolbox. The output of this block is then passed as input to the input port of the transmitter subsystem as the encoded bit stream for the LTE transceiver model.

The speech decoding sequence inverts the speech coding operations as follows:

1. The output bits of the LTE transceiver model are packed into bytes using the Bit-to-Integer block of the DSP System Toolbox.
2. The resulting bytes are processed through the G.711 PCM coder block from the DSP System Toolbox.
3. The resulting G.711 PCM samples are converted to a floating-point data type and normalized to a range of values between -1 and 1.
4. The To Multimedia File block of the DSP System Toolbox is used to write the resulting samples on to the disk as an audio file.

8.6.2   Subjective Voice-Quality Testing

We can simulate the model for various conditions, including the SIMO, transmit-diversity, and spatial-multiplexing transmission modes. As we listen to the speech file output, we note that the quality of the voice signal depends on the parameters of the transceiver and the channel model. For example, if the fading delay spread, as reflected in the fading-channel path-delay parameter, is within 4.6 μs (as prescribed by the standard) then the SNR in recovering speech will be within representative values and the perceived speech quality will be reasonably good. Similarly, by using modes with improved link quality, such as transmit diversity instead of SISO (Single Input Single Output), we can obtain a better voice quality.

8.7   Chapter Summary

In this chapter, we composed a system model for the LTE PHY model. We integrated the first four modes of downlink transmission into the system model comprising the transmitter, the channel model, and the receiver. Then we simulated the system model in order to quantitatively assess the performance of the overall system. We studied the effects of various transmission modes, channel models, link SNRs, channel estimation techniques, MIMO receiver algorithms, and channel delay spreads on the overall performance. We also performed experiments aimed at gauging the throughput of the LTE system model.

Next, we composed a Simulink model for the LTE transceiver model. We built the Simulink model step by step by integrating the MATLAB functions for the transmitter, receiver, and channel model incrementally within the system. We then automated the way we specify the LTE system model parameters by developing parameter dialogs into our Simulink model. Finally, we added a speech coder and decoder to the Simulink model in order to enable a qualitative assessment of the system performance.

References

[1] Jafarkhani, H. (2005) *Space-time Coding; Theory and Practice*, Cambridge University Press, New York.
[2] 3GPP Evolved Universal Terrestrial Radio Access (E-UTRA); User Equipment (UE) Radio Transmission and Reception (Version 11.4.0), March 2013. TS 36.101.

9

Simulation

So far we have provided a functional description of the LTE (Long Term Evolution) PHY (Physical Layer) standard and its implementation in MATLAB®. To verify whether this functional model will meet the requirements of the standardization process, we need to perform large-scale simulations. Like many other standards, the LTE standard has a mode-based specification. This means we need to perform a series of simulations in order to ensure that all possible combinations of modes, including modulation, coding, and MIMO (Multiple Input Multiple Output) modes, are exercised. The combined effects of using large simulation data sets and the computationally complex nature of the LTE standard will inevitably result in a familiar challenge: exceedingly long simulation times and the necessity to accelerate the speed of simulations.

The simulations can be performed on a software model or on a physical hardware prototype. Most designers find it useful to first run a computer model of the standard to verify various technical aspects related to the system performance before proceeding to a hardware prototype. When talking about accelerating the execution of a software model, it is natural to start with a baseline or initial version. The optimizations that lead to acceleration of the simulation speed of the baseline algorithm may or may not alter the functional accuracy of the model. To be true to a standard implementation, in this book we only highlight optimizations that preserve the numerical accuracy of the baseline algorithms. As such, optimizations examined here highlight various ways of implementing the same functionality more efficiently. In this chapter we discuss in detail many optimizations in MATLAB and Simulink that result in substantial acceleration of the simulation speed.

9.1 Speeding Up Simulations in MATLAB

When we model and simulate a communications system, our focus and priorities may be different at different stages of the workflow. In the early stages of development, we might focus on accuracy in expressing the mathematical model. At this stage we want to use visualization and debugging features of the MATLAB environment to ensure that the sequence of operations in the MATLAB function and scripts is correct. This stage of functional verification is sometimes referred to as unit testing and involves testing a limited set of data for which the

Understanding LTE with MATLAB®: From Mathematical Modeling to Simulation and Prototyping, First Edition.
Houman Zarrinkoub.
© 2014 John Wiley & Sons, Ltd. Published 2014 by John Wiley & Sons, Ltd.

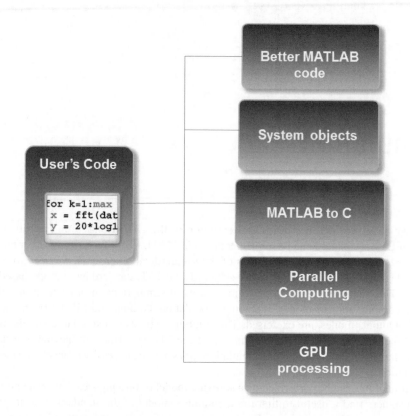

Figure 9.1   Simulation acceleration methods in MATLAB

correct response is known. Unit testing helps make sure that the mathematical model correctly implements the design. After satisfying the unit-testing requirements, most designers execute the same simulation model with a large amount of data within a simulation loop. Identifying the bottlenecks of design in large-scale testing helps us focus on the portions in which optimization efforts provide the most return. We can optimize the baseline model and resolve design bottlenecks in one of two ways (see Figure 9.1):

- **MATLAB code optimization**: Involves changing the MATLAB program code for a more efficient implementation. This includes steps that: (i) ensure constant parameters are only computed once during initializations, (ii) reduce parameter validation overhead, (iii) use variables that are preallocated in order to avoid overhead of dynamic memory allocations, and (iv) use more efficient algorithms implemented with System objects.
- **Use of acceleration features**: Involves applying such techniques as: (i) converting MATLAB code to a compiled C code, (ii) exploiting multiple cores or clusters for parallel processing, or (iii) using MATLAB features that are optimized for Graphics Processing Unit (GPU) processing.

9.2   Workflow

In this chapter, we start with a baseline MATLAB program. Following a series of code optimizations, we then successively accelerate the speed of simulation. At each step, the algorithm generates the same numerical outputs. The only difference between steps is the introduction of a more efficient programming technique.

Both numerical and timing results provided throughout the book depend on the platform where MATLAB is installed, and the type of operating system, C/C++ compiler or GPU that is used. Results in this book for non-GPU experiments are obtained by running MATLAB on a laptop computer with the following specifications:

- **Hardware:** Intel Dual-Core i7-2620M CPU @ 2.70 GHz with 8 GB of RAM
- **Operating system:** 64-bit Windows 7 Enterprise (Service Pack 1)
- **C/C++ compiler:** Microsoft Visual Studio 2010 with Microsoft Windows SDK v7.1.

The GPU experiments use NVIDIA Tesla GPU Accelerators installed on a desktop computer with an Intel Quad-core i7 CPU with 12 GB of RAM and the same operating system and C/C++ compiler as mentioned above.

9.3   Case Study: LTE PDCCH Processing

We use a simplified version of the signal processing applied to the Physical Downlink Control Channel (PDCCH) of the LTE standard in this chapter as a case study. We have already showcased this algorithm in Chapter 7. As Figure 9.2 illustrates, processing the PDCCH signal in the transmitter side involves the following operations: Cyclic Redundancy Check (CRC) generation, tail-biting convolutional encoding, rate matching, scrambling, Quadrature Phase Shift Keying (QPSK) modulation, and transmit-diversity MIMO encoding. Channel modeling consists of a combination of a two-by-two MIMO channel and an Additive White Gaussian Noise (AWGN) channel. We perform the inverse operations at the receiver, including transmit-diversity MIMO combination, QPSK demodulation, descrambling, rate dematching, Vitetbi decoding, and CRC detection. To reduce the complexity of the algorithm, in this section we

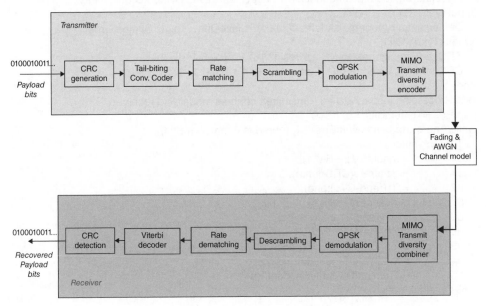

Figure 9.2   A simplified PDCCH signal processing chain

will update it with two modifications: (i) omission of the frequency-domain transformations involving Orthogonal Frequency Division Multiplexing (OFDM) resource-grid formation and signal generation; and (ii) use of hard-decision demodulation at the receiver.

9.4   Baseline Algorithm

The following baseline function shows the first implementation of the PDCCH processing chain. The sequence of operations characterizing the PDCCH algorithm already described is implemented with a series of functions. Some of the functions, such as *convenc* and *vitdec*, and objects, such as *modem.pskmod* and *crc.generator*, are available in the Communications System Toolbox. Others, such as *TransmitDiversityEncoder1* and *MIMOFadingChan*, are user-defined and are composed using a combination of basic MATLAB functions and constructs.

Algorithm

MATLAB function

```
function [ber, bits]=zPDCCH_v1(EbNo, maxNumErrs, maxNumBits)
%% Constants
FRM=2048;
M=4; k=log2(M); codeRate=1/3;
snr = EbNo + 10*log10(k) + 10*log10(codeRate);
trellis=poly2trellis(7, [133 171 165]);
L=FRM+24;C=6; Index=[L+1:(3*L/2) (L/2+1):L];
%% Initializations
persistent Modulator Demodulator CRCgen CRCdet
if isempty(Modulator)
    Modulator=modem.pskmod('M', 4, 'PhaseOffset', pi/4, 'SymbolOrder', 'Gray',
'InputType', 'Bit');
    Demodulator= modem.pskdemod('M', 4, 'PhaseOffset', pi/4, 'SymbolOrder', 'Gray',
'OutputType', 'Bit');
    CRCgen = crc.generator([1 1 zeros(1, 16) 1 1 0 0 0 1 1]);
    CRCdet = crc.detector   ([1 1 zeros(1, 16) 1 1 0 0 0 1 1]);
end
%% Processing loop modeling transmitter, channel model and receiver
numErrs = 0; numBits = 0; nS=0;
while ((numErrs < maxNumErrs) && (numBits < maxNumBits))
    % Transmitter
    u        = randi([0 1], FRM,1);
    u1       = generate(CRCgen,u);
    u2       = u1((end-C+1):end);
    [~, state] = convenc(u2,trellis);
    u3       = convenc(u1,trellis,state);
    u31      = fcn_RateMatcher(u3, L, codeRate);
    u32      = fcn_Scrambler(u31, nS);
    u4       = modulate(Modulator, u32);
    u5       = TransmitDiversityEncoder1(u4);
    % Channel model
```

```
[u6, h6]    = MIMOFadingChan(u5);
u7          = awgn(u6,snr);
% Receiver
u8          = TransmitDiversityCombiner1(u7, h6);
u9          = demodulate(Demodulator,u8);
u91         = fcn_Descrambler(u9, nS);
u92         = fcn_RateDematcher(u91, L);
uA          = [u999;u999];
uB          = vitdec(uA ,trellis,34,'trunc','hard');
uC          = uB(Index);
y           = detect(CRCdet, uC );
numErrs     = numErrs + sum(y~=u);
numBits     = numBits + FRM;
nS          = nS + 2; nS = mod(nS, 20);
end
%% Clean up & collect results
ber = numErrs/numBits;
bits=numBits;
```

Let us start by running this baseline algorithm to establish a benchmark for performance. The following MATLAB script (*zPDCCH_v1_test*) executes this algorithm within a for loop. In each iteration, the script calls the baseline algorithm with given Signal-to-Noise Ratio (SNR) values and computes the Bit Error Rate (BER). It also uses a combination of MATLAB *tic* and *toc* functions to measure the time needed to complete the loop iterations.

Algorithm

MATLAB script: zPDCCH_v1_test

```
MaxSNR=8;
MaxNumBits=1e5;
fprintf(1,'\nVersion 1: Baseline algorithm\n\n');
tic;
for snr=1:MaxSNR
   fprintf(1,'Iteration number %d\r',snr);
   ber= zPDCCH_v1 (snr, MaxNumBits, MaxNumBits);
end
time_1=toc;
fprintf(1,'Version 1: Time to complete %d iterations = %6.4f (sec)\n',
MaxSNR, time_1);
```

When we execute the MATLAB script, messages regarding the version of the algorithm, the iteration that is being executed, and the final tally of elapsed time will print in the command prompt. The results are shown in Figure 9.3. In this case, processing of 1 million bits in each of the eight iterations of the baseline algorithm takes about 411.30 seconds to complete.

```
Version 1: Baseline algorithm

Iteration number 1
Iteration number 2
Iteration number 3
Iteration number 4
Iteration number 5
Iteration number 6
Iteration number 7
Iteration number 8
Version 1: Time to complete 8 iterations = 411.3030 (sec)
```

Figure 9.3   Baseline algorithm: time taken to execute eight iterations

We use this measure as a yard stick and try to improve the performance using the code optimizations discussed later. Before proceeding to any code optimization, it is important to identify the code bottlenecks. These are the portions of the algorithm that contribute most to its computational complexity and take up the most processing time. We will now use some MATLAB tools to identify the bottlenecks in our algorithm.

9.5   MATLAB Code Profiling

MATLAB provides a variety of tools to help assess and optimize the performance of code. MATLAB Profiler shows where code is spending its time. It can be applied to the baseline algorithm by performing the following three commands:

Algorithm

MATLAB script

```
profile on;
ber= zPDCCH_v1 (snr, MaxNumBits, MaxNumBits);
profile viewer;
```

Calling the *profile viewer* command brings up the MATLAB Profiler report as illustrated in Figure 9.4. MATLAB Profiler provides a summary report of statistics on the overall execution of a code, including a list of all functions called, the number of times each function was called, and the total time spent in each function. It can also provide timing information about each function, such as information on the lines of code that use the most processing time.

Once bottlenecks have been identified, we can focus on improving the performance of these particular sections. For example, in this profile summary the function *TransmitDiversity-Combiner* takes 4.385 seconds of the 7.262 seconds it takes to run the entire function. That qualifies the *TransmitDiversityCombiner* function as one of the bottlenecks of our baseline algorithm.

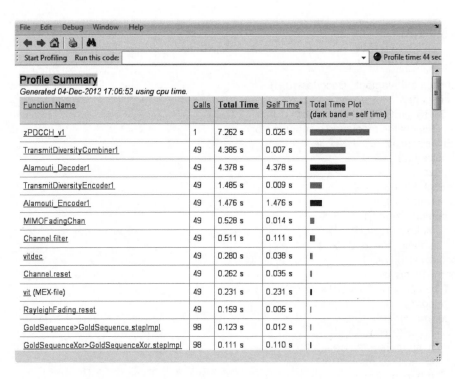

Figure 9.4 Profile summary report for the baseline algorithm

Algorithm

MATLAB function

```
function y = TransmitDiversityCombiner1(in, chEst)
%#codegen
% Alamouti Transmit Diversity Combiner
% Scale
in = sqrt(2) * in;
% STBC Alamouti
y = Alamouti_Decoder1(in, chEst);
% Space-Frequency to Space-Time transformation
y(2:2:end) = -conj(y(2:2:end));
```

When we drill down through the function hyperlink in the profile summary report, we can find out exactly which lines of the *TransmitDiversityCombiner1* function take the most time. Of the three lines of code, *Alamouti_Decoder1* can be easily identified as the processing bottleneck. The following *Alamouti_Decoder1* function shows the first implementation of the Alamouti combining algorithm.

Algorithm

MATLAB function

```
function s = Alamouti_Decoder1(u,H)
%#codegen
% STBC_DEC STBC Combiner
%   Outputs the recovered symbol vector
LEN=size(u,1);
Nr=size(u,2);
BlkSize=2;
NoBlks=LEN/BlkSize;
% Initialize outputs
h=complex(zeros(1,2));
s=complex(zeros(LEN,1));
% Alamouti code for 2 Tx
indexU=(1:BlkSize);
for m=1:NoBlks
    t_hat=complex(zeros(BlkSize,1));
    h_norm=0.0;
    for n=1:Nr
        h(:)=H(2*m-1,:,n);
        h_norm=h_norm+real(h*h');
        r=u(indexU,n);
        r(2)=conj(r(2));
        shat=[conj(h(1)), h(2); conj(h(2)), -h(1)]*r;
        t_hat=t_hat+shat;
    end
    s(indexU)=t_hat/h_norm; % Maximum-likelihood combining
    indexU=indexU+BlkSize;
end
end
```

By following the hyperlink to the *Alamouti_Decoder1* function, we can see a more detailed line-by-line profile of execution time (Figure 9.5). This level of breakdown enables us to identify what feature of the code contributes most to its performance. In this case, the algorithm performs two nested for loops and computes every element of a vector one by one within a scalar programming routine. Vectorizing this code can lead to acceleration.

9.6 MATLAB Code Optimizations

In this section we discuss some typical code-optimization techniques in MATLAB. These techniques include vectorizing the code, preallocating data, separating initialization from

```
time    calls  line
               1 function s = Alamouti_Decoder1(u,H)
               2 %#codegen
               3 % STBC_DEC STBC Combiner
               4 %    Outputs the recovered symbol vector
        49     5 LEN=size(u,1);
        49     6 Nr=size(u,2);
        49     7 BlkSize=2;
        49     8 NoBlks=LEN/BlkSize;
               9 % Initialize outputs
        49    10 h=complex(zeros(1,2));
< 0.01  49    11 s=complex(zeros(LEN,1));
              12 % Alamouti code for 2 Tx
        49    13 indexU=(1:BlkSize);
        49    14 for m=1:NoBlks
0.24    76146 15     t_hat=complex(zeros(BlkSize,1));
0.08    76146 16     h_norm=0.0;
0.05    76146 17     for n=1:Nr
0.52    152292 18        h(:)=H(2*m-1,:,n);
0.55    152292 19        h_norm=h_norm+real(h*h');
0.31    152292 20        r=u(indexU,n);
0.07    152292 21        r(2)=conj(r(2));
1.65    152292 22        shat=[conj(h(1)), h(2); conj(h(2)), -h(1)]*r;
0.21    152292 23        t_hat=t_hat+shat;
0.03    152292 24     end
0.60    76146 25     s(indexU)=t_hat/h_norm; % Maximum-likelihood combining
0.05    76146 26     indexU=indexU+BlkSize;
0.01    76146 27 end
        49    28 end
```

Figure 9.5   Line-by-line processing time in the *Alamouti_Decoder1* function

in-loop processing, and using System objects. To illustrate these techniques, we continue updating and optimizing the PDCCH processing algorithm.

9.6.1 Vectorization

Vectorization is one of the most important code-optimization techniques in MATLAB. In vectorization, we convert a code from using loops to using matrix and vector operations. Since MATLAB uses processor-optimized libraries for matrix and vector computations, we can often gain performance improvement by vectorizing our code.

The second version of the PDCCH algorithm is optimized based on vectorization. The only difference between this version of the algorithm and the baseline is the use of the *TransmitDiversityCombine2* function instead of *TransmitDiversityCombine1*. This function is the second version of the transmit-diversity combiner function and uses the *Alamouti_Decoder2* function, a vectorized version of *Alamouti_Decoder1*. When we examine the *Alamouti_Decoder2* function, we can see that the nested double for loop is modified to a single for loop and that operations are more vectorized in the single loop. These changes are illustrated in the following function:

MATLAB function

function [ber, bits]=zPDCCH_v1(...)	function [ber, bits]=zPDCCH_v2(...)
.	.
.	.

u5 = TransmitDiversityDecoder1(u4);	u5 = TransmitDiversityDecoder2(u4);
.	.
.	.
end	end

```
function y = TransmitDiversityCombiner1(in,
    chEst)
%#codegen
% Alamouti Transmit Diversity Combiner
% Scale
in = sqrt(2) * in;
% STBC Alamouti
```

```
function y = TransmitDiversityCombiner2(in,
    chEst)
%#codegen
% Alamouti Transmit Diversity Combiner
% Scale
in = sqrt(2) * in;
% STBC Alamouti
```

```
y =  Alamouti_Decoder1(in, chEst);
```

```
y =  Alamouti_Decoder2(in, chEst);
```

```
% Space-Frequency to Space-
    Time transformation
y(2:2:end) = -conj(y(2:2:end));
```

```
% Space-Frequency to Space-
    Time transformation
y(2:2:end) = -conj(y(2:2:end));
```

```
function s = Alamouti_Decoder1(u,H)
LEN=size(u,1);
Nr=size(u,2);
BlkSize=2;
NoBlks=LEN/BlkSize;
% Initialize outputs
h=complex(zeros(1,2));
s=complex(zeros(LEN,1));
% Alamouti code for 2 Tx
indexU=(1:BlkSize);
```

```
function s = Alamouti_Decoder2(u,H)
LEN=size(u,1);
BlkSize=2;
NoBlks=LEN/BlkSize;
T=[0 1;-1 0];
% Initialize outputs
s=complex(zeros(LEN,1));
% Alamouti code for 2 Tx
h=complex(zeros(BlkSize,BlkSize));
```

```
for m=1:NoBlks
   t_hat=complex(zeros(BlkSize,1));
   h_norm=0.0;
   for n=1:Nr
```

```
for m=1:NoBlks
```

```
      h(:)=H(2*m-1,:,n);
      h_norm=h_norm+real(h*h');
      r=u(indexU,n);
      r(2)=conj(r(2));
```

```
   indexU=(m-1)*BlkSize+(1:BlkSize);
   h(:)=H(2*m-1,:,:);
   h_norm=sum(h(:).*conj(h(:)));
   r=u(indexU,:);
   r(2,:)=conj(r(2,:));
   H1=conj(h);
   H2=T*h;
   M=[H1(:,1),H2(:,1),H1(:,2),H2(:,2)];
```

```
    shat=[conj(h(1)), h(2); conj(h(2)), -h(1)]*r;        s(indexU)=(M*r(:))/h_norm; % Maximum-
       t_hat=t_hat+shat;                                    likelihood combining
    end                                                  end
    s(indexU)=t_hat/h_norm; % Maximum-
    likelihood combining
    indexU=indexU+BlkSize;
end
end
```

In order to verify whether this optimization leads to a faster execution time, we run the following MATLAB script. This script is identical to the one we used for the baseline algorithm except that it calls the second version of the PDCCH algorithm (*zPDCCH_v2.m*). This time, processing 1 million bits in eight iterations takes about 326.50 seconds to complete (Figure 9.6).

```
Version 2: Vectorization

Iteration number 1
Iteration number 2
Iteration number 3
Iteration number 4
Iteration number 5
Iteration number 6
Iteration number 7
Iteration number 8
Version 2: Time to complete 8 iterations = 326.5071 (sec)
```

Figure 9.6   Second version of the algorithm: time taken to execute eight iterations

Algorithm

MATLAB script: zPDCCH_v2_test

```
MaxSNR=8;
MaxNumBits=1e5;
fprintf(1,'\nVersion 1: Baseline algorithm\n\n');
tic;
```

```
for snr=1:MaxSNR
    fprintf(1,'Iteration number %d\r',snr);
    ber= zPDCCH_v2 (snr, MaxNumBits, MaxNumBits);
end
time_2=toc;
fprintf(1,'Version 1: Time to complete %d iterations = %6.4f (sec)\n', MaxSNR, time_2);
```

The second version of the algorithm performs a single for loop with a number of iterations specified by the *NoBlks* variable, which relates to the first dimension of the function input. The first dimension is a rather large number: 3108 in this algorithm. Frequent iteration based on the first dimension and performance of vectorized operations on smaller-sized vectors does not take optimal advantage of vectorization.

The third version of the algorithm is designed to vectorize along the first dimension of the input function. In this version, we iterate only twice (along the second dimension) and perform vectorized operations on large vectors and matrices along the first dimension. It features better optimization based on vectorization of the code with large vectors and matrices. The only difference between this version and the second is the use of *TransmitDiversityCombine3* instead of *TransmitDiversityCombine2*.

MATLAB function

---

function [ber, bits]=zPDCCH_v2(...)	**function [ber, bits]=zPDCCH_v3(...)**
.	.
.	.
u5 = TransmitDiversityDecoder2(u4);	**u5 = TransmitDiversityDecoder3(u4);**
.	.
.	.
end	end

function y = TransmitDiversityCombiner2(in, chEst)	function y = TransmitDiversityCombiner3(in, chEst)
%#codegen	%#codegen
% Alamouti Transmit Diversity Combiner	% Alamouti Transmit Diversity Combiner
% Scale	% Scale
in = sqrt(2) * in;	in = sqrt(2) * in;
% STBC Alamouti	% STBC Alamouti

y = Alamouti_Decoder2(in, chEst);	y = Alamouti_Decoder3(in, chEst);

% Space-Frequency to Space-
   Time transformation
	% Space-Frequency to Space-
	   Time transformation

y(2:2:end) = -conj(y(2:2:end));	y(2:2:end) = -conj(y(2:2:end));

function s = Alamouti_Decoder2(u,H) LEN=size(u,1);	function y = Alamouti_Decoder3(u,Ch) %#codegen

```
BlkSize=2;
function s = Alamouti_Decoder2(u,H)
LEN=size(u,1);
BlkSize=2;
NoBlks=LEN/BlkSize;
T=[0 1;-1 0];
% Initialize outputs
s=complex(zeros(LEN,1));
% Alamouti code for 2 Tx
h=complex(zeros(BlkSize,BlkSize));

for m=1:NoBlks

   indexU=(m-1)*BlkSize+(1:BlkSize);
   h(:)=H(2*m-1,:,:);
   h_norm=sum(h(:).*conj(h(:)));
   r=u(indexU,:);
   r(2,:)=conj(r(2,:));
   H1=conj(h);
   H2=T*h;
   M=[H1(:,1),H2(:,1),H1(:,2),H2(:,2)];
   s(indexU)=(M*r(:))/h_norm; % Maximum-
   likelihood combining
end
```

```
% STBC_DEC STBC Combiner
LEN=size(u,1);
BlkSize=2;
NoBlks=LEN/BlkSize;
Nr=size(u,2);
idx1=1:BlkSize:LEN;
idx2=idx1+1;
% Initalize outputs
s=complex(zeros(LEN,Nr));
mynorm=complex(zeros(LEN,BlkSize));
vec_u=complex(zeros(NoBlks,BlkSize));
% Alamouti code for 2 Tx
H=complex(zeros(NoBlks,BlkSize));

for n=1:Nr

   vec_u(:,1)     = u(idx1,n);
   vec_u(:,2)     = conj(u(idx2,n));
   H(:)           = Ch(1:BlkSize:end,:,n);
   conjH          = conj(H);
   cn1            = [conjH(:,1), H(:,2)];
   s(idx1,n)      = sum(cn1.*vec_u,2);
   mynorm(idx1,n) = sum(H.*conj(H),2);
   cn2            = [conjH(:,2), -H(:,1)];
   s(idx2,n)      = sum(cn2.*vec_u,2);
end;
nn=sum(mynorm,2);
nn(idx2)=nn(idx1);
y=sum(s,2)./nn;
end
```

In order to verify whether this optimization leads to a faster execution time, we run the following MATLAB script (*zPDCCH_v3_test*). The third version of the algorithm takes about 175.84 seconds to process 1 million bits in eight iterations, as illustrated in Figure 9.7.

```
Version 3: Vectorization along larger dimension

Iteration number 1
Iteration number 2
Iteration number 3
Iteration number 4
Iteration number 5
Iteration number 6
Iteration number 7
Iteration number 8
Version 3: Time to complete 8 iterations = 175.8478 (sec)
```

Figure 9.7   Third version of the algorithm: time taken to execute eight iterations

Profile Summary

Generated 11-Dec-2012 14:23:55 using cpu time.

Function Name	Calls	Total Time	Self Time*	Total Time Plot (dark band = self time)
zPDCCH_v3	1	2.759 s	0.026 s	████████████
TransmitDiversityEncoder1	49	1.485 s	0.006 s	██████
Alamouti_Encoder1	49	1.479 s	1.479 s	█████
MIMOFadingChan	49	0.385 s	0.005 s	██
Channel.filter	49	0.378 s	0.057 s	██
vitdec	49	0.270 s	0.024 s	██
vit (MEX-file)	49	0.237 s	0.237 s	■
Channel.reset	49	0.206 s	0.025 s	■
GoldSequence>GoldSequence.stepImpl	98	0.124 s	0.006 s	∎
RayleighFading.reset	49	0.121 s	0.006 s	∎

Figure 9.8   Profile summary report for the third version of the algorithm

Algorithm

MATLAB script: zPDCCH_v3_test

```
MaxSNR=8;
MaxNumBits=1e5;
fprintf(1,'\nVersion 3: Better vectorized algorithm\n\n');
tic;
for snr=1:MaxSNR
    fprintf(1,'Iteration number %d\r',snr);
    ber= zPDCCH_v3 (snr, MaxNumBits, MaxNumBits);
end
time_3=toc;
fprintf(1,'Version 3: Time to complete %d iterations = %6.4f (sec)\n', MaxSNR, time_3);
```

By profiling the third version of the algorithm (*zPDCCH_v3*), we identify the function *TransmitDiversityEncoder1* as the next bottleneck. This function calls the first version of the Alamouti encoder function (*Alamouti_Encoder1*). The MATLAB profiler commands and resulting report are shown in Figure 9.8.

Algorithm

MATLAB script

```
profile on;
ber= zPDCCH_v3(snr, MaxNumBits, MaxNumBits);
profile viewer;
```

```
time    calls  line
                      1 function y= Alamouti_Encoder1(u)
                      2 % STBCENC Space-Time Block Encoder
                      3 % Outputs the Space-Time block encoded matrix
          49          4 Tx=2;
          49          5 LEN=size(u,1);
          49          6 idx1=1:Tx:LEN-1;
          49          7 idx2=idx1+1;
                      8 % Alamouti Space-Time Block Encoder
                      9 %    G = [   s1        s2 ]
                     10 %        [ -s2*       s1*]
  < 0.01  49         11 y=[];
  < 0.01  49         12 for n=1:LEN/Tx
    0.55  76146      13      G=[         u(idx1(n))          u(idx2(n));...
                     14               -conj(u(idx2(n)))  conj(u(idx1(n)))];
    0.73  76146      15      y=[y;G];
    0.02  76146      16 end
```

Figure 9.9   Profiling *Alamouti_Encoder1*

By following the hyperlink to the *Alamouti_Encoder1* function in the profile report, we can see a more detailed line-by-line profile of the execution time (Figure 9.9). Note that we initialize output matrix y with an empty matrix. In iterations of the for loop, we then grow the size of matrix y by appending a 2×2 Alamouti matrix to its end. In successive iterations, we must allocate new memory and copy the existing matrix into the new one. We can thus identify preallocation as a feature of the code that can be improved. Next we will discuss preallocation as a MATLAB code-optimization feature.

9.6.2   Preallocation

Preallocation refers to the initialization of an array of known size at the beginning of a computation. It helps prevent dynamic resizing of an array while a code is executing, especially when using for and while loops. Since arrays require contiguous blocks of memory, repeated resizing of them often compells MATLAB to spend time looking for larger contiguous blocks and then moving the array into them. By preallocating arrays, we can avoid these unnecessary memory operations and improve overall execution time.

The fourth version of the PDCCH algorithm is optimized based on preallocation. This version uses the function *TransmitDiversityEncoder2* instead of *TransmitDiversityEncoder1*; this function uses the second version of the transmit diversity encoder function, which in turn uses *Alamouti_Encoder2*, a preallocated version of *Alamouti_Encoder1*.

Algorithm

MATLAB function

```
function [ber, bits]=zPDCCH_v4(EbNo, maxNumErrs, maxNumBits)
%% Constants
FRM=2048;
```

```
M=4; k=log2(M); codeRate=1/3;
snr = EbNo + 10*log10(k) + 10*log10(codeRate);
trellis=poly2trellis(7, [133 171 165]);
L=FRM+24;C=6; Index=[L+1:(3*L/2) (L/2+1):L];
%% Initializations
persistent Modulator Demodulator CRCgen CRCdet
if isempty(Modulator)
    Modulator    = modem.pskmod('M', 4, 'PhaseOffset', pi/4, 'SymbolOrder', 'Gray',
'InputType', 'Bit');
    Demodulator = modem.pskdemod('M', 4, 'PhaseOffset', pi/4, 'SymbolOrder', 'Gray',
'OutputType', 'Bit');
    CRCgen       = crc.generator([1 1 zeros(1, 16) 1 1 0 0 0 1 1]);
    CRCdet       = crc.detector  ([1 1 zeros(1, 16) 1 1 0 0 0 1 1]);
end
%% Processing loop
numErrs = 0; numBits = 0; nS=0;
while ((numErrs < maxNumErrs) && (numBits < maxNumBits))
    % Transmitter
    u          = randi([0 1], FRM,1);                    % Generate bit payload
    u1         = generate(CRCgen,u);                     % CRC insertion
    u2         = u1((end-C+1):end);                      % Tail-biting convolutional coding
    [~, state] = convenc(u2,trellis);
    u3         = convenc(u1,trellis,state);
    u4         = fcn_RateMatcher(u3, L, codeRate);  % Rate matching
    u5         = fcn_Scrambler(u4, nS);             % Scrambling
    u6         = modulate(Modulator, u5);           % Modulation
    u7         = TransmitDiversityEncoder2(u6);     % MIMO Alamouti encoder
    % Channel
    [u8, h8]   = MIMOFadingChan(u7);                % MIMO fading channel
    sigpower = 10*log10(real(var(u8(:))));
    u9         = awgn(u8,snr,sigpower,'dB');
    % Receiver
    uA         = TransmitDiversityCombiner3(u9, h8); % MIMO Alamouti combiner
    uB         = demodulate(Demodulator,uA);         % Demodulation
    uC         = fcn_Descrambler(uB, nS);            % Descrambling
    uD         = fcn_RateDematcher(uC, L);           % Rate de-matching
    uE         = [uD;uD];                            % Tail-biting
    uF         = vitdec(uE ,trellis,34,'trunc','hard'); % Viterbi decoding
    uG         = uF(Index);
    y          = detect(CRCdet, uG );                % CRC detection
    numErrs  = numErrs + sum( y~=u );                % Update number of bit errors
    numBits  = numBits + FRM;
    nS         = nS + 2; nS = mod(nS, 20);
end
%% Clean up & collect results
ber = numErrs/numBits;
bits=numBits;
```

When we examine *Alamouti_Encoder2*, we can see that it first initializes the output
with information derived from the size of the input, then transforms the input and inserts
selected samples of it into the predetermined locations of the output matrix. Also, the
updated *Alamouti_Encoder2* function is not only preallocated but also vectorized, whereas
Alamouti_Encoder1 is a scalarized function. The main problem with *Alamouti_Encoder1* is
that it initializes the output to an empty matrix and then performs a for loop in which in each
iteration the output matrix grows in size. These types of frequent dynamic memory allocation
contribute to the degradation of performance.

MATLAB function

function [ber, bits]=zPDCCH_v3(...)	**function [ber, bits]=zPDCCH_v4(...)**
.	.
.	.

u7 = TransmitDiversityEncoder1(u6);	u7 = TransmitDiversityEncoder2(u6);
.	.
.	.
end	**end**

```
function y = TransmitDiversityEncoder1(in)
% Alamouti Transmit Diversity Encoder
% Space-Frequency to Space-
   Time transformation
in(2:2:end) = -conj(in(2:2:end));
% STBC Alamouti
```

```
function y = TransmitDiversityEncoder2(in)
% Alamouti Transmit Diversity Encoder
% Space-Frequency to Space-
   Time transformation
in(2:2:end) = -conj(in(2:2:end));
% STBC Alamouti
```

```
y = Alamouti_Encoder1(in);
```

```
y = Alamouti_Encoder2(in);
```

```
% Scale
y = y/sqrt(2);
```

```
% Scale
y = y/sqrt(2);
```

```
function y= Alamouti_Encoder1(u)
% Space-Time Block Encoder
Tx=2;
LEN=size(u,1);
idx1=1:Tx:LEN-1;
idx2=idx1+1;
% Alamouti Space-Time Block Encoder
%  G = [ s1     s2 ]
%      [ -s2*   s1*]
y=[];
for n=1:LEN/Tx
  G=[    u(idx1(n))    u(idx2(n));...
      -conj(u(idx2(n))) conj(u(idx1(n)))];
  y=[y;G];
end
```

```
function y= Alamouti_Encoder2(u)
% Space-Time Block Encoder
Tx=2;
LEN=size(u,1);
idx1=1:Tx:LEN-1;
idx2=idx1+1;
% Alamouti Space-Time Block Encoder
%  G = [ s1     s2 ]
%      [ -s2*   s1*]
y=complex(zeros(LEN,Tx));
y(idx1,1)=u(idx1);
y(idx1,2)=u(idx2);
y(idx2,1)=-conj(u(idx2));
y(idx2,2)=conj(u(idx1));
```

```
Version 4: Vectorization + Preallocation

Iteration number 1
Iteration number 2
Iteration number 3
Iteration number 4
Iteration number 5
Iteration number 6
Iteration number 7
Iteration number 8
Version 4: Time to complete 8 iterations = 82.7194 (sec)
```

Figure 9.10   Fourth version of the algorithm: time taken to execute eight iterations

By running the following MATLAB script, we can verify whether this optimization leads to a faster execution time. The results show that processing of 1 million bits in eight iterations takes about 82.71 seconds (Figure 9.10).

Algorithm

MATLAB script: zPDCCH_v4_test

```
MaxSNR=8;
MaxNumBits=1e5;
fprintf(1,'\nVersion 4: Vectorization + Preallocation\n\n');
tic;
for snr=1:MaxSNR
    fprintf(1,'Iteration number %d\r',snr);
    ber= zPDCCH_v4 (snr, MaxNumBits, MaxNumBits);
end
time_4=toc;
fprintf(1,'Version 4: Time to complete %d iterations = %6.4f (sec)\n',
MaxSNR, time_4);
```

This series of optimizations reveals a pattern. We started with a baseline algorithm that implemented the Alamouti encoder and combiner algorithms with the simplest MATLAB code. The baseline code can be considered the transcribed version of the mathematical formula of the algorithm, which can be obtained from any textbook describing space–time block coding. MATLAB codes based on scalar operations may not be sufficient to run the same algorithm with a faster execution time. In most cases, we need to alter the sequence of operations in order to leverage the vector-based character of the MATLAB language. This means implementing the same algorithm by vectorizing the code and preallocating data.

However, these extra optimizations lead to rewriting of the MATLAB code. We can either spend time optimizing our code or, if we have access to them, take advantage of the functionality available in various MATLAB toolboxes. MATLAB toolboxes are written in such a way as to be sensitive to simulation performance. All MATLAB toolbox functions are based on preallocation and vectorization. Furthermore, as discussed earlier, DSP and the Communications System Toolbox provide efficient algorithmic components as System objects. In the next section we will take advantage of some of the System objects in the Communications System Toolbox to obtain faster implementations of many components of this algorithm.

9.6.3   System Objects

System objects can be used to accelerate a MATLAB code, largely in the areas of signal processing and communications. System objects are MATLAB object-oriented implementations of algorithms available in MATLAB toolboxes such as the Communications System Toolbox. By using System objects, we decouple the declaration (System object creation) from the execution of an algorithm, resulting in more efficient loop-based calculations, since we can perform parameter handling and initializations only once. A System object can be created and configured outside the loop, and then the step method can be called inside it. A majority of System objects from the DSP and Communications System Toolbox are implemented as MATLAB Executables (MEXs). A MEX implementation of a code is essentially a compiled C code. This can also speed up simulation, since many algorithmic optimizations have been included in the MEX implementations of objects.

The fifth version of the PDCCH algorithm uses System objects of the Communications System Toolbox to implement the Alamouti encoder (*Alamouti_EncoderS* function) and the Alamouti combiner (*Alamouti_CombinerS* function).

Algorithm

MATLAB function

```
function [ber, bits]=zPDCCH_v5(EbNo, maxNumErrs, maxNumBits)
%% Constants
FRM=2048;
M=4; k=log2(M); codeRate=1/3;
snr = EbNo + 10*log10(k) + 10*log10(codeRate);
trellis=poly2trellis(7, [133 171 165]);
L=FRM+24;C=6; Index=[L+1:(3*L/2) (L/2+1):L];
%% Initializations
persistent Modulator Demodulator CRCgen CRCdet
if isempty(Modulator)
    Modulator    = modem.pskmod('M', 4, 'PhaseOffset', pi/4, 'SymbolOrder', 'Gray',
'InputType', 'Bit');
    Demodulator = modem.pskdemod('M', 4, 'PhaseOffset', pi/4, 'SymbolOrder', 'Gray',
'OutputType', 'Bit');
    CRCgen      = crc.generator([1 1 zeros(1, 16) 1 1 0 0 0 1 1]);
    CRCdet      = crc.detector   ([1 1 zeros(1, 16) 1 1 0 0 0 1 1]);
end
%% Processing loop
numErrs = 0; numBits = 0; nS=0;
while ((numErrs < maxNumErrs) && (numBits < maxNumBits))
```

```
% Transmitter
u          = randi([0 1], FRM,1);                      % Generate bit payload
u1         = generate(CRCgen,u);                        % CRC insertion
u2         = u1((end-C+1):end);                         % Tail-biting convolutional coding
[~, state] = convenc(u2,trellis);
u3         = convenc(u1,trellis,state);
u4         = fcn_RateMatcher(u3, L, codeRate);  % Rate matching
u5         = fcn_Scrambler(u4, nS);              % Scrambling
u6         = modulate(Modulator, u5);            % Modulation
u7         = TransmitDiversityEncoderS(u6);      % MIMO Alamouti encoder
% Channel
[u8, h8]   = MIMOFadingChan(u7);                 % MIMO fading channel
sigpower = 10*log10(real(var(u8(:))));
u9         = awgn(u8,snr,sigpower,'dB');
% Receiver
uA         = TransmitDiversityCombinerS(u9, h8); % MIMO Alamouti combiner
uB         = demodulate(Demodulator,uA);         % Demodulation
uC         = fcn_Descrambler(uB, nS);            % Descrambling
uD         = fcn_RateDematcher(uC, L);           % Rate de-matching
uE         = [uD;uD];                            % Tail-biting
uF         = vitdec(uE ,trellis,34,'trunc','hard'); % Viterbi decoding
uG         = uF(Index);
y          = detect(CRCdet, uG );                % CRC detection
numErrs  = numErrs + sum( y~=u );                % Update number of bit errors
numBits  = numBits + FRM;
nS         = nS + 2; nS = mod(nS, 20);
end
%% Clean up & collect results
ber = numErrs/numBits;
bits=numBits;
```

This version uses the functions *TransmitDiversityEncoderS* and *TransmitDiversityCombinerS*, which in turn use the Alamouti encoder and combiner functions implemented with System objects.

MATLAB function

```
function y = TransmitDiversityEncoderS(in)
%#codegen
% Alamouti Transmit Diversity Encoder
% Space-Frequency to Space-
   Time transformation
in = sqrt(2) * in;
% STBC Alamouti
y = Alamouti_EncoderS(in);
```

```
function y = TransmitDiversityCombinerS(in,
   chEst)
%#codegen
% Alamouti Transmit Diversity Combiner
% Scale
in = sqrt(2) * in;
% STBC Alamouti
y =  Alamouti_DecoderS(in, chEst);
```

```
% Scale                                    % Space-Frequency to Space-
y = y/sqrt(2);                                Time transformation
                                           y(2:2:end) = -conj(y(2:2:end));
```

```
function y = Alamouti_EncoderS(u)          function s = Alamouti_DecoderS(u,H)
% STBCENC Space-Time Block Encoder         %#codegen
% Outputs the Space-                       % STBC_DEC STBC Combiner
   Time block encoded matrix               persistent hTDDec
persistent hTDEnc;                         if isempty(hTDDec)
if isempty(hTDEnc)                              hTDDec= comm.OSTBCCombiner(...
    % Use same object for either scheme        'NumTransmitAnten-
    hTDEnc = comm.OSTBCEncoder                  nas',2,'NumReceiveAntennas',2);
    ('NumTransmitAntennas', 2);            end
end                                        s = step(hTDDec, u, H);
% Alamouti Space-Time Block Encoder
y = step(hTDEnc, u);
```

Note that we create the *comm.OSTBCEncoder* and *comm.OSTBCCombiner* System objects only the first time we enter the function. This is accomplished by denoting the System objects as MATLAB *persistent* variables. We then use the *isempty* function, which ensures that everything is performed only the first time the persistent variable is "empty," or in other words not initialized. Both of the Alamouti algorithms are then executed by calling the *step* functions of their corresponding System objects.

Let us now verify how, by using available System objects, we can avoid the preallocation and vectorization steps yet arrive at a faster execution time. Running the following MATLAB script will call the new fifth version of the algorithm, which uses System objects. The results (as illustrated in Figure 9.11) show that processing 1 million bits in eight iterations takes about 81.91 seconds. This execution time is close to that obtained with the fourth version of the algorithm. Note that we avoided all code updates by using available System object functionality in the toolbox.

```
Version 5: System objects for MIMO

Iteration number 1
Iteration number 2
Iteration number 3
Iteration number 4
Iteration number 5
Iteration number 6
Iteration number 7
Iteration number 8
Version 5: Time to complete 8 iterations = 81.9175 (sec)
```

Figure 9.11 Fifth version of the algorithm: time taken to execute eight iterations

Algorithm

MATLAB script: zPDCCH_v4_test

```
MaxSNR=8;
MaxNumBits=1e5;
fprintf(1,'\nVersion 5: Using System objects for MIMO \n\n');
tic;
for snr=1:MaxSNR
    fprintf(1,'Iteration number %d\r',snr);
    ber= zPDCCH_v5 (snr, MaxNumBits, MaxNumBits);
end
time_5=toc;
fprintf(1,'Version 5: Time to complete %d iterations = %6.4f (sec)\n', MaxSNR, time_5);
```

In order to keep track of the extent of acceleration using the various techniques discussed so far, we have included a helper MATLAB function (*Report_Timing_Results.m*). This function takes as input four parameters: the algorithm version, the time to process the baseline, the time to process the current version, and a text string describing the optimization technique. It returns a table that tracks the simulation times.

Algorithm

MATLAB function: Report_Timing_Results

```
function y=Report_Timing_Results(M,a,b,str)
persistent Results
if isempty(Results)
    Results={};
end
Results(M).name=str;
Results(M).elapsed_time=b;
Results(M).acceleration=a/b;
disp('------------------------------------------------------------------------------------');
disp('Versions of the Transceiver                    | Elapsed Time (sec)| Acceleration Ratio');
for m=1:M
fprintf(1,'%d. %-49s| %17.4f | %12.4f\n',m, Results(m).name, Results(m).elapsed_time,
Results(m).acceleration);
end
disp('------------------------------------------------------------------------------------');
y=Results;
end
```

By running this function, we can recall different versions and their execution times and compute the acceleration ratios compared to the baseline algorithm. The results indicate that the version that uses System objects accelerates the simulation by a factor of 5.02 (Figure 9.12).

```
Versions of the Transceiver                   | Elapsed Time (sec)| Acceleration Ratio
 1. Baseline                                   |           411.3030 |          1.0000
 2. Vectorization                              |           326.5071 |          1.2597
 3. Vectorization along larger dimension       |           175.8478 |          2.3390
 4. Vectorization + Preallocation              |            82.7194 |          4.9723
 5. System objects for MIMO                    |            81.9175 |          5.0209
```

Figure 9.12   Execution times and acceleration ratios for the first five versions of the algorithm

Algorithm

MATLAB script

```
Report_Timing_Results(1,time_1,time_1,'Baseline');
Report_Timing_Results(2,time_1,time_2,'Vectorization');
Report_Timing_Results(3,time_1,time_3,'Vectorization along larger dimension');
Report_Timing_Results(4,time_1,time_4,'Vectorization + Preallocation');
Report_Timing_Results(5,time_1,time_5,'System objects for MIMO');
```

Profiling the fifth version of the algorithm can help us identify the next target bottleneck for optimization. The MATLAB profiler commands and the report are shown in Figure 9.13.

Algorithm

MATLAB script

```
profile on;
ber= zPDCCH_v5(snr, MaxNumBits, MaxNumBits);
profile viewer;
```

Profile Summary

Generated 15-Dec-2012 17:03:43 using cpu time.

Function Name	Calls	Total Time	Self Time*	Total Time Plot (dark band = self time)
zPDCCH_v5	1	1.490 s	0.024 s	
MIMOFadingChan	49	0.503 s	0.009 s	
Channel.filter	49	0.493 s	0.057 s	
vitdec	49	0.299 s	0.046 s	
Channel.reset	49	0.281 s	0.046 s	
vit (MEX-file)	49	0.243 s	0.243 s	

Figure 9.13   Profile summary report for the fifth version of the algorithm

The profile summary identifies two algorithms – the MIMO channel model (*MIMOFadingChan.m*) and the Viterbi decoder (*vitdec* function) – as the next bottlenecks. These algorithms are based on two functions from the Communications System Toolbox: *mimochan* and *vitdec*. These functions, like all functions in MATLAB toolboxes, are vectorized and preallocated. However, replacing these two with the corresponding System objects will result in performance improvements. Using these System objects highlights the two mechanisms by which System objects achieve accelerations: avoidance of repeated parameter validations and use of a MATLAB MEX implementation.

9.6.3.1   MATLAB MEX Implementation

The function *MIMOFadingChan.m* represents the next bottleneck to be addressed in the fifth version of the PDCCH algorithm. Examining the following *MIMOFadingChan* function reveals that it uses the *mimochan* object of the Communications System Toolbox to perform a MIMO channel-filtering operation. Using a persistent variable in MATLAB, only the first time the function is entered is the *mimochan* object initialized. Each time we call the function, we execute the object's filter method to obtain both the filtered output of the channel model (*variable y*) and the channel gains (*variable h*).

In the sixth version of the algorithm (see Figure 9.14), we use an alternative implementation of MIMO channel filtering and employ the *comm.MIMOChannel* System object. Examining the *MIMOFadingChanS* function, we notice that only the first time the function is entered is the *comm.MIMOChannel* System object initialized. Executing the step method provides both the filtered output and the channel gains.

MATLAB function

---

function [ber, bits]=zPDCCH_v5(...)	**function [ber, bits]=zPDCCH_v6(...)**
.	.
.	.
.	.

[u8, h8] = MIMOFadingChan(u7);	[u8, h8] = MIMOFadingChanS(u7);
.	.
.	.

---

end	**end**

```
function [y, h] = MIMOFadingChan(in)
% MIMOFadingChan
numTx=2;
numRx=2;
chanSRate=(2048*15000);
Doppler=70;
PathDelays = 0;
PathGains  = 0;
persistent chanObj
if isempty(chanObj)
```

```
function [y, h] = MIMOFadingChanS(in)
% MIMOFadingChan
numTx=2;
numRx=2;
chanSRate=(2048*15000);
Doppler=70;
PathDelays = 0;
PathGains  = 0;
persistent chanObj
if isempty(chanObj)
```

```
chanObj = mimochan(numTx,numRx,
(1/chanSRate),Doppler,PathDelays,
   PathGains);
   chanObj.NormalizePathGains = 1;
   chanObj.StorePathGains = 1;
   chanObj.ResetBeforeFiltering = 1;
end
y        = filter(chanObj, in);
ChGains = chanObj.PathGains;
Len      = size(in,1);
h     = complex(zeros(Len,numTx,numRx));
h(:) = ChGains(:,1,:,:);
```

```
chanObj = comm.MIMOChannel
('SampleRate', chanSRate,
          'MaximumDopplerShift', Doppler,
          'PathDelays', PathDelays,
          'AveragePathGains', PathGains,
          'NumTransmitAntennas', numTx,...
          'TransmitCorrelationMatrix',
eye(numTx),...
          'NumReceiveAntennas', numRx,...
          'ReceiveCorrelationMatrix',
eye(numRx),...
          'PathGainsOutputPort', true,...
          'NormalizePathGains', true,...
          'NormalizeChannelOutputs', true);
end
[y, G] = step(chanObj, in);
Len        = size(in,1);
PathG      = com-
   plex(zeros(Len,numTx,numRx));
PathG(:) = G(:,1,:,:);
h          = PathG;
```

As we can readily see, there are many similarities between the *mimochan* object and the *comm.MIMOChannel* System object. The System object is based on a MATLAB MEX implementation (compiled C code) and integrates various optimizations. Therefore, we expect to see a performance improvement from the use of the *MIMOFadingChanS* function, which employs the *comm.MIMOChannel* System object. To verify this, we run the following MATLAB script, which recalls the performance improvements up to this point (Figure 9.15).

```
Version 6: System objects for MIMO & Channel modeling

Iteration number 1
Iteration number 2
Iteration number 3
Iteration number 4
Iteration number 5
Iteration number 6
Iteration number 7
Iteration number 8
Version 6: Time to complete 8 iterations = 59.9528 (sec)
```

Figure 9.14 Sixth version of the algorithm: time taken to execute eight iterations

```
Versions of the Transceiver                    | Elapsed Time (sec)| Acceleration Ratio
 1. Baseline                                    |          411.3030 |         1.0000
 2. Vectorization                               |          326.5071 |         1.2597
 3. Vectorization along larger dimension        |          175.8478 |         2.3390
 4. Vectorization + Preallocation               |           82.7194 |         4.9723
 5. System objects for MIMO                     |           81.9175 |         5.0209
 6. System objects for MIMO & Channel           |           59.9528 |         6.8604
```

Figure 9.15   Execution times and acceleration ratios for the first six versions of the algorithm

Algorithm

MATLAB script: zPDCCH_v6_test

```
MaxSNR=8;
MaxNumBits=1e5;
fprintf(1,'\nVersion 6: System objects for MIMO & Channel\n\n');
tic;
for snr=1:MaxSNR
    fprintf(1,'Iteration number %d\r',snr);
    ber= zPDCCH_v6 (snr, MaxNumBits, MaxNumBits);
end
time_6=toc;
fprintf(1,'Version 6: Time to complete %d iterations = %6.4f (sec)\n', MaxSNR, time_6);
Report_Timing_Results(6,time_1,time_6, System objects for MIMO & Channel);
```

Following the last five optimizations, we profile the sixth version of the algorithm. As illustrated by the profile summary report in Figure 9.13, the Viterbi decoder function *vitdec* from the Communications System Toolbox represents the bottleneck of the sixth version (Figure 9.16).

9.6.3.2   Avoiding Repeated Parameter Validations

The seventh version of the algorithm (Figure 9.17) optimizes the code by replacing the *vitdec* function that implements the Viterbi decoder with the corresponding System object,

Profile Summary
Generated 18-Dec-2012 13:26:20 using cpu time.

Function Name	Calls	Total Time	Self Time*	Total Time Plot (dark band = self time)
zPDCCH_v6	1	1.162 s	0.011 s	
vitdec	49	0.351 s	0.041 s	
vit (MEX-file)	49	0.310 s	0.310 s	

Figure 9.16   Profile summary report for the sixth version of the algorithm

```
Version 7: System objects for MIMO & Channel & Viterbi

Iteration number 1
Iteration number 2
Iteration number 3
Iteration number 4
Iteration number 5
Iteration number 6
Iteration number 7
Iteration number 8
Version 7: Time to complete 8 iterations = 58.8660 (sec)
```

Figure 9.17   Seventh version of the algorithm: time taken to execute eight iterations

comm. ViterbiDecoder. Using this System object can enable acceleration by avoiding repeated parameter validations. Since System objects decouple declaration (System object creation) from execution, parameter handling and initializations occur only once outside the while loop. However, in the *vitdec* function, every time the function is called within the loop, parameters such as trellis structure and termination and decision methods are checked for validity and appropriate intermediate variables are created before the main function is called.

This type of parameter-handling overhead is necessary when we are experimenting with different modes of a function and interacting with it at the command line. However, when function parameters are fixed and already determined and the function is being executed in a loop, avoiding extra parameter handling – as System objects are designed to do – can improve the simulation performance.

MATLAB function

```
function [ber, bits]=zPDCCH_v6(...)       function [ber, bits]=zPDCCH_v7(...)
.                                         .
.                                         .
                                          Viterbi=comm.ViterbiDecoder(
while ((numErrs < maxNumErrs) &&          'TrellisStructure', trellis, 'InputFormat','Hard',
   (numBits < maxNumBits))                'TerminationMethod','Truncated');
.                                         .
```
```
uF       = vitdec(uE,trellis,34,'trunc','hard');    while ((numErrs < maxNumErrs) &&
   % Viterbi decoding                                  (numBits < maxNumBits))
                                                     .
```
```
.                                         uF       = step(Viterbi, uE); % Viterbi decoding
.
end                                       .
                                          .
                                          end
```

```
Versions of the Transceiver                      | Elapsed Time (sec)| Acceleration Ratio
  1. Baseline                                     |          411.3030 |          1.0000
  2. Vectorization                                |          326.5071 |          1.2597
  3. Vectorization along larger dimension         |          175.8478 |          2.3390
  4. Vectorization + Preallocation                |           82.7194 |          4.9723
  5. System objects for MIMO                      |           81.9175 |          5.0209
  6. System objects for MIMO & Channel            |           59.9528 |          6.8604
  7. System objects for MIMO & Channel & Viterbi  |           58.8660 |          6.9871
```

Figure 9.18   Execution times and acceleration ratios for the first seven versions of the algorithm

We can verify this optimization by running the following MATLAB script, which calls the seventh version of the algorithm and recalls the collective performance improvements (Figure 9.18).

Algorithm

MATLAB script: zPDCCH_v7_test

```
MaxSNR=8;
MaxNumBits=1e5;
fprintf(1,'\nVersion 7: System objects for MIMO & Channel & Viterbi\n\n');
tic;
for snr=1:MaxSNR
    fprintf(1,'Iteration number %d\r',snr);
    ber= zPDCCH_v7 (snr, MaxNumBits, MaxNumBits);
end
time_7=toc;
fprintf(1,'Version 7: Time to complete %d iterations = %6.4f (sec)\n', MaxSNR,
time_7);
Report_Timing_Results(7,time_1,time_7,'System objects for MIMO & Channel
& Viterbi');
```

```
Version 8: Using All available System objects

Iteration number 1
Iteration number 2
Iteration number 3
Iteration number 4
Iteration number 5
Iteration number 6
Iteration number 7
Iteration number 8
Version 8: Time to complete 8 iterations = 48.9014 (sec)
```

Figure 9.19   Eighth version of the algorithm: time taken to execute eight iterations

9.6.3.3 Using All Available System Objects

In the eighth version (Figure 9.19), we use all the System objects pertinent to this algorithm. In addition to the System objects used so far, we also implement the modulator, the demodulator, two convolutional encoders (used in tail-biting encoding), and CRC generation and detection functionalities.

Algorithm

MATLAB function

```
function [ber, bits]=zPDCCH_v8(EbNo, maxNumErrs, maxNumBits)
%% Constants
FRM=2048;
M=4; k=log2(M); codeRate=1/3;
snr = EbNo + 10*log10(k) + 10*log10(codeRate);
trellis=poly2trellis(7, [133 171 165]);
L=FRM+24;C=6; Index=[L+1:(3*L/2) (L/2+1):L];
%% Initializations
persistent Modulator AWGN DeModulator BitError ConvEncoder1 ConvEncoder2 Viterbi
CRCGen CRCDet
if isempty(Modulator)
    Modulator    = comm.QPSKModulator('BitInput',true);
    AWGN         = comm.AWGNChannel('NoiseMethod', 'Variance', 'VarianceSource',
'Input port');
    DeModulator =  comm.QPSKDemodulator('BitOutput',true);
    BitError     = comm.ErrorRate;
    ConvEncoder1=comm.ConvolutionalEncoder('TrellisStructure', trellis,
'FinalStateOutputPort', true, ...
        'TerminationMethod','Truncated');
    ConvEncoder2 =  comm.ConvolutionalEncoder('TerminationMethod','Truncated',
'InitialStateInputPort', true,...
        'TrellisStructure', trellis);
    Viterbi=comm.ViterbiDecoder('TrellisStructure', trellis,
'InputFormat','Hard','TerminationMethod','Truncated');
    CRCGen = comm.CRCGenerator('Polynomial',[1 1 zeros(1, 16) 1 1 0 0 0 1 1]);
    CRCDet = comm.CRCDetector   ('Polynomial',[1 1 zeros(1, 16) 1 1 0 0 0 1 1]);
end
%% Processing loop modeling transmitter, channel model and receiver
numErrs = 0; numBits = 0; nS=0;
results=zeros(3,1);
while ((numErrs < maxNumErrs) && (numBits < maxNumBits))
    % Transmitter
    u       = randi([0 1], FRM,1);            % Generate bit payload
    u1      = step(CRCGen, u);                % CRC insertion
    u2      = u1((end-C+1):end);              % Tail-biting convolutional coding
```

```
[~, state]  = step(ConvEncoder1, u2);
u3          = step(ConvEncoder2, u1,state);
u4          = fcn_RateMatcher(u3, L, codeRate);  % Rate matching
u5          = fcn_Scrambler(u4, nS);             % Scrambling
u6          = step(Modulator, u5);               % Modulation
u7          = TransmitDiversityEncoderS(u6);     % MIMO Alamouti encoder
  % Channel
[u8, h8]    = MIMOFadingChanS(u7);               % MIMO fading channel
noise_var = real(var(u8(:)))/(10.^(0.1*snr));
u9          = step(AWGN, u8, noise_var);         % AWGN
% Receiver
uA          = TransmitDiversityCombinerS(u9, h8);% MIMO Alamouti combiner
uB          = step(DeModulator, uA);             % Demodulation
uC          = fcn_Descrambler(uB, nS);           % Descrambling
uD          = fcn_RateDematcher(uC, L);          % Rate de-matching
uE          = [uD;uD];                           % Tail-biting
uF          = step(Viterbi, uE);                 % Viterbi decoding
uG          = uF(Index);
y           = step(CRCDet, uG );                 % CRC detection
results     = step(BitError, u, y);              % Update number of bit errors
numErrs    = results(2);
numBits    = results(3);
nS          = nS + 2; nS = mod(nS, 20);
end
%% Clean up & collect results
ber = results(1); bits= results(3);
reset(BitError);
```

By running the following MATLAB script, which calls the eighth version of the algorithm, and recalling the collective performance improvements (Figure 9.20), we can verify that using all pertinent System objects makes a positive difference to simulation speed.

```
Versions of the Transceiver                      | Elapsed Time (sec)| Acceleration Ratio
 1. Baseline                                     |        411.3030 |           1.0000
 2. Vectorization                                |        326.5071 |           1.2597
 3. Vectorization along larger dimension         |        175.8478 |           2.3390
 4. Vectorization + Preallocation                |         82.7194 |           4.9723
 5. System objects for MIMO                      |         81.9175 |           5.0209
 6. System objects for MIMO & Channel            |         59.9528 |           6.8604
 7. System objects for MIMO & Channel & Viterbi  |         58.8660 |           6.9871
 8. System objects for all                       |         48.9014 |           8.4109
```

Figure 9.20 Execution times and acceleration ratios for the first eight versions of the algorithm

Algorithm

MATLAB script: zPDCCH_v8_test

```
MaxSNR=8;
MaxNumBits=1e5;
fprintf(1,'\nVersion 8: Using All available System objects\n\n');
tic;
for snr=1:MaxSNR
    fprintf(1,'Iteration number %d\r',snr);
    ber= zPDCCH_v8(snr, MaxNumBits, MaxNumBits);
end
time_8=toc;
fprintf(1,'Version 8: Time to complete %d iterations = %6.4f (sec)\n', MaxSNR, time_8);
Report_Timing_Results(8,time_1,time_8,'System objects for all');
```

So far we have shown how writing better MATLAB code can result in faster simulation. We have also showcased the use of System objects from the Communications System Toolbox as a way of accelerating the simulation speed of an algorithm (in most cases). Another benefit of using System objects is that they support MATLAB-to-C code generation with MATLAB Coder. This feature is one of three additional acceleration features that will be discussed next.

9.7   Using Acceleration Features

The techniques described so far focus on ways of optimizing MATLAB programs. Beside code optimization, performance improvements can be gained from the use of additional computing power or by retargeting a design to compiled C code. MATLAB parallel-computing products take advantage of multicore processors, computer clusters, and GPUs. MATLAB Coder provides the ability to automatically convert a MATLAB code to C code, which can be compiled to provide faster simulations. In the next section we take advantage of these features to further accelerate simulation speed.

9.7.1   MATLAB-to-C Code Generation

Replacing parts of a MATLAB code with automatically generated MEX (function) may speed up simulations. Using MATLAB Coder, we can generate readable and portable C code and compile it into a MEX function that replaces the appropriate parts of an existing MATLAB algorithm. The amount of acceleration will depend on the algorithm. The best way of determining acceleration is to use MATLAB Coder to generate a MEX function and test the speed-up firsthand. If the algorithm contains single-precision data types, fixed-point data types, loops with states, or code that cannot be vectorized, we are likely to see speed-ups. Much of the

MATLAB language and many toolboxes, including the Communications System Toolbox, support code generation.

In this step, we generate a MEX function for the eighth version of the PDCCH algorithm. The MEX function generated will be the ninth version in our sequence of acceleration steps. This process involves using a single MATLAB command (*codegen*) available in MATLAB Coder. The following MATLAB script shows how to call the *codegen* command to convert the function *zPDCCH_v8.m* to C code and compile it into a MEX function. If the name of the output MEX function is not specified, the default will be the name of the MATLAB function followed by a *_mex* suffix, in this case *zPDCCH_v8_mex*.

Algorithm

MATLAB script: zPDCCH_v8_codegen

```
MaxSNR=8;
MaxNumBits=1e5;
fprintf(1,'\n\nGenerating MEX function for zPDCCH_v8.m \r');
codegen -args { MaxSNR, MaxNumBits, MaxNumBits } zPDCCH_v8.m
fprintf(1,'Done.\r');
```

By running the following MATLAB script we can verify whether this optimization leads to a faster execution time. The results indicate that processing 1 million bits in eight iterations takes about 37.18 seconds (Figures 9.21 and 9.22).

Algorithm

MATLAB script: zPDCCH_v9_test

```
MaxSNR=8;
MaxNumBits=1e6;
fprintf(1,'\nVersion 9: MATLAB to C code generation (MEX)\n\n');
tic;
for EbNo=1:MaxSNR
   fprintf(1,'Iteration number %d\r',EbNo);
   ber= zPDCCH_v8_mex(snr, MaxNumBits, MaxNumBits);
end
time_9=toc;
fprintf(1,'Version 9: Time to complete %d iterations = %6.4f (sec)\n', MaxSNR, time_9);
```

The results show the simulation time of the MEX version of the algorithm. Note that when the System-object algorithm is compiled into a MEX function, the MEX version of the algorithm runs faster than either earlier version.

```
Version 9: MATLAB to C code generation (MEX)

Iteration number 1
Iteration number 2
Iteration number 3
Iteration number 4
Iteration number 5
Iteration number 6
Iteration number 7
Iteration number 8
Version 9: Time to complete 8 iterations = 31.2941 (sec)
```

Figure 9.21   Ninth version of the algorithm: time taken to execute eight iterations

Versions of the Transceiver	Elapsed Time (sec)	Acceleration Ratio
1. Baseline	411.3030	1.0000
2. Vectorization	326.5071	1.2597
3. Vectorization along larger dimension	175.8478	2.3390
4. Vectorization + Preallocation	82.7194	4.9723
5. System objects for MIMO	81.9175	5.0209
6. System objects for MIMO & Channel	59.9528	6.8604
7. System objects for MIMO & Channel & Viterbi	58.8660	6.9871
8. System objects for all	48.9014	8.4109
9. Version 8 + MATLAB to C code generation (MEX)	31.4722	13.0688

Figure 9.22   Execution times and acceleration ratios for the first nine versions of the algorithm

This behavior is expected because one of the advantages of using MATLAB-to-C code generation is simulation acceleration. Although the algorithm that uses System objects is highly optimized, code generation can accelerate simulation by locking down the sizes and data types of variables inside the function. This process makes the execution more efficient because it removes the overhead of the interpreted language that checks for size and data type in every line of code. If an algorithm contains MATLAB functions that have implicitly multithread computations, functions that call IPP or BLAS libraries, built-in functions optimized for execution in MATLAB on a PC (such as Fast Fourier Transforms, FFTs), or algorithms that can vectorize the code, we are not likely to see any speed-ups.

9.7.2   Parallel Computing

Using the Parallel Computing Toolbox, we can run multiple MATLAB workers (MATLAB computational engines) on a desktop multicore machine. Simulations can be sped up by dividing computations across multiple MATLAB workers. This approach allows more control over the parallelism than is available in the built-in multithreading found in MATLAB and it is often used for applications that involve parameter sweeps and Monte Carlo simulations.

Additionally, we can scale parallel applications that use MATLAB workers to a computer, cluster, or grid.

The Parallel Computing Toolbox also offers high-level programming constructs such as the *parfor* command. Using parfor, we can accelerate for loops by dividing loop iterations for simultaneous executions across a number of MATLAB workers. To use parfor, the loop iterations must be independent, with none depending on on any of the others. If we want to accelerate dependent or state-based loops, we should consider either optimization of the body of the for loop or generation of C code instead. Since there is a communications cost involved in a parfor loop, there might be no advantage to using one when we have only a small number of simple calculations.

In this step, we call the MEX function representing the ninth version of the PDCCH algorithm within a parfor loop. Before doing so, we must access multiple cores in our computer. The *matlabpool* command (or the *parpool* command in more recent releases of MATLAB) allows us to access various cores on the computer and assigns each a MATLAB worker.

Algorithm

MATLAB script: zPDCCH_vA_test

```
isOpen = matlabpool('size') > 0;
if ~isOpen
    fprintf(1,'Parallel Computing Toolbox is starting ...\n');
    matlabpool;
end
```

At this stage we can run the parfor loop instead of the for loop and take advantage of parallel computing. The MATLAB script for these operations is as follows.

Algorithm

MATLAB script: zPDCCH_vA_test

```
MaxSNR=8;
MaxNumBits=1e6;
fprintf(1,'\nVersion 10: Parallel computing (parfor) + MEX \n\n');
tic;
parfor snr=1:MaxSNR
    fprintf(1,'Iteration number %d\r',snr);
    ber= zPDCCH_v9(snr, MaxNumBits, MaxNumBits);
end
time_A=toc;
fprintf(1,'Version 10: Time to complete %d iterations = %6.4f (sec)\n', MaxSNR, time_A);
```

```
Version 10: Parallel computing (parfor) + MEX

Iteration number 6
Iteration number 5
Iteration number 4
Iteration number 7

Iteration number 3
Iteration number 2
Iteration number 1
Iteration number 8

Version 10: Time to complete 8 iterations = 18.3899 (sec)
```

Figure 9.23   Tenth version of the algorithm: time taken to execute eight iterations

```
Versions of the Transceiver                          | Elapsed Time (sec)| Acceleration Ratio
 1. Baseline                                         |        411.3030 |         1.0000
 2. Vectorization                                    |        326.5071 |         1.2597
 3. Vectorization along larger dimension             |        175.8478 |         2.3390
 4. Vectorization + Preallocation                    |         82.7194 |         4.9723
 5. System objects for MIMO                          |         81.9175 |         5.0209
 6. System objects for MIMO & Channel                |         59.9528 |         6.8604
 7. System objects for MIMO & Channel & Viterbi      |         58.8660 |         6.9871
 8. System objects for all                           |         48.9014 |         8.4109
 9. Version 8 + MATLAB to C code generation (MEX)    |         31.4722 |        13.0688
10. Version 8 + MEX + Parallel computing (parfor)    |         18.3899 |        22.3657
```

Figure 9.24   Execution times and acceleration ratios for the first 10 versions of the algorithm

We observe that with the combination of System objects from the Communications System Toolbox, MATLAB-to C-code generation, and parallel processing, we can reduce the processing time to about 18.38 seconds (Figure 9.23). This corresponds to a 22.36-times acceleration compared to the original baseline and a 3.45-times acceleration compared to the fourth version, which employs the basic MATLAB programming guidelines (see Figure 9.24). As with many performance benchmarks, the extent of acceleration depends on the algorithm, the platform where MATLAB is installed, the C/C++ compiler used to create the MEX function, and the number of cores available in the computer.

9.8   Using a Simulink Model

So far we have updated our MATLAB programs for better performance. The same process can be applied to the algorithms represented by Simulink models. Simulink allows us to represent

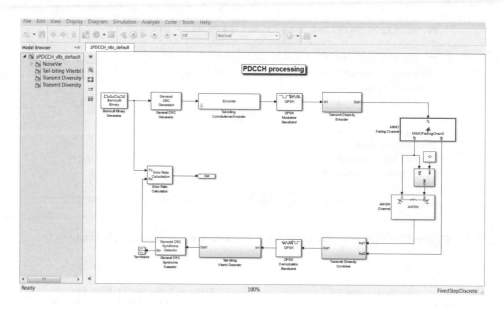

Figure 9.25 Simulink model representing the PDCCH algorithm

a design as a block diagram. Such a graphical representation naturally captures the architecture and hierarchy and makes them easier to understand.

The Communications System Toolbox provides algorithms either as System objects for use in MATLAB or as blocks for use in a Simulink. The eighth version of the PDCCH algorithm, for example, uses many System objects from the Communications System Toolbox. In this section we implement the same algorithm as a Simulink model. We will first verify that it has the same numerical performance as the MATLAB program and then examine various Simulink optimization techniques that can substantially accelerate the simulation speed.

9.8.1 Creating the Simulink Model

Figure 9.25 shows the Simulink model (*zPDCCH_v8_default.xls*) representing the eighth version of the PDCCH algorithm. The process of transforming the MATLAB program to Simulink is made easier by the fact that the MATLAB implementation uses System objects. From the block library of the Communications System Toolbox we can access such blocks as convolutional encoders, Viterbi decoders, and so on. The System objects and blocks from a given system toolbox are numerically identical and have the same properties. Therefore, we can easily set the properties of the blocks in Simulink by copying them from the System objects. For algorithms such as transmit diversity, which involves System objects and some basic MATLAB code, we compose a subsystem that represents the same operations with Simulink blocks. For an algorithm component that cannot be easily implemented with a few blocks or a subsystem, we can use the MATLAB function block in Simulink to directly turn a MAT-LAB function into a Simulink block. We use this approach to implement the MIMO fading channel block.

9.8.2   Verifying Numerical Equivalence

We use *bertool* to verify that the implementations in MATLAB and Simulink produce the same numerical results. *bertool* performs the following operations:

- Iterates through a set of *Eb/N0* values
- Executes the Simulink model or MATLAB function for each value
- Signals the simulation stopping criteria to the Error Rate Calculation block using two parameters: maximum number of errors and maximum number of bits
- Records the BER value of the current iteration and displays it on the BER curve.

As illustrated in Figure 9.26, we process the eight version of the algorithm in both MATLAB (*zPDCCH_v8.m*) and Simulink (*zPDCCH_v8_default.xls*). By iterating with SNR values from 0 to 4, with a resolution of ½, and by setting both maximum number of bits and maximum number of errors to 10 million, we can compare the BER curve as a function of SNR. As Figure 9.27 illustrates, the numerical results are very similar. The reason for the small discrepancy at higher SNR values is that we have chosen to simulate with only 10 million bits for each SNR value. At high SNR values with BERs around 1e-6 to 1e-7, a few error bits will affect the performance results. Running simulations with larger numbers of bits can make the numerical results identical.

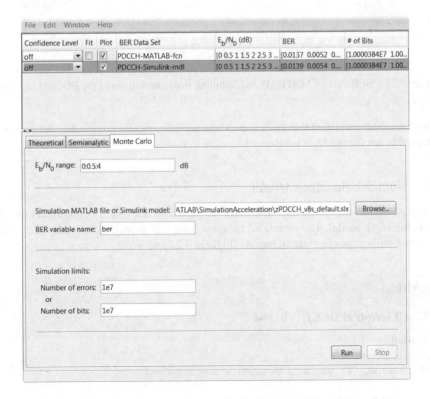

Figure 9.26   *bertool* iterating PDCCH MATLAB and Simulink models

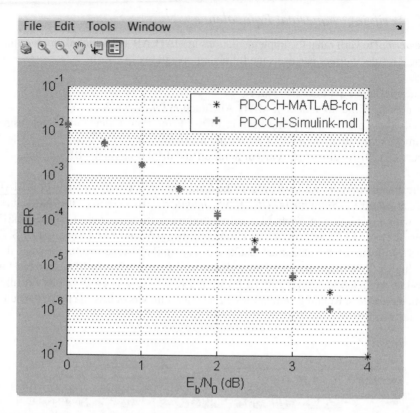

Figure 9.27   BER curves: MATLAB and Simulink implementations of the PDCCH algorithm

Now that we have verified that both algorithms are numerically compatible, let us compare the elapsed times for the MATLAB and Simulink versions.

9.8.3   Simulink Baseline Model

In this step, we run the following MATLAB script, which uses the *sim* command to run the baseline Simulink model. The simulated baseline Simulink model takes about 84.59 seconds to process 1 million bits in eight iterations (Figures 9.28 and 9.29).

Algorithm

MATLAB script: zPDCCH_vB_test

```
MaxSNR=8;
MaxNumBits=1e6;
fprintf(1,'\nVersion 11: Version 8 Simulink normal mode'\n\n');
```

```
tic;
for snr=1:MaxSNR
   fprintf(1,'Iteration number %d\r',snr);
   sim('zPDCCH_v8s_default');
end
time_11=toc;
fprintf(1,'Version 11: Time to complete %d iterations = %6.4f (sec)\n', MaxSNR, time_11);
```

9.8.4   Optimizing the Simulink Model

We can accelerate the simulation speed of a Simulink model via multiple methods, including turning off visualizations and debugging, and introducing the acceleration features already discussed for MATLAB programs, including C code generation and parallel computing. We will discuss all of these techniques in this section, starting from the baseline model of the PDCCH algorithm in Simulink.

```
Version 11: Version 8 Simulink normal mode

Iteration number 1
Iteration number 2
Iteration number 3
Iteration number 4
Iteration number 5
Iteration number 6
Iteration number 7
Iteration number 8
Version 11: Time to complete 8 iterations = 84.5959 (sec)
```

Figure 9.28   Eleventh version of the algorithm: time taken to execute eight iterations

```
Versions of the Transceiver                          | Elapsed Time (sec)| Acceleration Ratio
 1. Baseline                                         |        411.3030 |       1.0000
 2. Vectorization                                    |        326.5071 |       1.2597
 3. Vectorization along larger dimension             |        175.8478 |       2.3390
 4. Vectorization + Preallocation                    |         82.7194 |       4.9723
 5. System objects for MIMO                          |         81.9175 |       5.0209
 6. System objects for MIMO & Channel                |         59.9528 |       6.8604
 7. System objects for MIMO & Channel & Viterbi      |         58.8660 |       6.9871
 8. System objects for all                           |         48.9014 |       8.4109
 9. Version 8 + MATLAB to C code generation (MEX)    |         31.4722 |      13.0688
10. Version 8 + MEX + Parallel computing (parfor)    |         18.3899 |      22.3657
11. Version 8 Simulink normal mode                   |         84.5959 |       4.8620
```

Figure 9.29   Execution times and acceleration ratios for the first 11 versions of the algorithm

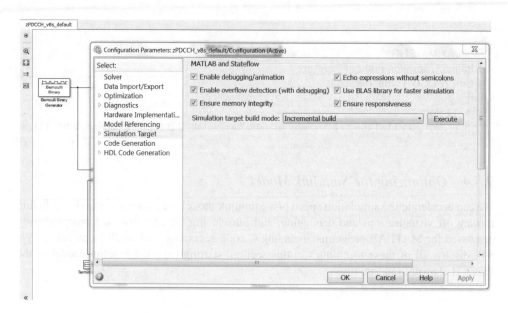

Figure 9.30   Model configuration parameters: default – normal mode

9.8.4.1   Simulation Configurations

The most straightforward method of acceleration involves turning off visualizations and debugging features during simulation. Simulink executes a model in multiple modes, including the normal one, accelerator mode, and rapid accelerator mode. In the normal mode (the default mode of simulation), the default configuration parameters are selected to enhance debugging and help with incremental building up of a valid simulation model. This is reflected in the model configuration parameters accessible from the Simulation menu in every model. Figure 9.30 illustrates the default configuration parameters of the *zPDCCH_v8_default.slx* model, as found in the Simulation Target tab.

As we can see, properties related to debugging, such as "Enable debugging/animation," "Enable overflow detection," and "Echo expressions without semicolon," are by default turned on. Some run-time checks that help designers identify semantic problems, such as "Ensure memory integrity" and "Ensure responsiveness" are also on by default. Obviously, these checks represent simulation overhead, and by turning them off we may achieve a level of acceleration. Figure 9.31 illustrates a new profile of Simulation Target parameters that unchecks many of these features.

To gauge how much of an improvement can be made by this type of optimization, we run the following MATLAB script, which iterates through simulation of the more optimized simulation model in the normal mode. In this case, the simulation time is reduced from about 84 down to 44 seconds (Figure 9.32). This may be considered rather a substantial improvement and brings the simulation speed of our Simulink model on par with the eighth version of the MATLAB algorithm (Figure 9.33).

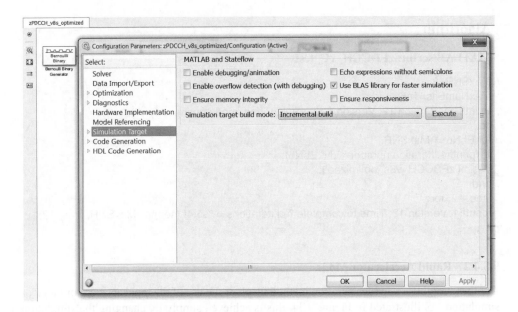

Figure 9.31 Model configuration parameters: simulation target

```
Version 12: Version 8 Simulink normal mode optimized

Iteration number 1
Iteration number 2
Iteration number 3
Iteration number 4
Iteration number 5
Iteration number 6
Iteration number 7
Iteration number 8
Version 12: Time to complete 8 iterations = 44.3369 (sec)
```

Figure 9.32 Twelfth version of the algorithm: time taken to execute eight iterations

```
Versions of the Transceiver                          | Elapsed Time (sec)| Acceleration Ratio
 1. Baseline                                          |      411.3030 |        1.0000
 2. Vectorization                                     |      326.5071 |        1.2597
 3. Vectorization along larger dimension              |      175.8478 |        2.3390
 4. Vectorization + Preallocation                     |       82.7194 |        4.9723
 5. System objects for MIMO                           |       81.9175 |        5.0209
 6. System objects for MIMO & Channel                 |       59.9528 |        6.8604
 7. System objects for MIMO & Channel & Viterbi       |       58.8660 |        6.9871
 8. System objects for all                            |       48.9014 |        8.4109
 9. Version 8 + MATLAB to C code generation (MEX)     |       31.4722 |       13.0688
10. Version 8 + MEX + Parallel computing (parfor)     |       18.3899 |       22.3657
11. Version 8 Simulink normal mode                    |       84.5959 |        4.8620
12. Version 8 Simulink normal mode optimized          |       44.3369 |        9.2768
```

Figure 9.33 Execution times and acceleration ratios for the first 12 versions of the algorithm

Algorithm

MATLAB script: zPDCCH_vC_test

```
MaxSNR=8;
MaxNumBits=1e6;
fprintf(1,'\nVersion 12: Version 8 Simulink normal mode optimized\n\n');
tic;
for EbNo=1:MaxSNR
    fprintf(1,'Iteration number %d\r',EbNo);
    sim('zPDCCH_v8s_optimized');
end
time_12=toc;
fprintf(1,'Version 12: Time to complete %d iterations = %6.4f (sec)\n', MaxSNR, time_12);
```

9.8.4.2   Rapid Accelerator Mode

Another straightforward acceleration method involves the use of the rapid accelerator mode of simulation. As illustrated in Figure 9.34, this is achieved simply by changing the simulation mode to rapid accelerator. Rapid accelerator mode creates a MEX version of the model and

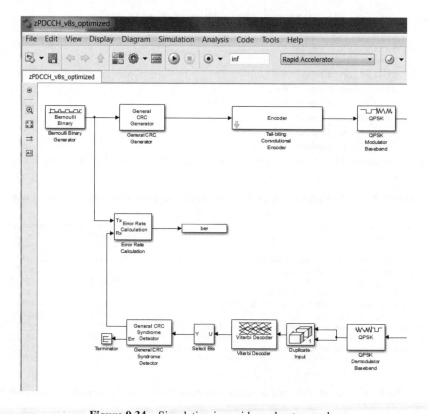

Figure 9.34   Simulation in rapid accelerator mode

executes the resulting compiled code. In this respect, rapid accelerator is analogous to generation of the MEX file for a MATLAB function.

The first time the simulation is run, the model is compiled and the MEX function is generated. In the case of the model *zPDCCH_v8s_optimized.slx*, the messages shown in Figure 9.35 appear in the MATLAB workspace and the MEX file *zPDCCH_v8s_optimized_sfun* is generated.

By running the following MATLAB script, we iterate through simulation of the optimized model in the rapid accelerator mode. The results indicate that the script takes about 40 seconds to complete (Figure 9.36). Comparison with the MATLAB function that used MATLAB-to-C code generation (the ninth version of the algorithm) shows an advantage in terms of simulation speed in MATLAB after MEX file generation (Figure 9.37). In the next step, we will introduce a simple fix to alleviate this discrepancy.

```
### Building the rapid accelerator target for model: zPDCCH_v8s_optimized
### Successfully built the rapid accelerator target for model: zPDCCH_v8s_optimized
```

Figure 9.35   Compiling a Simulink model in rapid accelerator mode

```
Version 13: Version 8 Simulink rapid accelerator

Iteration number 1
Iteration number 2
Iteration number 3
Iteration number 4
Iteration number 5
Iteration number 6
Iteration number 7
Iteration number 8
Version 13: Time to complete 8 iterations = 40.5559 (sec)
```

Figure 9.36   Thirteenth version of the algorithm: time taken to execute eight iterations

Versions of the Transceiver	Elapsed Time (sec)	Acceleration Ratio
1. Baseline	411.3030	1.0000
2. Vectorization	326.5071	1.2597
3. Vectorization along larger dimension	175.8478	2.3390
4. Vectorization + Preallocation	82.7194	4.9723
5. System objects for MIMO	81.9175	5.0209
6. System objects for MIMO & Channel	59.9528	6.8604
7. System objects for MIMO & Channel & Viterbi	58.8660	6.9871
8. System objects for all	48.9014	8.4109
9. Version 8 + MATLAB to C code generation (MEX)	31.4722	13.0688
10. Version 8 + MEX + Parallel computing (parfor)	18.3899	22.3657
11. Version 8 Simulink normal mode	84.5959	4.8620
12. Version 8 Simulink normal mode optimized	44.3369	9.2768
13. Version 8 Simulink rapid accelerator	40.5559	10.1416

Figure 9.37   Execution times and acceleration ratios for the first 13 versions of the algorithm

Algorithm

MATLAB script: zPDCCH_vD_test

```
MaxSNR=8;
MaxNumBits=1e6;
fprintf(1,'\nVersion 13: Version 8 Simulink rapid accelerator\n\n');
tic;
for EbNo=1:MaxSNR
    fprintf(1,'Iteration number %d\r',EbNo);
    sim('zPDCCH_v8s_optimized','SimulationMode','rapid');
end
time_13=toc;
fprintf(1,'Version 13: Time to complete %d iterations = %6.4f (sec)\n', MaxSNR, time_13);
```

9.8.4.3   Optimized Rapid Accelerator

In rapid accelerator mode, Simulink regenerates the MEX file every time the Simulink model is changed. The time it takes for Simulink to determine whether the rapid-accelerator executable is up to date is significantly less than that required to generate code. We can take advantage of this characteristic when we wish to test design tradeoffs. For example, we can generate the rapid-accelerator target code once and use it to simulate a model with a series of SNR settings. This is an especially efficient way of using this mode because this type of change does not result in the target code being regenerated. The target code is generated the first time the model runs, but on subsequent runs the Simulink code only takes the time necessary to verify that the target is up to date. We can even bypass this recurring update check by running the *sim* command with a property that turns off update checking:

Algorithm

sim(model_name,'SimulationMode','rapid','RapidAcceleratorUpToDateCheck', 'off')

To verify the effect of this type of optimized rapid accelerator simulation on speed, we call the following MATLAB testbench. As the results indicate, this optimization almost doubles the speed of simulation. At this stage, the code generated in Simulink runs faster than the MATLAB code (ninth version) we examined earlier (Figures 9.38 and 9.39).

Algorithm

MATLAB script: zPDCCH_vE_test

```
MaxSNR=8;
MaxNumBits=1e6;
```

```
fprintf(1,'\nVersion 14: Version 8 Simulink rapid accelerator optimized\n\n');
tic;
for EbNo=1:MaxSNR
   fprintf(1,'Iteration number %d\r',EbNo);
   sim('zPDCCH_v8s_optimized','SimulationMode','rapid',
'RapidAcceleratorUpToDateCheck', 'off');
end
time_14=toc;
fprintf(1,'Version 14: Time to complete %d iterations = %6.4f (sec)\n',
MaxSNR, time_14);
```

9.8.4.4 Parallel Computing

In this step, we combine parallel computing with the rapid accelerator mode in Simulink. To
do so, we call the rapid accelerator target code of our Simulink model within a parfor loop.
First we verify that the *matlabpool* command is called and that access to multiple cores in our

```
Version 14: Version 8 Simulink rapid accelerator optimized

Iteration number 1
Iteration number 2
Iteration number 3
Iteration number 4
Iteration number 5
Iteration number 6
Iteration number 7
Iteration number 8
Version 14: Time to complete 8 iterations = 22.2629 (sec)
```

Figure 9.38   Fourteenth version of the algorithm: time taken to execute eight iterations

```
Versions of the Transceiver                      | Elapsed Time (sec)| Acceleration Ratio
 1. Baseline                                      |        411.3030 |        1.0000
 2. Vectorization                                 |        326.5071 |        1.2597
 3. Vectorization along larger dimension          |        175.8478 |        2.3390
 4. Vectorization + Preallocation                 |         82.7194 |        4.9723
 5. System objects for MIMO                       |         81.9175 |        5.0209
 6. System objects for MIMO & Channel             |         59.9528 |        6.8604
 7. System objects for MIMO & Channel & Viterbi   |         58.8660 |        6.9871
 8. System objects for all                        |         48.9014 |        8.4109
 9. Version 8 + MATLAB to C code generation (MEX) |         31.4722 |       13.0688
10. Version 8 + MEX + Parallel computing (parfor) |         18.3899 |       22.3657
11. Version 8 Simulink normal mode                |         84.5959 |        4.8620
12. Version 8 Simulink normal mode optimized      |         44.3369 |        9.2768
13. Version 8 Simulink rapid accelerator          |         40.5559 |       10.1416
14. Version 8 Simulink rapid accelerator optimized|         22.2629 |       18.4748
```

Figure 9.39   Execution times and acceleration ratios for the first 14 versions of the algorithm

computer has been established. The parfor command divides loop iterations for simultaneous executions across a number of MATLAB workers (in this case, two). The following program shows how the parfor loop calls the Simulink model running in the optimized rapid accelerator mode. This is quite similar to the way in which we parallelized our MATLAB code in the tenth version earlier.

Algorithm

MATLAB script: zPDCCH_vF_test

```
MaxSNR=8;
MaxNumBits=1e6;
fprintf(1,'\nVersion 15: Version 8 Simulink rapid accel. optimized + parfor\n\n');
tic;
parfor EbNo=1:MaxSNR
    fprintf(1,'Iteration number %d\r',EbNo);
    sim('zPDCCH_v8s_optimized','SimulationMode','rapid',
'RapidAcceleratorUpToDateCheck', 'off');
end
time_15=toc;
fprintf(1,'Version 15: Time to complete %d iterations = %6.4f (sec)\n',
MaxSNR, time_15);
```

The results also indicate that the combination of compilation of the Simulink model in rapid accelerator mode and parallel computing substantially accelerates the simulation. We first ran the simulation model in default normal mode in 85.29 seconds. The combined optimizations available in the fifteenth version result in a simulation time of just 12.31 seconds (Figure 9.40). This corresponds to a 7-times acceleration compared to the eleventh version and a 33-times acceleration compared to the MATLAB-code baseline (Figure 9.41).

```
Version 15: Version 8 Simulink rapid accel. optimized + parfor

Iteration number 6
Iteration number 5
Iteration number 4
Iteration number 8

Iteration number 3
Iteration number 2
Iteration number 1
Iteration number 7

Version 15: Time to complete 8 iterations = 12.3180 (sec)
```

Figure 9.40   Fifteenth version of the algorithm: time taken to execute eight iterations

```
Versions of the Transceiver                      | Elapsed Time (sec)| Acceleration Ratio
  1. Baseline                                     |          411.3030 |          1.0000
  2. Vectorization                                |          326.5071 |          1.2597
  3. Vectorization along larger dimension         |          175.8478 |          2.3390
  4. Vectorization + Preallocation                |           82.7194 |          4.9723
  5. System objects for MIMO                      |           81.9175 |          5.0209
  6. System objects for MIMO & Channel            |           59.9528 |          6.8604
  7. System objects for MIMO & Channel & Viterbi  |           58.8660 |          6.9871
  8. System objects for all                       |           48.9014 |          8.4109
  9. Version 8 + MATLAB to C code generation (MEX)|           31.4722 |         13.0688
 10. Version 8 + MEX + Parallel computing (parfor)|           18.3899 |         22.3657
 11. Version 8 Simulink normal mode               |           85.2923 |          4.8223
 12. Version 8 Simulink normal mode optimized     |           44.3369 |          9.2768
 13. Version 8 Simulink rapid accelerator         |           40.5559 |         10.1416
 14. Version 8 Simulink rapid accelerator optimized|          22.2629 |         18.4748
 15. Version 8 Simulink rapid accel. optim. + parfor|         12.3180 |         33.3903
```

Figure 9.41   Execution times and acceleration ratios for the first 15 versions of the algorithm.

9.9   GPU Processing

GPUs were originally developed to accelerate graphical applications but are now increasingly applied to a range of scientific calculations. MATLAB has functionality that takes advantage of the power of GPUs. Computations can be performed on CUDA (Compute Unified Device Architecture)-enabled NVIDIA GPUs directly from MATLAB to accelerate algorithms. FFT, Inverse Fast Fourier Transform (IFFT), and linear algebraic operations are among more than 100 built-in MATLAB functions that can be executed directly on the GPU by providing an input argument of the type *GPUArray*, a special MATLAB array type provided by the Parallel Computing Toolbox. These GPU-enabled functions operate differently depending on the data type of the arguments passed to them. Similarly, toolboxes such as the Neural Networks Toolbox, the Communications System Toolbox, the Signal Processing Toolbox, and the Phased Array System Toolbox also provide GPU-accelerated algorithms.

As a rule of thumb, an application may be a good fit for a GPU if it is computationally intensive and massively parallel. This translates to two criteria. First, the time it takes for the application to run on the GPU should be significantly greater than the time it takes to transfer the same amount of data between the CPU and the GPU during application execution. Second, the best GPU performance will be seen when all of the cores are kept busy, exploiting the inherent parallel nature of the GPU. Vectored MATLAB calculations on larger arrays and GPU-enabled toolbox functions fit into this category.

With access to GPUs, we can tap into their power to dramatically improve the simulation speed of an algorithm in MATLAB, especially if the data are sufficiently large. Algorithms optimized for GPUs are a perfect fit for mobile communication systems, since most leverage large data sizes and involve repeated operations performed independently for multiple users.

9.9.1   Setting up GPU Functionality in MATLAB

Running the following MATLAB examples on a supported GPU requires use of the Parallel Computing Toolbox and Communications System Toolbox. The following commands help verify whether the proper licenses are held:

Algorithm

license('test','distrib_computing_toolbox');
license('test','communication_toolbox');

If the answer provided by MATLAB to both of these commands is 1, which stands for true, then the correct product licenses are held. To verify whether a GPU device is properly installed and ready to be used within MATLAB, the following command can be entered.

Algorithm

parallel.gpu.GPUDevice.isAvailable

If MATLAB returns a value of 1, the GPU is ready for use in MATLAB.

9.9.2   GPU-Optimized System Objects

The Communication System Toolbox has many specialized algorithms that support GPU processing. The Parallel Computing Toolbox can be used to execute many communications algorithms directly on the GPU. The following Communications System Toolbox System objects are GPU-optimized:

- *comm.gpu.AWGNChannel*
- *comm.gpu.BlockDeinterleaver*
- *comm.gpu.BlockInterleaver*
- *comm.gpu.ConvolutionalDeinterleaver*
- *comm.gpu.ConvolutionalEncoder*
- *comm.gpu.ConvolutionalInterleaver*
- *comm.gpu.LDPCDecoder*
- *comm.gpu.PSKDemodulator*
- *comm.gpu.PSKModulator*
- *comm.gpu.TurboDecoder*
- *comm.gpu.ViterbiDecoder.*

As can be seen, not all System objects are optimized for GPU processing. Those listed here correspond to computationally intensive algorithms encountered in many communication systems. They have an easy-to-use syntax: *.gpu* is added to the object name. With minor changes like this applied to the code, when a MATLAB® application is run on a supported GPU the simulation is usually accelerated.

9.9.3   Using a Single GPU System Object

The following MATLAB function shows the first GPU optimization applied to the eighth version of our PDCCH algorithm. The only change to the MATLAB code is the use of the *comm.gpu.ViterbiDecoder* System object instead of *comm.ViterbiDecoder*.

Algorithm

MATLAB function: zPDCCH_vG

```
function [ber, bits]=zPDCCH_vG(EbNo, maxNumErrs, maxNumBits)
%% Constants
FRM=2048;
M=4; k=log2(M); codeRate=1/3;
snr = EbNo + 10*log10(k) + 10*log10(codeRate);
trellis=poly2trellis(7, [133 171 165]);
L=FRM+24;C=6; Index=[L+1:(3*L/2) (L/2+1):L];
%% Initializations
persistent Modulator AWGN DeModulator BitError ConvEncoder1 ConvEncoder2 Viterbi
CRCGen CRCDet
if isempty(Modulator)
    Modulator     = comm.QPSKModulator('BitInput',true);
    AWGN          = comm.AWGNChannel('NoiseMethod', 'Variance', 'VarianceSource',
'Input port');
    DeModulator = comm.QPSKDemodulator('BitOutput',true);
    BitError      = comm.ErrorRate;
    ConvEncoder1=comm.ConvolutionalEncoder('TrellisStructure', trellis,
'FinalStateOutputPort', true, ...
        'TerminationMethod','Truncated');
    ConvEncoder2 = comm.ConvolutionalEncoder('TerminationMethod','Truncated',
'InitialStateInputPort', true,...
        'TrellisStructure', trellis);
    Viterbi=comm.gpu.ViterbiDecoder('TrellisStructure', trellis,
'InputFormat','Hard','TerminationMethod','Truncated');
    CRCGen = comm.CRCGenerator('Polynomial',[1 1 zeros(1, 16) 1 1 0 0 0 1 1]);
    CRCDet = comm.CRCDetector  ('Polynomial',[1 1 zeros(1, 16) 1 1 0 0 0 1 1]);
end
%% Processing loop modeling transmitter, channel model and receiver
numErrs = 0; numBits = 0; nS=0;
results=zeros(3,1);
while ((numErrs < maxNumErrs) && (numBits < maxNumBits))
    % Transmitter
    u         = randi([0 1], FRM,1);              % Generate bit payload
    u1        = step(CRCGen, u);                  % CRC insertion
    u2        = u1((end-C+1):end);                % Tail-biting convolutional coding
    [~, state] = step(ConvEncoder1, u2);
    u3        = step(ConvEncoder2, u1,state);
    u4        = fcn_RateMatcher(u3, L, codeRate); % Rate matching
```

```
u5          = fcn_Scrambler(u4, nS);              % Scrambling
u6          = step(Modulator, u5);                % Modulation
u7          = TransmitDiversityEncoderS(u6);      % MIMO Alamouti encoder
  % Channel
[u8, h8]    = MIMOFadingChanS(u7);                % MIMO fading channel
noise_var = real(var(u8(:)))/(10.^(0.1*snr));
u9          = step(AWGN, u8, noise_var);          % AWGN
% Receiver
uA          = TransmitDiversityCombinerS(u9, h8);% MIMO Alamouti combiner
uB          = step(DeModulator, uA);              % Demodulation
uC          = fcn_Descrambler(uB, nS);            % Descrambling
uD          = fcn_RateDematcher(uC, L);           % Rate de-matching
uE          = [uD;uD];                            % Tail-biting
uF          = step(Viterbi, uE);                  % Viterbi decoding
uG          = uF(Index);
y           = step(CRCDet, uG );                   % CRC detection
results     = step(BitError, u, y);               % Update number of bit errors
numErrs     = results(2);
numBits     = results(3);
nS          = nS + 2; nS = mod(nS, 20);
end
%% Clean up & collect results
ber = results(1); bits= results(3);
reset(BitError);
```

By running the following MATLAB script, which calls the first GPU-optimized version of the algorithm, and by recalling the collective performance improvements, we can see the effect on one of the algorithm's bottlenecks (the Viterbi decoder) of using a GPU. Note that all other functionality is performed on the CPU (Figures 9.42 and 9.43).

```
Version 16: Version 8 + Viterbi decoder on GPU

Iteration number 1
Iteration number 2
Iteration number 3
Iteration number 4
Iteration number 5
Iteration number 6
Iteration number 7
Iteration number 8
Version 16: Time to complete 8 iterations = 26.5780 (sec)
```

Figure 9.42 Sixteenth version of the algorithm: time taken to execute eight iterations.

```
--------------------------------------------------------------------------------
Versions of the Transceiver                      | Elapsed Time (sec)| Acceleration Ratio
 1. Baseline                                     |        411.3030 |        1.0000
 2. Vectorization                                |        326.5071 |        1.2597
 3. Vectorization along larger dimension         |        175.8478 |        2.3390
 4. Vectorization + Preallocation                |         82.7194 |        4.9723
 5. System objects for MIMO                      |         81.9175 |        5.0209
 6. System objects for MIMO & Channel            |         59.9528 |        6.8604
 7. System objects for MIMO & Channel & Viterbi  |         58.8660 |        6.9871
 8. System objects for all                       |         48.9014 |        8.4109
 9. Version 8 + MATLAB to C code generation (MEX)|         31.4722 |       13.0688
10. Version 8 + MEX + Parallel computing (parfor)|         18.3899 |       22.3657
11. Version 8 Simulink normal mode               |         85.2923 |        4.8223
12. Version 8 Simulink normal mode optimized     |         44.3369 |        9.2768
13. Version 8 Simulink rapid accelerator         |         40.5559 |       10.1416
14. Version 8 Simulink rapid accelerator optimized|        22.2629 |       18.4748
15. Version 8 Simulink rapid accel. optim. + rarfor |      12.3180 |       33.3904
16. Version 8 + Viterbi decoder on GPU           |         26.5780 |       15.4753
--------------------------------------------------------------------------------
```

Figure 9.43   Execution times and acceleration ratios for the first 16 versions of the algorithm

Algorithm

MATLAB function

```
fprintf(1,'\nVersion 16: Version 8 + Viterbi decoder on GPU\n\n');
tic;
for snr = 1:MaxSNR
ber= zPDCCH_vG(snr, MaxNumBits, MaxNumBits);
end
time_16=toc;
fprintf(1,'Version 16: Time to complete %d iterations = %6.4f (sec)\n', MaxSNR, time_16);
```

9.9.4   Combining Parallel Processing with GPUs

In this version of the algorithm, we combine GPU processing with parallel processing. To parallelize the algorithm, we use a function from the Parallel Processing Toolbox called spmd. The *spmd* function, which implements a "single program, multiple data" construct, executes MATLAB code on several MATLAB workers simultaneously. The general form of a *spmd* statement is as follows:

Algorithm

```
spmd;
<MATLAB statements>
end;
```

In order to execute the statements in parallel, we must first open a pool of MATLAB workers using the *matlabpool* function introduced earlier. Inside the body of the *spmd* statement, each MATLAB worker has a unique identifier denoted by a variable called labindex, while a variable called numlabs gives the total number of workers executing the block in parallel.

The following function shows our final version of the PDCCH algorithm. It combines the use of the *spmd* function to parallelize in-loop processing with the use of GPU-optimized System objects for the Viterbi decoder.

Algorithm

MATLAB function

```
function [ber, bits]=zPDCCH_vH(EbNov, maxNumErrs, maxNumBits)
%% Constants
wkrs = 2;
spmd(wkrs)
FRM=2048;
M=4; k=log2(M); codeRate=1/3;
snrv = EbNov + 10*log10(k) + 10*log10(codeRate);
bits=zeros(size(EbNov));errs=bits;
trellis=poly2trellis(7, [133 171 165]);
L=FRM+24;C=6; Index=[L+1:(3*L/2) (L/2+1):L];
s = RandStream.create('mrg32k3a', 'NumStreams', wkrs, 'CellOutput', true, 'Seed', 1);
RandStream.setGlobalStream(s{labindex});
Modulator    = comm.QPSKModulator('BitInput',true);
AWGN         = comm.AWGNChannel('NoiseMethod', 'Variance', 'VarianceSource',
'Input port');
DeModulator = comm.QPSKDemodulator('BitOutput',true);
BitError     = comm.ErrorRate;
ConvEncoder1=comm.ConvolutionalEncoder('TrellisStructure', trellis,
'FinalStateOutputPort', true, 'TerminationMethod','Truncated');
ConvEncoder2 = comm.ConvolutionalEncoder('TerminationMethod','Truncated',
'InitialStateInputPort', true,'TrellisStructure', trellis);
Viterbi=comm.gpu.ViterbiDecoder('TrellisStructure', trellis,
'InputFormat','Hard','TerminationMethod','Truncated');
CRCGen = comm.CRCGenerator('Polynomial',[1 1 zeros(1, 16) 1 1 0 0 0 1 1]);
CRCDet = comm.CRCDetector   ('Polynomial',[1 1 zeros(1, 16) 1 1 0 0 0 1 1]);
%end
for n=1:numel(snrv),
    %% Processing loop modeling transmitter, channel model and receiver
    numErrs = 0; numBits = 0; nS=0;
    results=zeros(3,1);
    while ((numErrs < maxNumErrs/numlabs) && (numBits < maxNumBits/numlabs))
        % Transmitter
        u       = randi([0 1], FRM,1);               % Generate bit payload
        u1      = step(CRCGen, u);                    % CRC insertion
        u2      = u1((end-C+1):end);                  % Tail-biting convolutional coding
        [~, state] = step(ConvEncoder1, u2);
        u3      = step(ConvEncoder2, u1,state);
```

```
u4        = fcn_RateMatcher(u3, L, codeRate);  % Rate matching
u5        = fcn_Scrambler(u4, nS);             % Scrambling
u8        = step(Modulator, u5);               % Modulation
u7        = TransmitDiversityEncoderS(u8);     % MIMO Alamouti encoder
% Channel
[u8, h8]  = MIMOFadingChanS(u7);               % MIMO fading channel
noise_var = real(var(u8(:)))/(10.^(0.1*snrv(n)));
u9        = step(AWGN, u8, noise_var);         % AWGN
% Receiver
uA        = TransmitDiversityCombinerS(u9, h8);% MIMO Alamouti combiner
uB        = step(DeModulator, uA);             % Demodulation
uC        = fcn_Descrambler(uB, nS);           % Descrambling
uD        = fcn_RateDematcher(uC, L);          % Rate de-matching
uE        = [uD;uD];                           % Tail-biting
uF        = step(Viterbi, uE);                 % Viterbi decoding
uG        = uF(Index);
y         = step(CRCDet, uG );                 % CRC detection
results   = step(BitError, u, y);              % Update number of bit errors
numErrs   = results(2);
numBits   = results(3);
nS        = nS + 2; nS = mod(nS, 20);
  end
  %% Clean up & collect results
  bits(n)= results(3);
  errs(n) = results(2);
  reset(BitError);
end
end
totbits = zeros(1, numel(EbNov));
toterrs = zeros(1, numel(EbNov));
for n=1:wkrs,
   totbits = totbits + bits{n};
   toterrs = toterrs + errs{n};
end
ber = toterrs./totbits;
```

The iterations over *Eb/N0* values are brought inside the function and the input values for the *Eb/N0* are stored in a vector rather than a single value. Furthermore, inside the body of the *spmd* function we compute the same operations on independent workers. To make the BER results valid, we must ensure that the random number generators used to produce the random bits and values added as AWGN noise are not related. This is achieved by the following two lines of code, which generate different random number streams for different workers:

Algorithm

s = RandStream.create('mrg32k3a', 'NumStreams', wkrs, 'CellOutput', true, 'Seed', 1);
RandStream.setGlobalStream(s{labindex});

We also subdivide the number of errors and number of bits by the number of workers, such that each worker will process its equal share of total bits. At the end of the *spmd* statement, each variable contains different values, computed independently in each worker. In the for loop, after the *spmd* statement, we add all the bits and BER values computed over each worker to find the total BER.

The following calling function executes this version of the algorithm to illustrate the effect of the combination of parallel and GPU processing on the simulation time. The results indicate that this version produces the same numerical results as all other versions of the algorithm in the least amount of time (Figures 9.44 and 9.45).

Algorithm

MATLAB function

```
fprintf(1,'\nVersion 17: Version 8 + Viterbi decoder on GPU + spmd\n\n');
tic;
ber= zPDCCH_vH(1:snr, MaxNumBits, MaxNumBits);
time_17=toc;
fprintf(1,'Version 17: Time to complete %d iterations = %6.4f (sec)\n', MaxSNR, time_17);
```

9.10   Case Study: Turbo Coders on GPU

In this section we present a case study to show how GPUs can be used to accelerate turbo-coding applications in MATLAB. Turbo coding is an essential part of the LTE standard. Because of the iterative nature of its algorithm, the turbo decoder is computationally intensive and thus an ideal candidate for GPU acceleration. Using a GPU-optimized System object for turbo decoding, we can accelerate BER simulations.

We first repeat the steps highlighted in the previous section, but using a turbo decoder instead of a Viterbi decoder. Then we show and resolve some of the pitfalls associated with GPU processing in MATLAB. The main one relates to the excessive transfer of data between the CPU and the GPU. This is a well-known issue with GPU processing and can slow the simulation. Finally, we show how increasing the size of the data running on the GPU can further accelerate the simulation.

```
Version 17: Version 8 + Viterbi decoder on GPU + spmd

Iteration number 1
Iteration number 2
Iteration number 3
Iteration number 4
Iteration number 5
Iteration number 6
Iteration number 7
Iteration number 8
Version 17: Time to complete 8 iterations = 15.2433 (sec)
```

Figure 9.44   Seventeenth version of the algorithm: time taken to execute eight iterations

```
--------------------------------------------------------------------------------
Versions of the Transceiver                        | Elapsed Time (sec)| Acceleration Ratio
 1. Baseline                                        |           411.3030 |           1.0000
 2. Vectorization                                   |           326.5071 |           1.2597
 3. Vectorization along larger dimension            |           175.8478 |           2.3390
 4. Vectorization + Preallocation                   |            82.7194 |           4.9723
 5. System objects for MIMO                         |            81.9175 |           5.0209
 6. System objects for MIMO & Channel               |            59.9528 |           6.8604
 7. System objects for MIMO & Channel & Viterbi     |            58.8660 |           6.9871
 8. System objects for all                          |            48.9014 |           8.4109
 9. Version 8 + MATLAB to C code generation (MEX)   |            31.4722 |          13.0688
10. Version 8 + MEX + Parallel computing (parfor)   |            18.3899 |          22.3657
11. Version 8 Simulink normal mode                  |            85.2923 |           4.8223
12. Version 8 Simulink normal mode optimized        |            44.3369 |           9.2768
13. Version 8 Simulink rapid accelerator            |            40.5559 |          10.1416
14. Version 8 Simulink rapid accelerator optimized  |            22.2629 |          18.4748
15. Version 8 Simulink rapid accel. optim. + rarfor |            12.3180 |          33.3904
16. Version 8 + Viterbi decoder on GPU              |            26.5780 |          15.4753
17. Version 8 + Viterbi decoder on GPU + spmd       |            15.2433 |          26.9825
--------------------------------------------------------------------------------
```

Figure 9.45 Execution times and acceleration ratios for the first 17 versions of the algorithm

9.10.1 Baseline Algorithm on a CPU

As a baseline, we use the turbo-coding algorithm developed in Chapter 8. The following function implements the algorithm for CPU processing. The input data size is 2432 bits per frame. The trellis structure and the interleaver used for turbo coding are the ones specified by the LTE standard. The transmitter is composed of a turbo encoder followed by a QPSK modulator. To simplify the code, we perform channel modeling by adding AWGN to the modulated symbols. In the receiver we perform soft-decision demodulation followed by turbo decoding.

Algorithm

MATLAB function: zTurboExample_gpu0

```
function [ber, bits]=zTurboExample_gpu0(EbNo, maxNumErrs, maxNumBits)
FRM=2432;
Indices = lteIntrlvrIndices(FRM);
M=4;k=log2(M);
R= FRM/(3* FRM + 4*3);
snr = EbNo + 10*log10(k) + 10*log10(R);
noiseVar = 10.^(-snr/10);
numIter=6; trellis =  poly2trellis(4, [13 15], 13);
persistent hTEnc Modulator AWGN DeModulator hTDec hBER
if isempty(Modulator)
   hTEnc = comm.TurboEncoder('TrellisStructure',trellis , 'InterleaverIndices', Indices);
   Modulator     = comm.PSKModulator(4, ...
       'BitInput', true, 'PhaseOffset', pi/4, 'SymbolMapping', 'Custom', ...
       'CustomSymbolMapping', [0 2 3 1]);
   AWGN          = comm.AWGNChannel('NoiseMethod', 'Variance', 'VarianceSource',
'Input port');
   DeModulator = comm.PSKDemodulator(...
```

```
          'ModulationOrder', 4, ...
          'BitOutput', true, ...
          'PhaseOffset', pi/4, 'SymbolMapping', 'Custom', ...
          'CustomSymbolMapping', [0 2 3 1],...
          'DecisionMethod', 'Approximate log-likelihood ratio', ...
          'VarianceSource', 'Input port');
     % Turbo Decoder
     hTDec  = comm.TurboDecoder('TrellisStructure', trellis,'InterleaverIndices', Indices,
'NumIterations', numIter);
     % BER measurement
     hBER = comm.ErrorRate;
end
%% Processing loop
Measures = zeros(3,1); %initialize BER output
while (( Measures(2)< maxNumErrs) && (Measures(3) < maxNumBits))
     data = randi([0 1], FRM, 1);
     % Encode random data bits
     yEnc = step(hTEnc, data);
     % Add noise to real bipolar data
     modout = step(Modulator, yEnc);
     rData = step(AWGN, modout,noiseVar );
     % Convert to log-likelihood ratios for decoding
     llrData = step( DeModulator, rData, noiseVar);
     % Turbo Decode
     decData = step(hTDec, -llrData);
     % Calculate errors
     Measures = step(hBER, data, decData);
end
bits = Measures(3);
ber= Measures(1);
reset(hBER);
```

We can apply the profiler to find the bottlenecks in this algorithm by performing the following commands. The results clearly show that the turbo decoder is the main bottleneck, taking 24.485 of the 26.503 seconds required to process 1 million bits of data (Figure 9.46).

Algorithm

MATLAB script

```
EbNo=0; maxNumErrs=1e6;maxNumBits=1e6;
profile on;
ber= zTurboExample_gpu0(EbNo, maxNumErrs, maxNumBits);
profile viewer;
```

Profile Summary
Generated 16-Jan-2013 19:03:39 using cpu time.

Function Name	Calls	**Total Time**	Self Time*	Total Time Plot (dark band = self time)
zTurboExample_gpu0	1	26.503 s	0.748 s	▬▬▬▬▬▬
TurboDecoder>TurboDecoder.stepImpl	412	24.485 s	24.426 s	▬▬▬▬▬
TurboEncoder>TurboEncoder.stepImpl	412	0.440 s	0.331 s	|
AWGNChannel>AWGNChannel.stepImpl	412	0.339 s	0.282 s	|

Figure 9.46   Profiler results for the baseline turbo-coding algorithm

```
Version 1: Everything on CPU

Iteration number 1
Iteration number 2
Iteration number 3
Iteration number 4
Iteration number 5
Iteration number 6
Iteration number 7
Version 1: Time to complete 7 iterations = 311.2196 (sec)
-----------------------------------------------------------------------------------
Versions of the Transceiver                     | Elapsed Time (sec)| Acceleration Ratio
  1. Everything on CPU                           |       311.2196 |       1.0000
-----------------------------------------------------------------------------------
```

Figure 9.47   Execution times and acceleration ratios for the baseline turbo-coding algorithm

To establish a starting yardstick for the turbo-coding execution time, we run the following MATLAB script. We iterate through $Eb/N0$ values of 0 to 1.2 dB, in seven 0.2 dB steps. It takes about 311 seconds to process 1 million bits in each of these seven iterations (Figure 9.47).

Algorithm

MATLAB script

```
MaxSNR=7;
Snrs=0:0.2:1.2;
MaxNumBits=1e6;
N=1;
fprintf(1,'\nVersion 1: Everything on CPU\n\n')
tic;
for idx = 1:MaxSNR
    fprintf(1,'Iteration number %d\r',idx);
    EbNo=Snrs(idx);
    ber= zTurboExample_gpu0(EbNo, MaxNumBits, MaxNumBits);
end
time_CPU=toc;
fprintf(1,'Version 1: Time to complete %d iterations = %6.4f (sec)\n', MaxSNR, time_CPU);
Report_Timing_Results(N,time_CPU,time_CPU,'Everything on CPU');
```

9.10.2   Turbo Decoder on a GPU

In the second version of the turbo-coding example, we execute the identified bottleneck, the turbo decoder, on the GPU, while the rest of the algorithm executes on the CPU. The only modification needed to achieve this is the use of the *comm.gpu.TurboDecoder* System object instead of *comm.TurboDecoder*. The following function shows the only line of code in which the baseline algorithm (*zTurboExample_gpu0*) differs from the second version (*zTurboExample_gpu1*).

MATLAB function

```
function [ber, bits] =                          function [ber, bits] =
   zTurboExample_gpu0(EbNo, maxNumErrs,           zTurboExample_gpu1(EbNo, maxNumErrs,
   maxNumBits)                                     maxNumBits)

hTDec  = comm.TurboDecoder                       hTDec  = comm.gpu.TurboDecoder
   ('TrellisStructure', trellis,'InterleaverIndices',  ('TrellisStructure', trellis,'InterleaverIndices',
   Indices, 'NumIterations', numIter);              Indices, 'NumIterations', numIter);
   .                                                .
   .                                                .
   .                                                .
end                                              end
```

By running the following MATLAB script, we can assess whether there is an advantage to running the turbo decoder on the GPU. It now takes about 75 seconds to process 1 million bits in each of the seven iterations: a four-times acceleration (Figure 9.48).

Algorithm

MATLAB script

```
MaxSNR=7;
Snrs=0:0.2:1.2;
MaxNumBits=1e6;
N=2;
fprintf(1,'\nVersion 2: Turbo coding Only on GPU\n\n');
tic;
for idx = 1:MaxSNR
   fprintf(1,'Iteration number %d\r',idx);
   EbNo=Snrs(idx);
   ber= zTurboExample_gpu1(EbNo, MaxNumBits, MaxNumBits);
end
time_GPU1=toc;
fprintf(1,'Version 2: Time to complete %d iterations = %6.4f (sec)\n', MaxSNR,
time_GPU1);
Report_Timing_Results(N,time_CPU,time_GPU1,'Turbo coding Only on GPU');
```

```
Version 2: Turbo coding Only on GPU

Iteration number 1
Iteration number 2
Iteration number 3
Iteration number 4
Iteration number 5
Iteration number 6
Iteration number 7
Version 2: Time to complete 7 iterations = 75.5764 (sec)
------------------------------------------------------------------------------------
Versions of the Transceiver          | Elapsed Time (sec)| Acceleration Ratio
 1. Everything on CPU                 |          311.2196 |         1.0000
 2. Turbo coding Only on GPU          |           75.5764 |         4.1179
------------------------------------------------------------------------------------
```

Figure 9.48 Execution times and acceleration ratios for the second version of the turbo-coding algorithm

9.10.3 Multiple System Objects on GPU

In the third version of the turbo-coding example, we use multiple GPU-optimized System objects in addition to the turbo decoder used in the previous version. These include *comm.gpu.PSKModulator*, *comm.gpu.AWGNChannel*, *comm.gpu.PSKDemodulator*, and *comm.gpu.TurboDecoder*. As a result, a portion of the algorithm runs on the CPU and a portion on the GPU. This necessitates communication of data between the GPU and the CPU. The *gpuArray* function communicates data from the CPU to the GPU and the *gather* function communicates data from the GPU back to the CPU. The following MATLAB function implements the third version of the algorithm.

Algorithm

MATLAB function

```
function [ber, bits]=zTurboExample_gpu2(EbNo, maxNumErrs, maxNumBits)
FRM=2432;
Indices = lteIntrlvrIndices(FRM);
M=4;k=log2(M);
R= FRM/(3* FRM + 4*3);
snr = EbNo + 10*log10(k) + 10*log10(R);
noiseVar = 10.^(-snr/10);
numIter=6; trellis =  poly2trellis(4, [13 15], 13);
persistent hTEnc Modulator AWGN DeModulator hTDec hBER
if isempty(Modulator)
    hTEnc = comm.TurboEncoder('TrellisStructure',trellis , 'InterleaverIndices', Indices);
    Modulator     = comm.gpu.PSKModulator(4, ...
        'BitInput', true, 'PhaseOffset', pi/4, 'SymbolMapping', 'Custom', ...
        'CustomSymbolMapping', [0 2 3 1]);
```

```
    AWGN            =comm.gpu.AWGNChannel('NoiseMethod', 'Variance', 'VarianceSource',
'Input port');
    DeModulator = comm.gpu.PSKDemodulator(...
        'ModulationOrder', 4, ...
        'BitOutput', true, ...
        'PhaseOffset', pi/4, 'SymbolMapping', 'Custom', ...
        'CustomSymbolMapping', [0 2 3 1],...
        'DecisionMethod', 'Approximate log-likelihood ratio', ...
        'VarianceSource', 'Input port');
    % Turbo Decoder
    hTDec  = comm.gpu.TurboDecoder('TrellisStructure', trellis,'InterleaverIndices', Indices,
'NumIterations', numIter);
    % BER measurement
    hBER = comm.ErrorRate;
end
%% Processing loop
Measures = zeros(3,1); %initialize BER output
while (( Measures(2)< maxNumErrs) && (Measures(3) < maxNumBits))
    data = randi([0 1], numFrames*FRM, 1);
    % Encode random data bits
    yEnc = gpuArray(step(hTEnc, data));
    % Modulate the signal - send to GPU
    modout = step(Modulator, yEnc);
    % Add noise to data
    rData = step(AWGN, modout,noiseVar );
    % Convert to log-likelihood ratios for decoding
    llrData = step( DeModulator, rData, noiseVar);
    % Turbo Decode
    decData = step(hTDec, -llrData);
    % Calculate errors
    Measures = step(hBER, data, gather(decData));
end
bits = Measures(3);
ber= Measures(1);
reset(hBER);
end
```

By using more than one GPU-optimized System object, we hope to further accelerate the simulation. However, the overhead resulting from GPU–CPU communication may actually counterbalance any benefits from the use of System objects. The result of simulation with the following MATLAB script supports this hypothesis. The third version of the algorithm runs a little slower than the second version, completing the seven iterations in about 86 seconds. Note that the bulk of the advantage of running the bottleneck algorithm (turbo decoder) on the GPU is still preserved. We still benefit from a 3.5-times acceleration compared to the CPU version, as indicated by the results in Figure 9.49.

```
Version 3:  Four GPU algorithms + Single-frame

Iteration number 1
Iteration number 2
Iteration number 3
Iteration number 4
Iteration number 5
Iteration number 6
Iteration number 7
Version 3: Time to complete 7 iterations = 86.6899 (sec)
-------------------------------------------------------------------------------
Versions of the Transceiver                  | Elapsed Time (sec)| Acceleration Ratio
  1. Everything on CPU                        |          311.2196 |      1.0000
  2. Turbo coding Only on GPU                 |           75.5764 |      4.1179
  3. Four GPU algorithms + Single-frame       |           86.6899 |      3.5900
-------------------------------------------------------------------------------
```

Figure 9.49 Execution times and acceleration ratios for the third version of the turbo-coding algorithm

Algorithm

MATLAB function

```
MaxSNR=7;
Snrs=0:0.2:1.2;
MaxNumBits=1e6;
N=3;
fprintf(1,'\nVersion 3:  Four GPU algorithms + Single-frame\n\n');
tic;
for idx = 1:MaxSNR
    fprintf(1,'Iteration number %d\r',idx);
    EbNo=Snrs(idx);
    ber= zTurboExample_gpu2(EbNo, MaxNumBits, MaxNumBits);
end
time_GPU2=toc;
fprintf(1,'Version 3: Time to complete %d iterations = %6.4f (sec)\n', MaxSNR,
time_GPU2);
Report_Timing_Results(N,time_CPU,time_GPU2,'Four GPU algorithms + Single-frame');
```

9.10.4   Multiple Frames and Large Data Sizes

The fourth version of the algorithm compensates for the inefficiencies introduced by excessive GPU–CPU communications by concatenating the input data to run multiple frames of data in parallel on the GPU. This approach is beneficial for two reasons. First, the advantage of using the GPU is more pronounced when larger data sizes are used. Second, through concatenation of the data, even the System objects other than the turbo decoder have a fuller GPU buffer to process, which reduces the overhead of the *gpuArray* and *gather* functions. The following MATLAB function shows the fourth version of the algorithm.

Algorithm

MATLAB function

```
function [ber, bits]=zTurboExample_gpu3(EbNo, maxNumErrs, maxNumBits)
FRM=2432;
Indices = lteIntrlvrIndices(FRM);
M=4;k=log2(M);
R= FRM/(3* FRM + 4*3);
snr = EbNo + 10*log10(k) + 10*log10(R);
noiseVar = 10.^(-snr/10);
numIter=6; trellis =  poly2trellis(4, [13 15], 13);
numFrames = 30;                    %Run 30 frames in parallel
persistent hTEnc Modulator AWGN DeModulator hTDec hBER
if isempty(Modulator)
   hTEnc = comm.TurboEncoder('TrellisStructure',trellis , 'InterleaverIndices', Indices);
   Modulator     = comm.gpu.PSKModulator(4, ...
      'BitInput', true, 'PhaseOffset', pi/4, 'SymbolMapping', 'Custom', ...
                                      'CustomSymbolMapping', [0 2 3 1]);
   AWGN          =comm.gpu.AWGNChannel('NoiseMethod', 'Variance', 'VarianceSource',
'Input port');
   DeModulator =  comm.gpu.PSKDemodulator(...
      'ModulationOrder', 4, ...
      'BitOutput', true, ...
      'PhaseOffset', pi/4, 'SymbolMapping', 'Custom', ...
      'CustomSymbolMapping', [0 2 3 1],...
      'DecisionMethod', 'Approximate log-likelihood ratio', ...
      'VarianceSource', 'Input port');
   % Turbo Decoder with MultiFrame processing
   hTDec  = comm.gpu.TurboDecoder('TrellisStructure', trellis,'InterleaverIndices', Indices,
...
      'NumIterations', numIter, 'NumFrames', numFrames);
   % BER measurement
   hBER = comm.ErrorRate;
end
%% Processing loop
   Measures = zeros(3,1); %initialize BER output
   while (( Measures(2)< maxNumErrs) && (Measures(3) < maxNumBits))
      data = randi([0 1], numFrames*FRM, 1);
      % Encode random data bits
      yEnc = gpuArray(multiframeStep(hTEnc, data, numFrames));
      % Add noise to real bipolar data
      modout = step(Modulator, yEnc);
      rData = step(AWGN, modout,noiseVar );
      % Convert to log-likelihood ratios for decoding
       llrData = step( DeModulator, rData, noiseVar);
      % Turbo Decode
      decData = step(hTDec, -llrData);
```

```
        % Calculate errors
        Measures = step(hBER, data, gather(decData));
    end
bits = Measures(3);
ber= Measures(1);
reset(hBER);
end
function y = multiframeStep(h, x, nf)
    xr = reshape(x,[], nf);
    ytmp = step(h,xr(:,1));
    y = zeros(size(ytmp,1), nf, class(ytmp));
    y(:,1) = ytmp;
    for ii =2:nf,
        y(:,ii) = step(h,xr(:,ii));
    end
    y = reshape(y, [], 1);
end
```

In this version, the changes relative to the third version are as follows. A variable called numFrames is used, which indicates how many frames of data are concatenated in this simulation. We have chosen a value of 30 for the *numFrames* parameter. This parameter is applied to the turbo decoder to parallelize the decoding operation on the GPU. A function called multiframeStep is also defined; this performs turbo-encoder operations multiple times and concatenates the results.

The following MATLAB script iterates through the SNR values and records the elapsed time. Note that despite relatively modest changes to the algorithm, we obtain a substantial improvement in the simulation speed. Figure 9.50 shows that this version of the algorithm needs only 28 seconds to complete.

```
Version 4:   Four GPU algorithms + Multi-frame

Iteration number 1
Iteration number 2
Iteration number 3
Iteration number 4
Iteration number 5
Iteration number 6
Iteration number 7
Version 4: Time to complete 7 iterations = 28.0369 (sec)
-----------------------------------------------------------------------------
Versions of the Transceiver               | Elapsed Time (sec)| Acceleration Ratio
  1. Everything on CPU                     |         311.2196 |        1.0000
  2. Turbo coding Only on GPU              |          75.5764 |        4.1179
  3. Four GPU algorithms + Single-frame    |          86.6899 |        3.5900
  4. Four GPU algorithms + Multi-frame     |          28.0369 |       11.1004
-----------------------------------------------------------------------------
```

Figure 9.50 Execution times and acceleration ratios for the fourth version of the turbo-coding algorithm

Algorithm

MATLAB function

```
MaxSNR=7;
Snrs=0:0.2:1.2;
MaxNumBits=1e6;
N=4;
fprintf(1,'\nVersion 4:  Four GPU algorithms + Multi-frame\n\n');
tic;
for idx = 1:MaxSNR
    fprintf(1,'Iteration number %d\r',idx);
    EbNo=Snrs(idx);
    ber= zTurboExample_gpu3(EbNo, MaxNumBits, MaxNumBits);
end
time_GPU3=toc;
fprintf(1,'Version 3: Time to complete %d iterations = %6.4f (sec)\n', MaxSNR,
time_GPU3);
Report_Timing_Results(N,time_CPU,time_GPU3,'Four GPU algorithms + Multi-frame');
```

9.10.5   Using Single-Precision Data Type

Finally, using a single-precision floating-point data type can also accelerate the simulation. Since operations on the GPU are optimized for a single-precision data type, all we need to do is make sure all variables in the fifth version of the algorithm are of single precision. Small modifications such as casting the output of the functions and variables by calling the MATLAB *single* function achieve this task. The fifth version of the algorithm (*zTurboExample_gpu4*), featuring GPU and CPU processing in single-precision floating point, is as follows:

Algorithm

MATLAB function

```
function [ber, bits]=zTurboExample_gpu4(EbNo, maxNumErrs, maxNumBits)
FRM=2432;
Indices = single(lteIntrlvrIndices(FRM));
M=4;k=log2(M);
R= FRM/(3* FRM + 4*3);
snr = EbNo + 10*log10(k) + 10*log10(R);
noiseVar = single(10.^(-snr/10));
numIter=6; trellis =  poly2trellis(4, [13 15], 13);
numFrames = 30;                    %Run 30 frames in parallel
persistent hTEnc Modulator AWGN DeModulator hTDec hBER
if isempty(Modulator)
    hTEnc = comm.TurboEncoder('TrellisStructure',trellis , 'InterleaverIndices', Indices);
    Modulator     = comm.gpu.PSKModulator(4,'OutputDataType', 'single', ...
```

```
      'BitInput', true, 'PhaseOffset', pi/4, 'SymbolMapping', 'Custom', ...
                                     'CustomSymbolMapping', [0 2 3 1]);
    AWGN            =comm.gpu.AWGNChannel('NoiseMethod', 'Variance', 'VarianceSource',
'Input port');
    DeModulator =  comm.gpu.PSKDemodulator(...
      'ModulationOrder', 4, ...
      'BitOutput', true, ...
      'PhaseOffset', pi/4, 'SymbolMapping', 'Custom', ...
      'CustomSymbolMapping', [0 2 3 1],...
      'DecisionMethod', 'Approximate log-likelihood ratio', ...
      'VarianceSource', 'Input port');
    % Turbo Decoder with MultiFrame processing
    hTDec  = comm.gpu.TurboDecoder('TrellisStructure', trellis,'InterleaverIndices', Indices,
...
      'NumIterations', numIter, 'NumFrames', numFrames);
    % BER measurement
    hBER = comm.ErrorRate;
end
%% Processing loop
    Measures = zeros(3,1); %initialize BER output
    while (( Measures(2)< maxNumErrs) && (Measures(3) < maxNumBits))
      data = randi([0 1], numFrames*FRM, 1);
      % Encode random data bits
      yEnc = gpuArray(multiframeStep(hTEnc, data, numFrames));
      % Add noise to real bipolar data
      modout = step(Modulator, yEnc);
      rData = step(AWGN, modout,noiseVar );
      % Convert to log-likelihood ratios for decoding
       llrData = step( DeModulator, rData, noiseVar);
      % Turbo Decode
      decData = step(hTDec, -llrData);
      % Calculate errors
      Measures = step(hBER, data, gather(decData));
    end
bits = Measures(3);
ber= Measures(1);
reset(hBER);
end
function y = multiframeStep(h, x, nf)
  xr = reshape(x,[], nf);
  ytmp = step(h,xr(:,1));
  y = zeros(size(ytmp,1), nf, class(ytmp));
  y(:,1) = ytmp;
  for ii =2:nf,
    y(:,ii) = step(h,xr(:,ii));
  end
  y = reshape(y, [], 1);
end
```

```
Version 5:  Four GPU algorithms + Multi-frame + float

Iteration number 1
Iteration number 2
Iteration number 3
Iteration number 4
Iteration number 5
Iteration number 6
Iteration number 7
Version 5: Time to complete 7 iterations = 14.5888 (sec)
-------------------------------------------------------------------------------
Versions of the Transceiver              | Elapsed Time (sec)| Acceleration Ratio
  1. Everything on CPU                    |          311.2196 |      1.0000
  2. Turbo coding Only on GPU             |           75.5764 |      4.1179
  3. Four GPU algorithms + Single-frame   |           86.6899 |      3.5900
  4. Four GPU algorithms + Multi-frame    |           28.0369 |     11.1004
  5. Four GPU algorithms + Multi-frame + float |     14.5888 |     21.3328
-------------------------------------------------------------------------------
  '
```

Figure 9.51 Execution times and acceleration ratios for the fifth version of the turbo-coding algorithm

We can immediately see the effect of running the fifth version with the following MATLAB calling script. Note that it takes only about 14 seconds to process the same number of bits in the same number of iterations, doubling the speed of the fourth version of the algorithm (double-precision version). Overall, by combining all the acceleration techniques introduced, we observe a 21-times acceleration compared to the CPU version of the algorithm. The results are summarized in Figure 9.51.

Algorithm

MATLAB function

```
MaxSNR=7;
Snrs=0:0.2:1.2;
MaxNumBits=1e6;
N=5;
fprintf(1,'\nVersion 5:  Four GPU algorithms + Multi-frame + float\n\n');
tic;
for idx = 1:MaxSNR
    fprintf(1,'Iteration number %d\r',idx);
    EbNo=Snrs(idx);
    ber= zTurboExample_gpu4(EbNo, MaxNumBits, MaxNumBits);
end
time_GPU4=toc;
fprintf(1,'Version 4: Time to complete %d iterations = %6.4f (sec)\n', MaxSNR,
time_GPU4);
Report_Timing_Results(N,time_CPU,time_GPU4,'Four GPU algorithms + Multi-
frame + float');
```

9.11   Chapter Summary

In this chapter we introduced multiple techniques for speeding up simulations in MATLAB and Simulink. Throughout, we showcased a series of optimizations used to accelerate the simulation of the LTE control-channel-processing algorithm and a turbo-coding algorithm. We started with baseline implementations and through successive profiling and code updates introduced the following optimizations: (i) better MATLAB serial programming techniques (vectorization, preallocation), (ii) System objects, (iii) MATLAB-to-C code generation (MEX), (iv) parallel computing (parfor, spmd), (v) GPU-optimized System objects, and (vi) rapid accelerator simulation mode in Simulink. We went through detailed examples in MATLAB and Simulink and showed that the extent of acceleration can be further amplified by combining two or more of these simulation-acceleration techniques.

10

Prototyping as C/C++ Code

So far we have developed MATLAB® programs and Simulink models in order to simulate the LTE (Long Term Evolution) PHY (Physical Layer) in the MATLAB environment. At some stage in the workflow of a communications system design, we might need to produce a software component that cannot be directly simulated in MATLAB. For example, we might need to interface to an existing simulation environment based on a C/C++ software implementation. If we want to export the result of modeling and simulation in MATLAB to an external C/C++ programming environment, we essentially have two choices: we can either manually translate algorithms developed in MATLAB into a C or C++ implementation or we can take advantage of automatic MATLAB C-code generation.

By using MATLAB Coder, we can generate standalone C and C++ code from MATLAB code. The generated source code is portable and readable. MATLAB Coder supports a subset of MATLAB language features, including program control constructs, functions, and matrix operations. It can generate MATLAB executable (MEX) functions that let us accelerate computationally intensive portions of MATLAB code and verify its behavior. It can also generate C/C++ source code for integration with existing C code, creation of an executable prototype, or direct implementation on a Digital Signal Processor (DSP) or general-purpose CPU using a C/C++ compiler.

In this chapter we examine the process of generating standalone C and C++ code from MATLAB code using MATLAB Coder. We first present use cases, motivations, and requirements for C/C++ code generation and then examine the mechanics of code generation using two methods: (i) calling code-generation functions from the MATLAB command line and (ii) using the MATLAB Coder Project Application. We then elaborate on the extent of support for code generation in MATLAB, highlighting code-generation support by various System toolboxes and support for various data types, including fixed-point data, and for MATLAB programs employing variable-sized data. Finally, we present a full workflow for the integration of generated code from a MATLAB algorithm into an existing C/C++ testbench.

Understanding LTE with MATLAB®: From Mathematical Modeling to Simulation and Prototyping, First Edition. Houman Zarrinkoub.
© 2014 John Wiley & Sons, Ltd. Published 2014 by John Wiley & Sons, Ltd.

10.1 Use Cases

Before we tackle the subject of generating C code from MATLAB, let us first elucidate the reasons why engineers translate MATLAB code to C today:

- **Integration**: We may want to integrate our MATLAB algorithms into an existing C-based project or software, such as a custom simulator, as source code or libraries.
- **Prototyping**: We may need to create a standalone prototype or executable for testing purposes or in order to create proof-of concept demonstrations.
- **Acceleration**: We may want to wrap the C code as MEX files for execution back in MATLAB. This use case is essentially for accelerating the execution of portions of algorithms that are numerically intensive.
- **Implementation**: We may need to take the C code and implement it in embedded processors as part of a larger system design.

10.2 Motivations

With the automatic translation of an algorithm from MATLAB to C, we can save the time it takes to rewrite the program and debug the low-level C code. This can provide more time for development and tuning of our algorithms at a high level in MATLAB. As we update each version of our MATLAB code, we can then generate a MEX file automatically. We can use the MEX file and call it in MATLAB in order to verify that the compiled version of the code executes properly. The MEX file can also be used to speed up the code in most cases. We can also generate source code, executables, or libraries automatically. As a result, we can maintain one design in MATLAB and periodically get a C/C++ code as a byproduct. Having a single software reference in MATLAB makes it easier to make changes or to improve the performance. As will be discussed in this chapter, we can also leverage automated tools to help assess the readiness of the MATLAB code for code generation. These tools can guide us in the steps needed to successfully generate C code from MATLAB algorithms.

10.3 Requirements

In order to generate C/C++ code from MATLAB algorithms, we must install MATLAB Coder and use a C/C++ compiler. First, we set up the compiler. For most platforms, MathWorks supplies a default compiler with MATLAB. If an installation does not include a default compiler, we must obtain and install a supported C/C++ compiler. The MATLAB documentation contains a list of supported compilers by platform [1]. To set up an installed compiler, at the MATLAB command line enter:

Algorithm

```
>> mex –setup
```

This will show a list of installed compilers and allow one to be selected. Note that the choice of compiler is quite important, because the speed of simulation of a compiled MATLAB code depends on the type of compiler and the compiler options used.

Both numerical and timing results provided throughout the book depend on the platform where MATLAB is installed, and the type of operating system, C/C++ compiler or GPU that is used. Results in this book for non-GPU experiments are obtained by running MATLAB on a laptop computer with the following specifications:

- **Hardware:** Intel Dual-Core i7-2620M CPU @ 2.70 GHz with 8 GB of RAM
- **Operating system:** 64-bit Windows 7 Enterprise (Service Pack 1)
- **C/C++ compiler:** Microsoft Visual Studio 2010 with Microsoft Windows SDK v7.1.

10.4 MATLAB Code Considerations

In order to convert MATLAB code into efficient C/C++ code we must always consider the following MATLAB code attributes:

- **Data types**: C and C++ use static typing. MATLAB, on the other hand, is a language based on dynamic typing of data. To bridge this gap, MATLAB Coder requires a complete assignment of type to each variable for successful code generation. MATLAB Coder offers many ways of determining variable types before use. For each variable, three properties must be determined: the class (or data type), size (or dimension), and complexity (whether or not the variable is a complex number). Examples in the following sections show how easily these properties can be specified and MATLAB code can be translated to C/C++.
- **Array sizing**: Variable sizes (dimensions) in MATLAB can be either fixed or variable during a simulation. Variable-sized arrays and matrices are supported for code generation. We can define inputs, outputs, and local variables in MATLAB functions to represent data that vary in size at run time.
- **Memory allocations**: We can choose whether generated code uses static or dynamic memory allocation. With dynamic memory allocation, potentially less memory is used, at the expense of the time required to manage it. With static memory, we get the best speed, but with higher memory usage. Most MATLAB algorithms take advantage of the dynamic sizing features in MATLAB. As a result, dynamic memory allocation typically enables us to generate code from existing MATLAB code with few modifications. Dynamic memory allocation also allows some programs to compile even when upper bounds cannot be found. Static allocation reduces the memory footprint of the generated code and is therefore suitable for applications where there is a limited amount of available memory, such as embedded applications.
- **Speed**: In applications that involve real-time signal processing, algorithms must be fast enough to keep up with the rate of arriving data. One way to improve the speed of the generated code for a real-time application is to disable unnecessary run-time checks. Run-time checks include extra code that ensures array bound integrity and responsiveness and avoids occurrence of operations that produce non-numerical results, such as division by zero. We can disable these checks and accelerate the generated C code after we have verified that the algorithm is designed properly and, for example, that the array boundaries are respected in successive assignments.

10.5 How to Generate Code

In this section we will review the typical steps involved in automatic MATLAB to C code generation. Code can be generated by using either a MATLAB Coder Project or the *codegen* command called at the MATLAB command line.

10.5.1   Case Study: Frequency-Domain Equalization

We start with a simple example, which implements an algorithm that performs frequency-domain equalization. In this algorithm, the output y is the result of multiplication of each sample of the received input signal u by corresponding samples of the channel-gain signal *coefficients*.

Algorithm

MATLAB function: Equalizer.m

```
function y = Equalizer( u, coefficients)
%#codegen
% Equalizer using element-by-element multiplication
y = u.*coefficients;
end
```

First we compose a calling script or testbench, *call_Equalizer.m*, that defines the input arguments and calls the function to produce the output. By executing this script, we get an expected value for the output before code generation. We can use this value to verify that after code generation the output function has generated the correct result. In this case, purely for illustration purposes, we use an arbitrary function (cosine of angles in radians ranging from 1 to 100) to provide a set of coefficients (*coef*) that multiply (in other words, equalize) the samples of the received data (*u*).

Algorithm

MATLAB testbench: call_Equalizer.m

```
u=1:100;              % First input
coef=cos(u);          % Second input
y=Equalizer(u,coef);  % Function call
```

There is another important reason to create the testbench: the process of code generation depends critically on the definitions of function inputs. The code-generation engine needs to map a dynamically typed language (MATLAB) to a statically typed language (C/C++). In this process, the data type, size, and complexity of every variable in the generated C code must be determined. The only requirement for the user is to define the data type, size, and complexity of the function inputs. The code-generation engine will then infer all other internal variables from those of the input parameters and generate the C code.

10.5.2   Using a MATLAB Command

The simplest way to generate C code from a MATLAB function is by using the *codegen* command. When *codegen* is called, a MATLAB function name and a list of command arguments must be specified. One important command argument is the *−args*, which defines the size, class, and complexity of all MATLAB function inputs by leveraging existing example variables in the MATLAB workspace.

For example, in order to generate C code for the function *Equalizer.m* we should first run the calling function *call_Equalizer.m* to create the function input variables *u* and *coef* in the MATLAB workspace. Then we just type the *codegen* command line as:

Algorithm

>> *codegen −args {u , coef} Equalizer.m*

The function *Equalizer.m* is the MATLAB entry-point function from which we generate a MEX function, C/C++ library, or C/C++ executable code. By default, when no additional arguments are specified, the *codegen* command generates a MEX function. The default name given to the generated MEX function is the function name followed by the *_mex* suffix. In this example, a MEX function called *Equalizer_mex.mex<platform>* is generated in the same directory as the MATLAB entry-point function. The *<platform>* suffix refers to the operating system. For example, if MATLAB is installed on a 64 bit machine running the Windows operating system, the full name is *Equalizer_mex.mexw64*.

To generate a C/C++ library instead of the default MEX function, at the command line the *−config* option should be used:

Algorithm

>> *codegen −args {u , coef} Equalizer.m −config:lib −report*

The type of output generated can be specified by using the *−config* option. In this case, using the *−config:lib* option generates a static C/C++ library composed of C source files and header files. By default, these files are stored in a folder related to the directory where the MATLAB entry-point function *<fcn_name>* resides. Table 10.1 shows how different uses of the *−config* options map to different code-generation output types and the relative locations of generated files.

The *−report* option provides a convenient hyperlink to the generated files. When we click on the hyperlink, the Code Generation Report opens. The report contains the result of the code generation. There are three tabs: MATLAB Code, Call Stack, and C Code. In the MATLAB Code tab, MATLAB functions are shown. In the C Code tab, as illustrated in Figure 10.1, the generated C files are shown.

The C source code generated has the same name as the MATLAB entry-point function (in this example, *Equalizer.c*). As we can see, it implements the equalization efficiently as an

Table 10.1 Mapping of configuration options, generated output types, and file locations

−config option	Output type	Relative location of generated files
mex	MEX function	*codegen/mex/<fcn_name>*
lib	static C/C++ library	*codegen/lib/<fcn_name>*
dll	dynamic C/C++ library	*codegen/dll/<fcn_name>*
exe	static C/C++ executable	*codegen/exe/<fcn_name>*

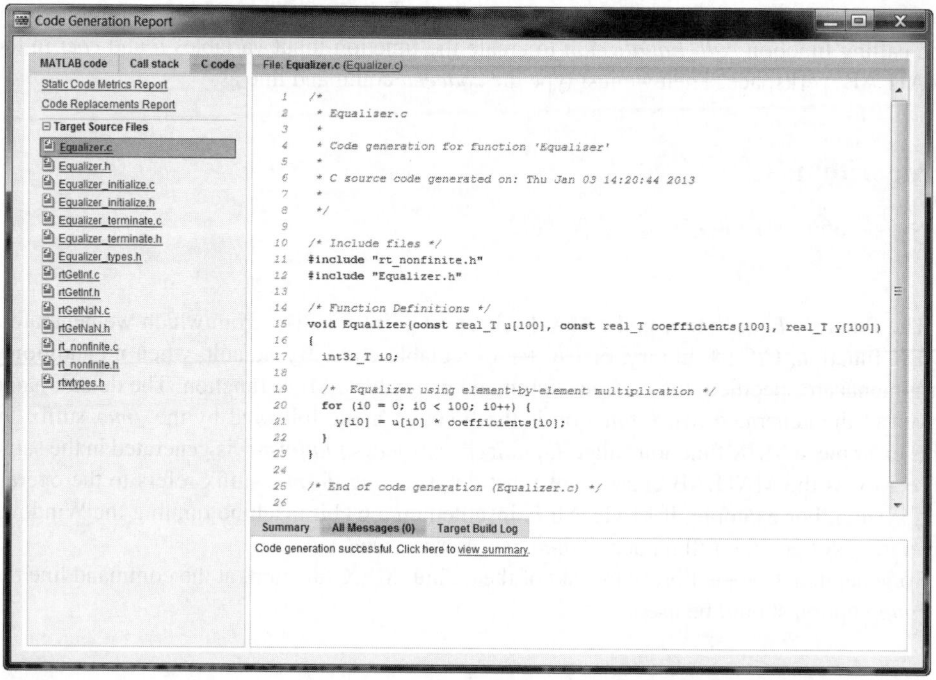

Figure 10.1   Code Generation Report showing generated C code

element-by-element multiplication using a for loop. Note that in the testbench, we assigned the function inputs a data type of double-precision floating point. As a result, the generated C code defines C-code variables as *real_T*, which is a MATLAB data type that corresponds to double-precision floating point.

The C code also features two header files. The first, *rt_nonfinite.h*, contains all the type definitions in MATLAB, such as *real_T*, and definitions for nonfinite MATLAB data such as *nan* and *inf*. The second, *Equalizer.h*, includes the function prototypes needed to include the source file in a C/C++ calling function. In the next subsection, we discuss the structure of generated C code and the corresponding files.

10.5.3   Using the MATLAB Coder Project

In this section we show how to generate C code using the MATLAB Coder Project (Figure 10.2). The MATLAB Coder Project is an example of a MATLAB application. It uses a Graphical User Interface (GUI) and is a handy tool in helping complete the code-generation process.

First, we create the project by typing the following MATLAB command:

Algorithm

```
>> coder –new MyEqualizer.prj
```

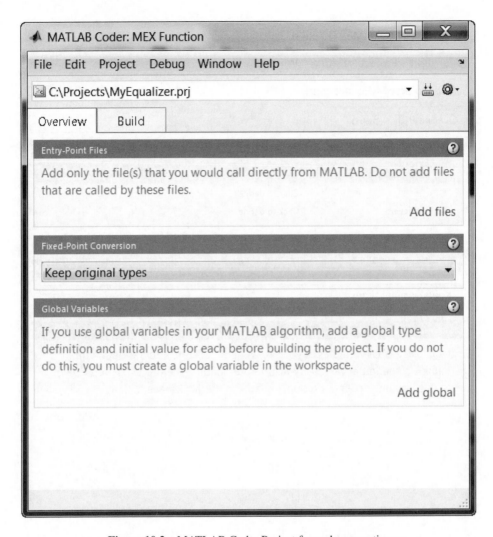

Figure 10.2   MATLAB Coder Project for code generation

This command starts a new MATLAB Coder Project, which we have called *MyEqualizer*. The Code Generation Project dialog box shown in Figure 10.3 opens, showing the path to the project in the directory structure. By default it is set to generate a MEX function. The next step is to add the MATLAB entry-point function to the project, either by dragging it to the section of the project called Entry-Point Files or by using the Add Files link, as illustrated in Figure 10.3.

At this stage, MATLAB Coder adds the file to the project. This function has two input parameters, *u* and *coefficients*, which appear below the file name. Note that so far the data type, size, and complexity properties of these input variables are still undefined. To compile this function, we specify the testbench so that MATLAB Coder can infer types for function input variables. By clicking on the Autodefine Types link, the Autodefine Entry-Point Input Types dialog box will appear (Figure 10.4). In the dialog box, we click the "+" button to add a test file to the project. Here, we add the testbench *call_Equalizer.m* as the test file.

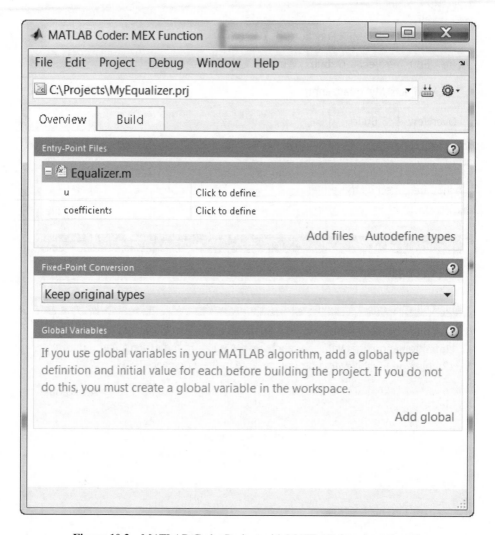

Figure 10.3   MATLAB Coder Project with MATLAB function selected

When we click on the Run button, the testbench executes. This enables MATLAB Coder to infer the size, data type, and complexity of each input variable of the MATLAB entry-point function. The results appear in a new dialog box, called Autodefine Input Types, as illustrated in Figure 10.5.

By clicking on the Use These Types button, we accept these properties and assign them to the input function parameters. As a last step, we click on the Build tab to select the output file name and output type and then click on the Build button to generate code (Figure 10.6).

By default, the output type is a MEX function. This means that following code generation, MATLAB Coder compiles the code as a MEX function that can only be called from within MATLAB environment.

The Verification section in the MATLAB Coder Project enables the generated MEX function to be run with the same testbench (calling script) used to define the data types. By comparing

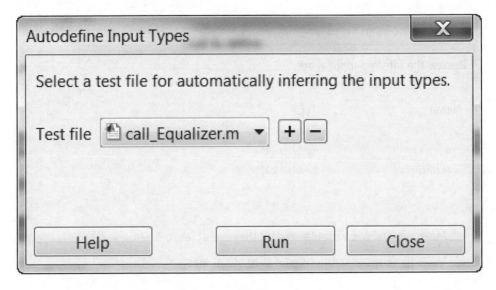

Figure 10.4   MATLAB Coder Project: dialog to select the testbench

the result of running the *Equalizer.m* function with the result of running the MEX function, we can verify that the MATLAB function and the generated MEX function are numerically identical.

We can obtain the actual C source code generated by MATLAB Coder by changing the output type to either dynamic C/C++ library or static C/C++ library. In this example, we just change the output type of the project to static C/C++ library and click on the Build button, as shown in Figure 10.7. After the Build button is pressed, the code-generation Build dialog appears (Figure 10.8). As illustrated in the figure, this dialog shows the code-generation progress and illustrates any error or warning messages that might be generated during the code-generation process.

If code generation is successful, we can click on a hyperlink that will open the Code Generation Report and show the result of code generation. In this example, the Code Generation Report is identical to that shown in Figure 10.1.

10.6   Structure of the Generated C Code

The generated C code exhibits a predefined structure. By examining the C Code tab of the Code Generation Report, we can see that besides the C source file and header file, which bear the same name as the MATLAB entry-point function (*Equalizer.c* and *Equalizer.h*), other files are generated. The list of generated files for this example is illustrated in Figure 10.9.

Operations performed in a MATLAB function can be subdivided into three categories according to the stage of simulation:

- Initialization contains operations that take place only once, during the initialization phase, before the processing loop starts.

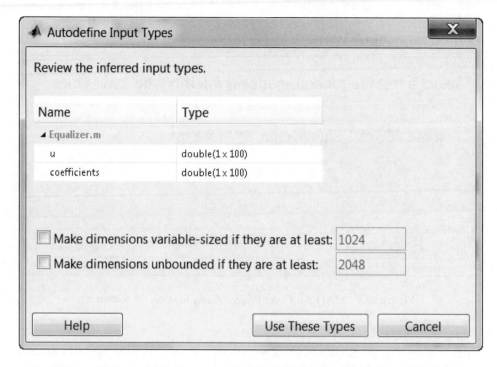

Figure 10.5   Autodefine Input Types dialog: inferring data types from testbench

- Function call (or in-loop processing) contains operations that are performed every time the function is called.
- Termination contains operations that are performed at the end of the simulation, in order to clean up the resources that were allocated during initialization and function calls.

The generated C code of a MATLAB function reflects the same structure for different types of operations. Note, for instance, in the Equalizer example:

- *Equalizer_initialize.c* and *Equalizer_initialize.h* correspond to the operations performed only during initialization.
- *Equalizer.c* and *Equalize.h* correspond to the main function-call operations performed every time.
- *Equalizer_terminate.c* and *Equalizer_terminate.h* correspond to the operations performed only during initializations.

For our simple case of element-wise operation performed in the Equalizer function, both the initialization and the termination C files *(Equalizer_initialize.c* and *Equalizer_terminate.c)* are empty and contain no operations. However, in more complex functions where variables that implement constants or stored data need to be initialized, the initialization functions in C are not empty and contain the initializations operations. Similarly, in functions where, for example, dynamic memory allocation is used to create a variable, the termination functions

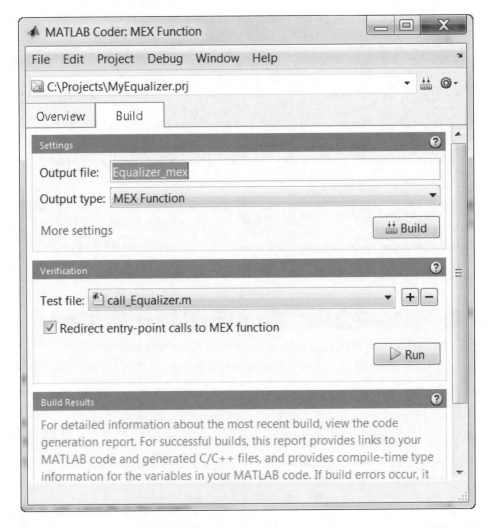

Figure 10.6   MATLAB Coder Project: using the Build tab to specify output type and file name

are not empty and contain typical *free()* operations in C that return the dynamically allocated memory back to the system resources.

Beside the generated C files related to operations performed in various stages, we also find six files that provide type definitions and operations for "nonfinite" numerical constructs in MATLAB. These are data types that are not natively defined in C. They are useful if an algorithm sometimes uses operations that allow variables to take on values such as *nan* (not-a-number) and *inf* (infinity). The list of generated files dedicated to "nonfinite" definitions includes *rt_nonfinite.c*, *rt_nonfinite.h*, *rtGetInf.c*, *rtGetInf.h*, *rtGetNaN.c*, and *rtGetNaN.h*.

The file *rtwtypes.h* contains all necessary type information and macros defining operation and data types supported in MATLAB. Depending on the MATLAB function, different types of other files are also generated. For a complete description of code generation and file partitioning, refer to the MATLAB documentation [2].

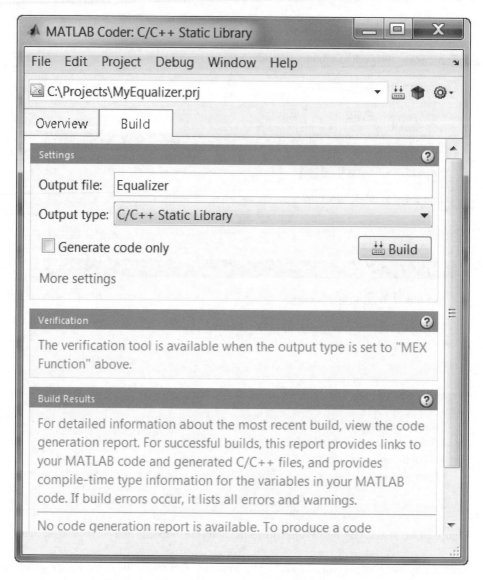

Figure 10.7   MATLAB Coder Project: choosing C/C++ source code as output type

10.7   Supported MATLAB Subset

A broad subset of the MATLAB language is supported for code generation. The supported language features include all of the standard matrix operations, various data types, and various program control constructs and structures. A comprehensive list of MATLAB language features that generate code is available in the MATLAB documentation [2] and includes the following: double-precision and single-precision floating-point, integer, and fixed-point arithmetic, complex numbers, characters, numeric classes, N-dimensional arrays, structures, matrix operations, arithmetic, relational, and logical operators, subscripting and function handles,

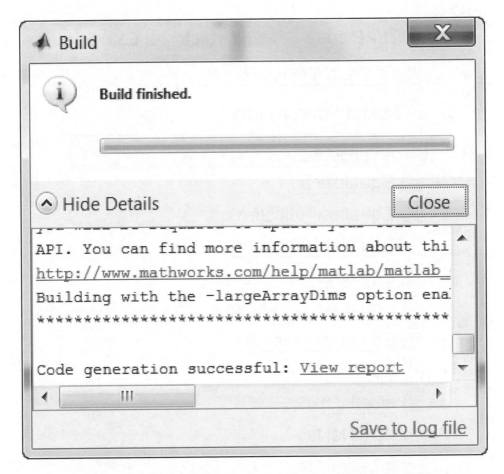

Figure 10.8 MATLAB Coder Project: Code Generation Report during the build process

persistent and global variables, program control statements (if, switch, for, and while loops), variable-sized data, variable-length input and output argument lists, a subset of MATLAB toolbox functions, and MATLAB classes.

More than 400 operators and functions from different toolboxes (including the Signal Processing Toolbox) can generate code. In the latest release of MATLAB (R2013a), more than 300 System objects that are part of the System toolboxes (DSP System Toolbox, Communications System Toolbox, and Computer Vision System Toolbox) are also supported for code generation. The following MATLAB language features are not supported for C/C++ code generation: anonymous and nested functions, cell arrays, Java, recursion, sparse matrices, and try/catch statements.

10.7.1 Readiness for Code Generation

MATLAB Coder provides automated tools to help assess the code-generation readiness of an algorithm. These tools can identify the portions of a MATLAB code that cannot be translated to C code. With a slew of detailed and targeted messages, these tools guide us through the steps

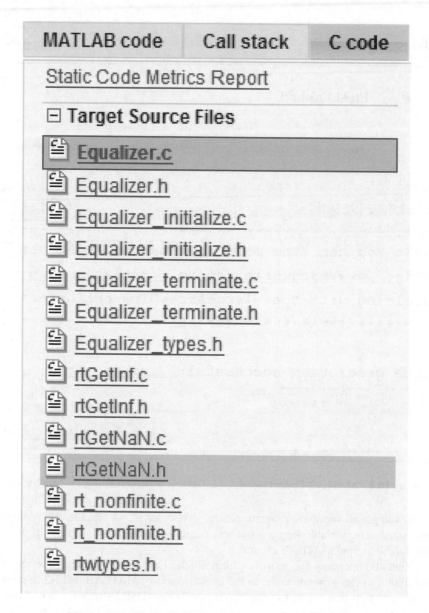

Figure 10.9 Code Generation Report: list of generated files

necessary to modify the unsupported portions so that C code can successfully be generated. Next we will discuss how to use two of these tools: the MATLAB Code Analyzer and the Code Readiness Report.

10.7.2   Case Study: Interpolation of Pilot Signals

In this section we use an example to illustrate how these tools help identify and correct code-generation issues. This example is a MATLAB function that finds the equalizer coefficients that are to be applied along all rows and columns – that is, along all subcarriers and all Orthogonal Frequency Division Multiplexing (OFDM) symbols in a subframe – by interpolating the coefficients found on a selected set of "pilot" symbols. This function (*Equalizer.m*) was developed in Section 5.16. The first version of the algorithm is shown here as *MyInterp0.m*.

Algorithm

MATLAB function: MyInterp0.m

```
function out = MyInterp0(y)
%#codegen
UpsampFactor=6;
out=interp(y,UpsampFactor);
```

When we edit functions and scripts in the MATLAB Editor, the Code Analyzer continuously checks the code as it is written. It shows warning and error messages about the code and allows functions to be modified. The messages update automatically and continuously to show whether the changes address the issues raised.

For example, in the function *MyInterp0.m*, if you end the line containing the call to the *interp* function with a "}" character instead of a ")" character, the Code Analyzer displays error messages in the MATLAB Editor (Figure 10.10). Note that the message indicator at the top of the message bar is red. At line four, the Code Analyzer displays a message regarding lack of code generation support for cell arrays. This message is correct, since cell arrays are denoted by { } rather than (). By modifying the code to call the function with the right syntax, these error messages will disappear and the message indicator will become green.

Let us see how the MATLAB Coder project handles code-generation issues for this function. Start by typing this command:

Algorithm

```
>> coder –new MyInterp
```

When *MyInterp0.m* is added as the MATLAB entry-point file, the following message appears in the project dialog box: "View code generation readiness issues." Upon clicking on this link, the Project Code Generation Readiness Report will appear as illustrated in Figure 10.11. The report identifies an unsupported function (*interp* function from the Signal Processing Toolbox)

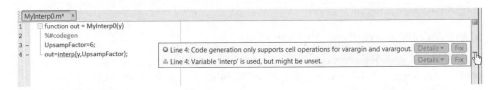

Figure 10.10   Code Analyzer reporting errors in the MATLAB editor

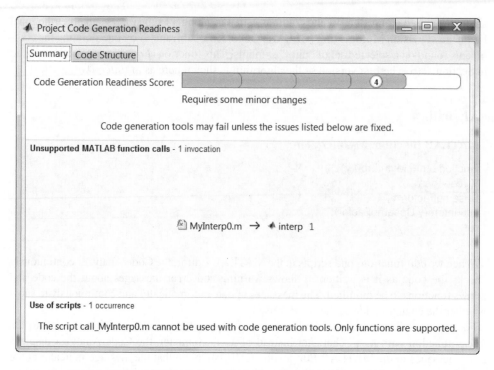

Figure 10.11   MATLAB Coder Project: code-generation readiness report

in our algorithm. When the unsupported function is replaced with one that uses the supported MATLAB functions and features, the readiness report will indicate that the code-generation issues are resolved and code generation can proceed.

10.8   Complex Numbers and Native C Types

In this section, we use the Equalizer example to show how to generate C code for algorithms that involve complex numbers. The Equalizer algorithm is written in such a way that the same operations are applied whether the input variables are scalars, vectors, or matrices of any size.

For example, to generate code when the input is a matrix, we do not need to change the algorithm: we only need to change the testbench that calls the function. The following testbench, *call_Equalizer2.m*, creates both input variables *u* and *coef* as complex matrices, with a dimension of 72 rows and 14 columns and a data type of single-precision floating point.

Algorithm

MATLAB calling script: call_Equalizer2.m

```
u=complex(single(randn(72,14)));                    % First input
coef= single(randn(72,14)) +1j * single(randn(72,14));   % Second input
y=Equalizer(u,coef);                                % Function call
```

When you run this testbench, variables (*u*, *coef*, and *y*) are created in the MATLAB workspace. By typing the MATLAB command *whos*, we can examine the sizes, classes (data types), and complexities of these variables (Figure 10.12).

We can repeat the steps highlighted in the previous section to generate the C code for the *Equalizer.m* function. First, by clicking on the Autodefine Types link in the MATLAB Coder Project, we use the new testbench to define the types and sizes for the input variables (Figure 10.13). At this point, we can accept the proposed data types as complex single-precision floating-point matrices with a 72×14 matrix size, as illustrated in Figure 10.14. Finally, by selecting the Build tab and choosing the C/C++ static library as the output type, we can generate the C code for the Equalizer function.

Figure 10.15 shows the output of code generation. Note that input variables in the C code are of a new type called *creal32_T*, signifying complex variables of single-precision floating-point type. All these type definitions, performed automatically by MATLAB Coder, can be found in the file known as *rtwtypes.h*. This file contains all the necessary type information and the macros defining complex variable operations. Essentially, every operation and every data type

```
>> whos
  Name          Size                  Bytes   Class     Attributes

  coef          72x14                  8064   single    complex
  u             72x14                  8064   single    complex
  y             72x14                  8064   single    complex
```

Figure 10.12   Examining the data types, sizes, and complexities of input variables to a function

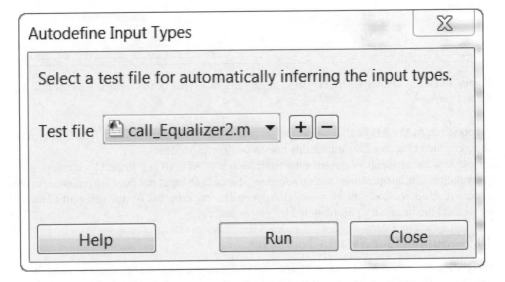

Figure 10.13   MATLAB Coder Project: selecting the testbench to test the generated MEX function

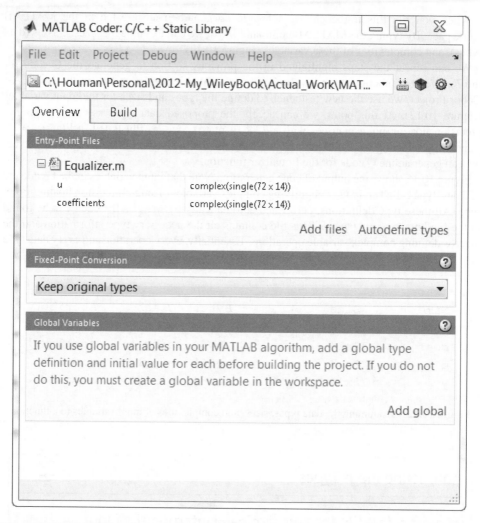

Figure 10.14   MATLAB Coder Project: changing the data types of the input arguments that use the test file

supported in MATLAB is also supported properly in the generated C code. The header files and generated C source files reflect this one-to-one correspondence.

Note how the element-wise matrix multiplication in MATLAB is reflected by element-wise array multiplication operations within a for loop. Since both input matrices are complex, in the generated C source code the element-wise operations are repeated for the real part (denoted by *.re*) and the imaginary part (denoted by *.im*) separately.

10.9   Support for System Toolboxes

A significant subset of MATLAB operators and functions support code generation. In addition, a majority of functions in the Signal Processing Toolbox and of System objects in the DSP and Communications System Toolboxes also support code generation.

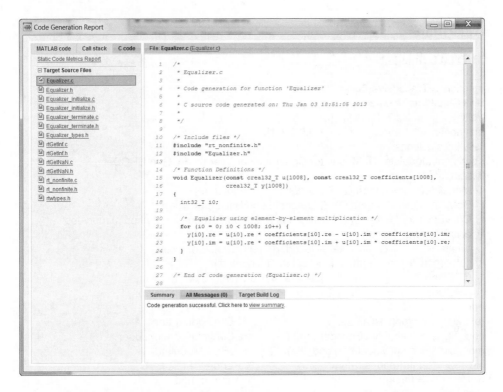

Figure 10.15 Code Generation Report: generated source code with complex data-type inputs

In this section we show examples of the use of System objects from System toolboxes for code generation. The benefits of using System toolboxes for code generation are twofold: first, System objects include many optimizations in the generated C code; second, by leveraging algorithms available in System toolboxes, more time is spent composing system components rather than recreating and optimizing algorithmic building blocks.

10.9.1   Case Study: FFT and Inverse FFT

The following function shows a simple example of a transceiver. In the transmitter, input bits are modulated and an Inverse Fast Fourier Transform (IFFT) operation is then applied to the modulated symbols before channel modeling with an Additive White Gaussian Noise (AWGN) channel. At the receiver, we first perform a Fast Fourier Transform (FFT) operation and then demodulate the signal to produce output bits. By comparing input and output bits, we compute the Bit Error Rate (BER). This example does not correspond to any particular known communications standard; we have chosen it for its simplicity and to demonstrate how to use System objects for code generation. This example uses multiple System objects from Communications System Toolbox, including: a modulator, a demodulator, a convolutional encoder, a Viterbi decoder, a Cyclic Redundancy Check (CRC) generator, a CRC detector, and an AWGN channel.

Algorithm

MATLAB function

```
function y=Transceiver0(u)
%% Constants
trellis=poly2trellis(7, [133 171]);
polynomial=[1 1 zeros(1, 16) 1 1 0 0 0 1 1];
%% Initializations
persistent Modulator DeModulator ConvEncoder Viterbi CRCGen CRCDet
if isempty(Modulator)
    Modulator      = comm.QPSKModulator('BitInput',true);
    DeModulator  = comm.QPSKDemodulator('BitOutput',true);
    ConvEncoder  = comm.ConvolutionalEncoder('TerminationMethod','Truncated',
'TrellisStructure', trellis);
    Viterbi          = comm.ViterbiDecoder('TrellisStructure', trellis,
'InputFormat','Hard','TerminationMethod','Truncated');
    CRCGen       = comm.CRCGenerator('Polynomial', polynomial);
    CRCDet       = comm.CRCDetector  ('Polynomial', polynomial);
end
tb          = step(CRCGen , u);            % CRC generator
cod_sig     = step(ConvEncoder , tb);      % Convolutional encoder
mod_sig     = step(Modulator, cod_sig);    % QPSK Modulator
sig         = ifft(mod_sig);               % Perform IFFT
rec         = fft(sig);                    % Perform FFT
demod       = step(DeModulator, rec);      % QPSK Demodulator
dec         = step(Viterbi , demod);       % Viterbi decoder
y           = step(CRCDet , dec);          % CRC detector
```

The process of code generation proceeds in the same way as in earlier examples and the results are shown in the Code Generation Report. More complex algorithms can be composed by using toolbox functionalities. This means that the size of the generated C file becomes rather large, and the entire generated C code cannot be shown here; Figure 10.16 shows the first few lines.

As a first optimization, we can turn off the support for nonfinite data types in order to reduce the amount of generated C code. The customization page in the MATLAB Coder Project has a Speed tab. Note that we can turn off the nonfinite data-type support by unchecking the relevant checkboxes as shown in Figure 10.17. A second observation is that the algorithm does not have variables that represent states and memory. Therefore, the generated C code does have many lines of code in the initialization function (*transceiver_initialize.c*) shown in Figure 10.18.

We would like to observe the effects of algorithms that contain variables representing states on the generated C code. To this end, we add a random bit generator to create the input bits. Since random number generators depend on maintaining the seed or states, there will be a good amount of initialization code within the generated C code.

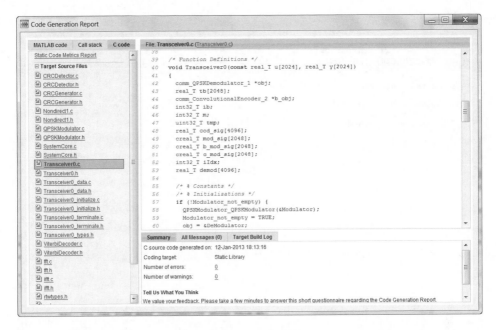

Figure 10.16 Code Generation Report: a few lines of the generated code

Algorithm

MATLAB function

```
function [u, y]=Transceiver1
%% Constants
trellis=poly2trellis(7, [133 171]);
polynomial=[1 1 zeros(1, 16) 1 1 0 0 0 1 1];
%% Initializations
persistent Modulator DeModulator ConvEncoder Viterbi CRCGen CRCDet
if isempty(Modulator)
    Modulator    = comm.QPSKModulator('BitInput',true);
    DeModulator  = comm.QPSKDemodulator('BitOutput',true);
    ConvEncoder  = comm.ConvolutionalEncoder('TerminationMethod','Truncated',
'TrellisStructure', trellis);
    Viterbi      = comm.ViterbiDecoder('TrellisStructure', trellis,
'InputFormat','Hard','TerminationMethod','Truncated');
    CRCGen       = comm.CRCGenerator('Polynomial', polynomial);
    CRCDet       = comm.CRCDetector    ('Polynomial', polynomial);
end
u          = randi([0 1], 2024,1);          % Random bits generation
tb         = step(CRCGen , u);              % CRC generator
cod_sig    = step(ConvEncoder , tb);        % Convolutional encoder
```

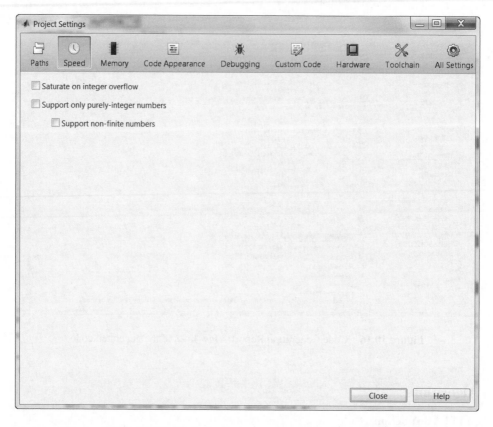

Figure 10.17   MATLAB Coder Project: options related to simulation speed

```
mod_sig   = step(Modulator, cod_sig);      % QPSK Modulator
sig       = ifft(mod_sig);                 % Perform IFFT
rec       = fft(sig);                      % Perform FFT
demod     = step(DeModulator, rec);        % QPSK Demodulator
dec       = step(Viterbi , demod);         % Viterbi decoder
y         = step(CRCDet , dec);             % CRC detector
```

In the updated function (*Transceiver1.m*), we generate the input bits using the MATLAB *randi* function. As a result, the function will have no inputs and two outputs. Generating code for this function illustrates how initializations involved in algorithms with states are handled in the initialization file *transceiver_initialize.c*. Note that the initialization function in C updates the variable *b_state*, which is defined in the generated C code as a static variable. Using a static variable in C is one way of representing a variable that maintains its value across multiple calls and implements a state variable (Figures 10.19 and 10.20).

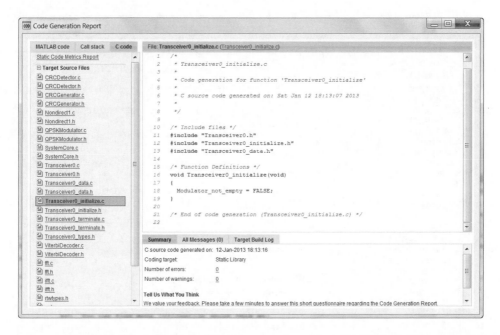

Figure 10.18 Code Generation Report: content of the initialization function

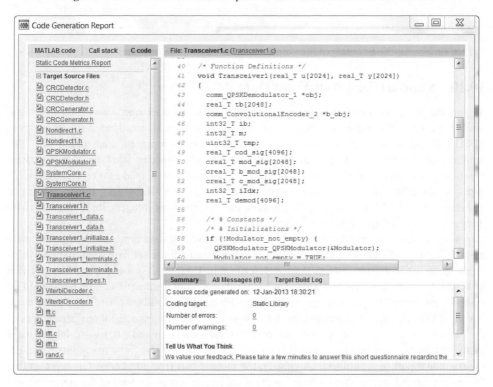

Figure 10.19 Generated code for *transceiver1.m*, showing the first few lines of generated C code

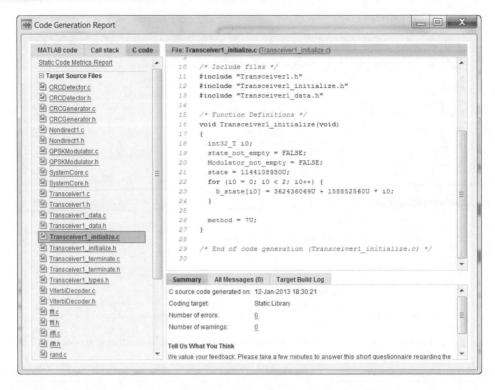

Figure 10.20   The new content of the initialization function – sets random-number-generator seeds

10.10   Support for Fixed-Point Data

So far, the functions developed in this chapter have performed operations on single- and double-precision floating-point data. In the last section we also introduced functions from MATLAB and some toolboxes that generate binary data to represent transmitted bits. The corresponding variables were represented efficiently by the Boolean data type. In many cases, the indices and quantized values used in functions are best represented as integers. MATLAB supports six different native integer types: uint8 (unsigned 8 bit integer), uint16 (unsigned 16 bit integer), uint32 (unsigned 32 bit integer), int8 (signed 8 bit integer), int16 (signed 16 bit integer), and int32 (signed 32 bit integer).

In some cases, we need to express data with a fixed-point data type. In fixed-point arithmetic, the range of values the data can take is drawn from a finite set. However, the real value is not necessarily a pure integer. As we described in Chapter 2, variables can be specified in MATLAB with fixed-point representation and fixed-point arithmetic can be performed by using Fixed-Point Designer (otherwise known as Fixed-Point Toolbox). Any fixed-point number in MATLAB can be expressed by the *fi* object. We need to specify three parameters of the object: (i) signed property (whether or not a variable is signed), (ii) word length (how many bits represent a number), and (iii) size of the fractional part (how many bits represent the fractional part of a number). Obviously, the integer part (the number of bits representing the integer part of the number) is equal to the word length minus the sum of the signed bit and the fractional bits.

10.10.1   Case Study: FFT Function

In this section, we update the last example (the function *Transceiver0.m*) to express the output
of the Quadrature Phase Shift Keying (QPSK) modulator with a fixed-point data type. Since
all operations before modulation use variables of the Boolean data type, if modulators are
converted to fixed-point and forward and inverse FFT operations have been performed, the
entire function will be based on fixed-point and integer data types. The first version of this
algorithm is as follows.

Algorithm

MATLAB function

```
function y=Transceiver0_fixed(u)
%% Constants
trellis=poly2trellis(7, [133 171]);
polynomial=[1 1 zeros(1, 16) 1 1 0 0 0 1 1];
%% Initializations
persistent Modulator DeModulator ConvEncoder Viterbi CRCGen CRCDet
if isempty(Modulator)
    Modulator      = comm.QPSKModulator('BitInput',true,'OutputDataType','Custom');
    DeModulator  = comm.QPSKDemodulator('BitOutput',true);
    ConvEncoder  = comm.ConvolutionalEncoder('TerminationMethod','Truncated',
'TrellisStructure', trellis);
    Viterbi         = comm.ViterbiDecoder('TrellisStructure', trellis,
'InputFormat','Hard','TerminationMethod','Truncated');
    CRCGen       = comm.CRCGenerator('Polynomial', polynomial);
    CRCDet       = comm.CRCDetector   ('Polynomial', polynomial);
end
tb           = step(CRCGen , u);                % CRC generator
cod_sig    = step(ConvEncoder , tb);           % Convolutional encoder
mod_sig    = step(Modulator, cod_sig);         % QPSK Modulator
sig          = ifft(mod_sig);                   % Perform IFFT
rec          = fft(sig);                         % Perform FFT
demod      = step(DeModulator, rec);           % QPSK Demodulator
dec         = step(Viterbi , demod);           % Viterbi decoder
y            = step(CRCDet , dec);             % CRC detector
```

Conversion of this function to fixed-point can be done in two ways: we can specify the output
data type of the modulator System object as fixed-point and specify its detail, or we can use
the *fi* object to create the fixed-point version of the modulator output after the (by default)
double-precision floating-point data have been created. The second approach maintains both
a floating-point and a fixed-point version of the same vector in the MATLAB workspace but
is not memory-efficient. After demodulation, since we use hard-decision decoding the output
data type is Boolean, again because hard-decision demodulation maps back to bits.

We can then run the testbench and generate code. When we examine the MATLAB Coder
Project, the readiness report indicates that the *ifft* and *fft* functions do not support fixed-point

data as their function inputs. To ensure that the entire function can perform fixed-point arithmetic, we must now find alternative algorithms that perform forward and inverse FFT operations and support fixed-point data types. The System objects *dsp.FFT* and *dsp.IFFT* from the DSP System Toolbox satisfy these requirements. This example actually clarifies one of the reasons for having redundant functionality in the Signal Processing Toolbox and DSP System Toolbox. The DSP System Toolbox has more implementation-oriented functionality and a more substantial support for fixed-point arithmetic, which is of interest to users concerned with hardware implementation. Support for fixed-point data types by *dsp.FFT* and *dsp.IFFT* in DSP System Toolbox is an obvious example of the mandate of the toolbox-supporting algorithm elaborations and shows the need for eventual hardware implementation. By replacing *fft* and *ifft* functions with *dsp.FFT* and *dsp.IFFT* System objects, respectively, from the DSP System Toolbox, the function can be made to handle fixed-point data.

Algorithm

MATLAB function

```
function y=Transceiver0_fixed2(u)
%% Constants
trellis=poly2trellis(7, [133 171]);
polynomial=[1 1 zeros(1, 16) 1 1 0 0 0 1 1];
%% Initializations
persistent Modulator DeModulator ConvEncoder Viterbi CRCGen CRCDet FFT IFFT
if isempty(Modulator)
   Modulator    = comm.QPSKModulator('BitInput',true,'OutputDataType','Custom');
   DeModulator  = comm.QPSKDemodulator('BitOutput',true, 'OutputDataType',
'Smallest unsigned integer');
   ConvEncoder  = comm.ConvolutionalEncoder('TerminationMethod','Truncated',
'TrellisStructure', trellis);
   Viterbi      = comm.ViterbiDecoder('TrellisStructure', trellis, 'InputFormat','Hard',...
      'TerminationMethod','Truncated', 'OutputDataType','logical');
   CRCGen       = comm.CRCGenerator('Polynomial', polynomial);
   CRCDet       = comm.CRCDetector  ('Polynomial', polynomial);
   FFT = dsp.FFT;
   IFFT = dsp.IFFT;
end
tb       = step(CRCGen , u);            % CRC generator
cod_sig  = step(ConvEncoder , tb);      % Convolutional encoder
mod_sig  = step(Modulator, cod_sig);    % QPSK Modulator
sig      = step(IFFT, mod_sig);         % Perform IFFT
rec      = step(FFT, sig);              % Perform FFT
demod    = step(DeModulator, rec);      % QPSK Demodulator
dec      = step(Viterbi , demod);       % Viterbi decoder
y        = step(CRCDet , dec);          % CRC detector
```

The MATLAB Coder Project will then successfully generate code for the function.

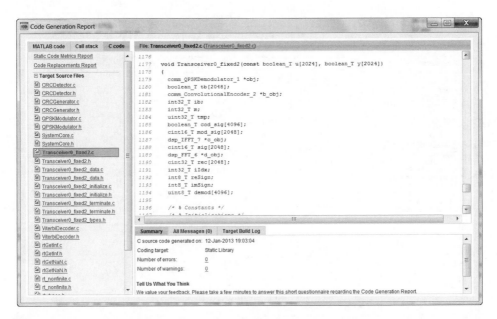

Figure 10.21 Code Generation Report: integer-based C code for the fixed-point version of the transceiver

Figure 10.21 shows the Code Generation Report of this function. The generated C code uses only integer data types. Note that no floating-point variables are present. Note also how for every arithmetic operation, including scaling, saturation, and wrapping, inline functions are defined.

For example, an inline function defining a multiword subtraction is illustrated in Figure 10.22. These functions are needed to perform arithmetic and logical operations on the integer data that implement the fixed-point arithmetic specified by the fixed-point data types. These low-level fixed-point operations are precisely the type of operation that, if performed manually, will add a lot of design time to our projects. By leveraging the Fixed-Point Toolbox and MATLAB Coder, we can save time and avoid many of the critical and tedious steps involved in converting a design from a floating-point to a fixed-point numerical representation.

10.11   Support for Variable-Sized Data

So far we have shown code generation from MATLAB functions in which the size of the input data is fixed. Fixed-sized code generation is quite straightforward; all that has to be done is to specify the size of each function input, and the generated C code is based on the same fixed data size.

In many situations, however, we need to generate code for a function whose input size changes during simulation. For example, in adaptive coding, as the coding rate changes, the output size of the channel coder changes, which means that the input size to the subsequent scrambler and modulator operations changes. Similarly, in adaptive modulation, even for a fixed number of bits at the input of the modulator, the output size can change depending on whether a QPSK, 16QAM, or 64QAM modulation scheme is used.

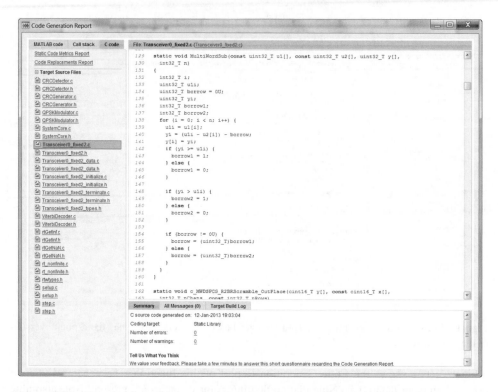

Figure 10.22 Sample of a multiword, integer-based routine developed automatically in fixed-point code generation

In this section, we show how C code can be generated from a function that can accommodate changes in variable size. When it comes to data sizes, we usually encounter three different code-generation modes: (i) data with fixed sizes, (ii) variable-sized data with an upper bound on the size, and (iii) unbounded variable-sized data. Each of these modes represents a tradeoff in terms of computational complexity, memory usage, and flexibility.

10.11.1   Case Study: Adaptive Modulation

We will illustrate here all three data-sizing modes and their effects on code generation, using an adaptive modulator as an example.

Algorithm

MATLAB function

```
function y=Modulator(u)
persistent QPSK
if isempty(QPSK)
```

```
QPSK       = comm.PSKModulator(4, 'BitInput', true, 'PhaseOffset', pi/4, ...
   'SymbolMapping', 'Custom', 'CustomSymbolMapping', [0 2 3 1]);
end
y=step(QPSK, u);
```

Let us start with a simple LTE QPSK modulator, examining the function (*Modulator.m*). This MATLAB function is very flexible in terms of input size. For example, for a QPSK modulator, as long as the input size is an even number, the function produces an output with half the size of the input.

Figure 10.23 illustrates how to we can execute the modulator function, first with an input bit vector of size 4200×1 and then with an input size of 256×1. We use the function MATLAB *whos* to examine the input and output sizes in each case. Without code generation, the function *Modulator.m* behaves in the way expected of MATLAB functions when it comes to changing the size of its input. As we change input size, the function generates an output whose size reflects the changes in input size. However, as we will see shortly, following code generation the generated MEX function may behave differently when the input size is changed.

10.11.2   Fixed-sized Code Generation

At a next step, we create a new code-generation project by typing the following command:

Algorithm

```
>> coder −new Modulator
```

```
>> u=randi([0 1], 4200, 1);
>> y=Modulator(u);
>> whos u y
   Name          Size              Bytes  Class     Attributes

   u            4200x1             33600  double
   y            2100x1             33600  double    complex

>> u2=randi([0 1], 256, 1);
>> y=Modulator(u2);
>> whos u2 y
   Name          Size              Bytes  Class     Attributes

   u2            256x1              2048  double
   y             128x1              2048  double    complex
```

Figure 10.23   Calling a modulator function with different input sizes

After adding the *Modulator* function to the project, we can use the following MATLAB testbench in the Autodefine Types link to specify the sizes and data types:

Algorithm

MATLAB script

```
u=randi([0 1], 4200,1);
y=Modulator(u);
```

After running this script, the AutoDefine Input Types tool correctly proposes 4200×1 as the input size to the function, as illustrated in Figure 10.24. We can identify two checkboxes in this window. These represent various options that determine the behavior of the generated MEX function when faced with changing input function sizes. If we leave both of the checkboxes unchecked, we are implementing a fixed-sized code generation. As we will see shortly, by clicking on the first checkbox we implement a variable-sized code generation with an upper bound. Finally, by clicking on the second checkbox, we implement an unbounded variable-sized code generation. Details of these cases will be discussed shortly.

To fully understand what we mean by "fixed-sized code generation," let us keep both of the checkboxes unchecked, proceed to the Build tab (Figure 10.25), and generate

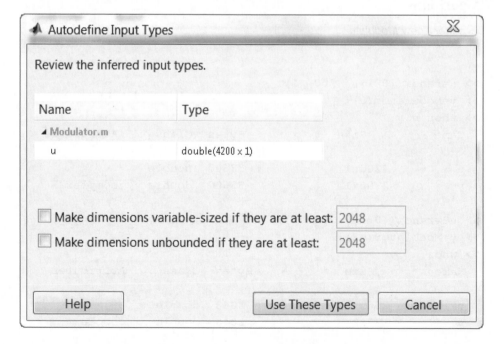

Figure 10.24   Options for fixed-sized code generation

Figure 10.25 MATLAB Coder Project: building a fixed-sized MEX function

the MEX function. We choose a default name of *Modulator_mex* for our output MEX function.

If we now call the generated MEX function with an input of any size other than 4200×1, the call produces error messages. In the following experiment, we first call the function with the correct input size and data type (a double vector with a 4200×1 size) and then with a double vector with a size of 256×1. The results and error messages are shown in Figure 10.26.

As we can see, the resulting fixed-sized code generation is not flexible and requires a particular size for the input. Alternatively, we can generate the MEX function of our modulator

```
>> u=randi([0 1], 4200, 1);
>> y=Modulator_mex(u);
>> whos u y
   Name          Size              Bytes  Class      Attributes

   u            4200x1             33600  double
   y            2100x1             33600  double     complex

>> u2=randi([0 1], 256, 1);
>> y=Modulator_mex(u2);
MATLAB expression 'u' is not of the correct size: expected [4200x1] found
[256x1].

Error in Modulator_mex
```

Figure 10.26   Calling the fixed-sized MEX function of the modulator

function to implement fixed-sized code generation by using the following command line script, which uses the *codegen* command:

Algorithm

MATLAB script

```
u=randi([0 1], 4200,1 );
codegen Modulator -args {u}
```

If we want to examine the C source code of the function, we can simply add two more options to the *codegen* command line, as follows:

Algorithm

MATLAB script

```
u=randi([0 1], 4200,1 );
codegen Modulator -args {u} −config:lib -report
```

The generated C source code can be examined by clicking on the View Report link the in MATLAB command line. When we open the Code Generation Report, as illustrated in Figure 10.27, we find that the C source file (*Modulator.c*) defines the modulator function with a constant real input array of 4200 elements.

Another way of ensuring a fixed-sized code generation is to specify the size of one variable based on the value of another. In this case, if the value is deemed constant then the

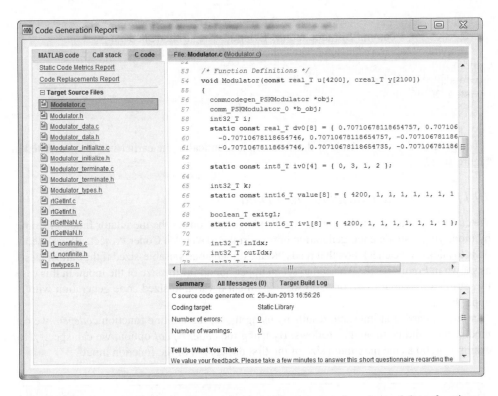

Figure 10.27   Code Generation Report: fixed-sized code generation of a modulator function

code-generation engine can easily infer the sizes of other variables. For example, in the *Modulator_fixedsize.m* function, the value of variable *N* determines the size of variable *u*, which corresponds to the input bits to the modulator. The *Modulator_fixedsize.m* function has no inputs and all variables are local to the body of the function. As a result, the MATLAB command that generates a fixed-sized code for the *Modulator_fixedsize* is simply:

Algorithm

```
>> codegen Modulator_fixedsize
```

Algorithm

MATLAB function

```
function y=Modulator_fixedsize
N=4200;
persistent QPSK
if isempty(QPSK)
```

```
QPSK      = comm.PSKModulator(4, 'BitInput', true, 'PhaseOffset', pi/4, ...
   'SymbolMapping', 'Custom', 'CustomSymbolMapping', [0 2 3 1]);
end
u=randi([0 1], N,1);
y=step(QPSK, u);
```

The generated C source code of this function is identical to the earlier version illustrated in Figure 10.27.

10.11.3 Bounded Variable-Sized Data

We can arrive at a bounded variable-sized code generation for the modulator function simply by modifying a single code-generation option. In the MATLAB Coder Project, all we need to do is to click on the check box that reads "Make dimensions variable-sized if they are at least," as shown in Figure 10.28. We need to set an upper bound for the size of the input. In this case, by setting a maximum size of 4200 we implement a variable-sized code generation with an upper bound.

In order to arrive at the same results by using the command line function *codegen*, we can modify the build command as follows. By using the *coder.typeof* option, we can specify, for example, 4200 as the maximum size of the first dimension of the function input.

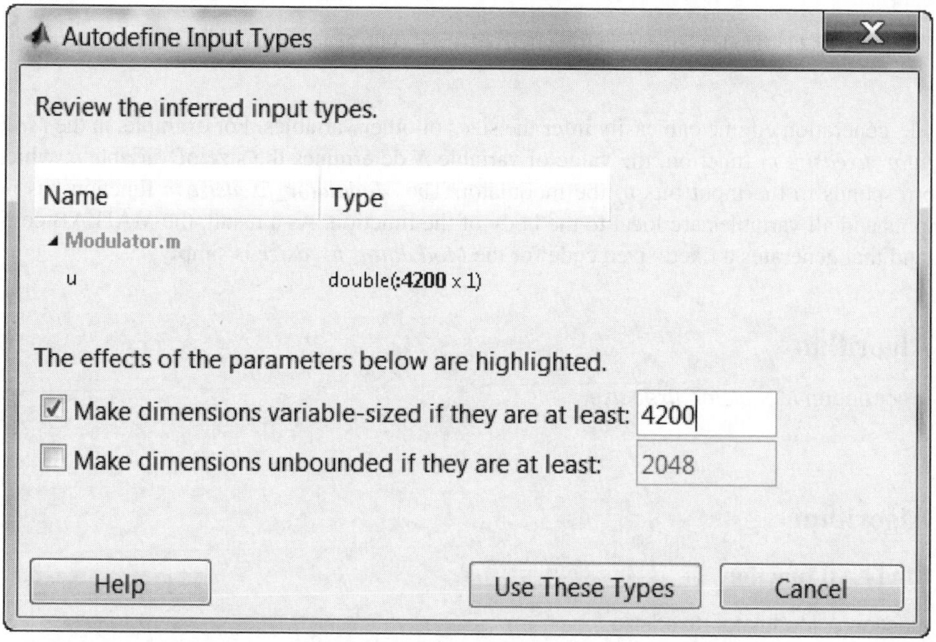

Figure 10.28 Options for bounded variable-sized code generation

Algorithm

MATLAB script

```
MaxSize = 4200;
u=randi([0 1], MaxSize,1);
codegen Modulator -args {coder.typeof(0,[MaxSize 1],1)}
```

The generated MEX function *Modulator_mex* can process input functions with sizes of up to 4200, as illustrated in Figure 10.29, by calling it in various scenarios. Note that the MEX function errors out whenever the input dimension exceeds 4200.

An alternative way of generating a variable-sized C code with an upper bound involves using the *assert* function. To illustrate this case, let us now bring the bit-generation function *randi* inside the modulator function. We call the modified function *Modulator_varsize_bounded*. The input variable N determines the size of the modulator input and output. To generate a bounded variable-sized C code for this function, we use the *assert* function to provide an upper limit for the value of variable N.

```
>> u=randi([0 1], 4200, 1);
>> y=Modulator_mex(u);
>> whos u y
  Name          Size            Bytes  Class      Attributes

  u            4200x1           33600  double
  y            2100x1           33600  double     complex

>> u2=randi([0 1], 256, 1);
>> y=Modulator_mex(u2);
>> whos u y
  Name          Size            Bytes  Class      Attributes

  u            4200x1           33600  double
  y             128x1            2048  double     complex

>> u3=randi([0 1], 123456, 1);
>> y=Modulator_mex(u3);
MATLAB expression 'u' is not of the correct size: expected [:4200x1] found
[123456x1].

Error in Modulator_mex
```

Figure 10.29   Calling the bounded variable-sized MEX function of a modulator

Algorithm

MATLAB function

```
function y=Modulator_varsize_bounded(N)
assert(N<=2400);
persistent QPSK
if isempty(QPSK)
    QPSK        = comm.PSKModulator(4, 'BitInput', true, 'PhaseOffset', pi/4, ...
        'SymbolMapping', 'Custom', 'CustomSymbolMapping', [0 2 3 1]);
end
u=randi([0 1], N,1);
y=step(QPSK, u);
```

The *Modulator_varsize_bounded.m* function has only one input (*N*) that determines the size of every variable inside the function. As a result, the MATLAB command that generates a bounded variable-sized code for the *Modulator_varsize_bounded* is:

Algorithm

>> codegen –args {N} Modulator_varsize_bounded

10.11.4   Unbounded Variable-Sized Data

To arrive at an unbounded variable-sized code generation for the modulator function, we need simply modify another code-generation option. In the MATLAB Coder Project, as illustrated in Figure 10.30, when we set the parameters of the Autodefine Input Types dialog we must click on the checkbox that reads "Make dimensions unbounded if they are at least." By setting a minimum size in the edit box that appears in this line, we signal to the code-generation engine that any input variable larger than the given size will be regarded as unbounded variable-sized data. As a result, the type of variable *u* reads as *double(:inf x 1)*, meaning that the first dimension of the variable is unbounded.

Alternatively, the *codegen* MATLAB command that generates a MEX function supporting unbounded input sizes is as follows:

Algorithm

>> codegen Modulator -args {coder.typeof(0,[inf 1],1)}

To verify proper operation, we run the same MATLAB script as in the last section, and we can see that no matter what the input size, the MEX function can produce the correct modulated outputs (Figure 10.31).

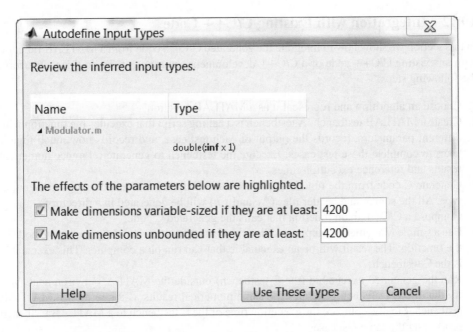

Figure 10.30 Options for unbounded variable-sized code generation

```
>> u=randi([0 1], 4200, 1);
>> y=Modulator_mex(u);
>> whos u y
  Name          Size              Bytes   Class      Attributes

  u             4200x1            33600   double
  y             2100x1            33600   double      complex

>> u2=randi([0 1], 256, 1);
>> y=Modulator_mex(u2);
>> whos u2 y
  Name          Size              Bytes   Class      Attributes

  u2            256x1              2048   double
  y             128x1              2048   double      complex

>> u3=randi([0 1], 123456, 1);
>> y=Modulator_mex(u3);
>> whos u3 y
  Name          Size              Bytes   Class      Attributes

  u3            123456x1         987648   double
  y             61728x1          987648   double      complex
```

Figure 10.31 Calling the unbounded variable-sized MEX function of a modulator

10.12   Integration with Existing C/C++ Code

In this section we show how to integrate the generated C/C++ code from a MATLAB function with an existing C/C++ code or a C/C++ development environment. To do this we perform the following steps:

1. Choose an algorithm and represent it as a MATLAB function.
2. Create a MATLAB testbench. A testbench is a calling script that executes the function with different parameters, records the output of each test case, and records how much time it takes to complete these test cases. Execute the testbench to generate reference numerical results and reference execution times.
3. Generate C code from the function. Choose static C library as the code-generation output type. All the source and header files (*.c and *.h) will be generated in a directory.
4. Compose a C/C++ main function that calls the generated C code.
5. Use a simple Makefile to compile and link the C main function and the generated C code of the function. The result will be an executable that can run on a computer. This executable is the C testbench.
6. Run the generated executable (the C testbench) outside the MATLAB environment. Verify that the C testbench generates the same numerical results as the reference MATLAB testbench. Finally, compare the execution time of the C testbench to a MATLAB testbench processing the same test cases.

10.12.1   Algorithm

To start the process of integrating MATLAB code with an external C code, we first need to choose an algorithm. We have selected a simplified form of the Physical Downlink Control Channel (PDCCH) processing algorithm [3]. In Chapter 9 we examined 17 different versions of the PDCCH algorithm. In this section we have chosen version 9, the MEX function of the eighth version of the algorithm, which incorporates all available System objects in the Communications System Toolbox. Version 9 has been shown to simulate faster than the first eight versions in the absence of parallel multicore processing. The MATLAB function that captures the eighth version of the algorithm is as follows:

Algorithm

MATLAB function

```
function [ber, bits]=zPDCCH_v8(EbNo, maxNumErrs, maxNumBits)
%% Constants
FRM=2048;
M=4; k=log2(M); codeRate=1/3;
snr = EbNo + 10*log10(k) + 10*log10(codeRate);
trellis=poly2trellis(7, [133 171 165]);
L=FRM+24;C=6; Index=[L+1:(3*L/2) (L/2+1):L];
%% Initializations
```

```
persistent Modulator AWGN DeModulator BitError ConvEncoder1 ConvEncoder2 Viterbi
CRCGen CRCDet
if isempty(Modulator)
    Modulator     = comm.QPSKModulator('BitInput',true);
    AWGN          = comm.AWGNChannel('NoiseMethod', 'Variance', 'VarianceSource',
'Input port');
    DeModulator =  comm.QPSKDemodulator('BitOutput',true);
    BitError      = comm.ErrorRate;
    ConvEncoder1=comm.ConvolutionalEncoder('TrellisStructure', trellis,
'FinalStateOutputPort', true, ...
        'TerminationMethod','Truncated');
    ConvEncoder2 =  comm.ConvolutionalEncoder('TerminationMethod','Truncated',
'InitialStateInputPort', true,...
        'TrellisStructure', trellis);
    Viterbi=comm.ViterbiDecoder('TrellisStructure', trellis,
'InputFormat','Hard','TerminationMethod','Truncated');
    CRCGen = comm.CRCGenerator('Polynomial',[1 1 zeros(1, 16) 1 1 0 0 0 1 1]);
    CRCDet = comm.CRCDetector   ('Polynomial',[1 1 zeros(1, 16) 1 1 0 0 0 1 1]);
end
%% Processing loop modeling transmitter, channel model and receiver
numErrs = 0; numBits = 0; nS=0;
results=zeros(3,1);
while ((numErrs < maxNumErrs) && (numBits < maxNumBits))
    % Transmitter
    u          = randi([0 1], FRM,1);               % Generate bit payload
    u1         = step(CRCGen, u);                    % CRC insertion
    u2         = u1((end-C+1):end);                  % Tail-biting convolutional coding
    [~, state] = step(ConvEncoder1, u2);
    u3         = step(ConvEncoder2, u1,state);
    u4         = fcn_RateMatcher(u3, L, codeRate);  % Rate matching
    u5         = fcn_Scrambler(u4, nS);              % Scrambling
    u6         = step(Modulator, u5);               % Modulation
    u7         = TransmitDiversityEncoderS(u6);      % MIMO Alamouti encoder
    % Channel
    [u8, h8]   = MIMOFadingChanS(u7);               % MIMO fading channel
    noise_var = real(var(u8(:)))/(10.^(0.1*snr));
    u9         = step(AWGN, u8, noise_var);          % AWGN
    % Receiver
    uA         = TransmitDiversityCombinerS(u9, h8);% MIMO Alamouti combiner
    uB         = step(DeModulator, uA);             % Demodulation
    uC         = fcn_Descrambler(uB, nS);            % Descrambling
    uD         = fcn_RateDematcher(uC, L);           % Rate de-matching
    uE         = [uD;uD];                            % Tail-biting
    uF         = step(Viterbi, uE);                 % Viterbi decoding
    uG         = uF(Index);
    y          = step(CRCDet, uG );                  % CRC detection
    results    = step(BitError, u, y);               % Update number of bit errors
```

```
   numErrs   = results(2);
   numBits   = results(3);
   nS           = nS + 2; nS = mod(nS, 20);
end
%% Clean up & collect results
ber = results(1); bits= results(3);
reset(BitError);
```

In order to manage the files and directories associated with C-code generation properly, we create a new directory in our computer and place all the MATLAB files in it. In this example, we create a directory called *C:\Examples\PDCCH* and copy all the files needed to run the eighth version of the algorithm to it. The MATLAB script performing these tasks is as follows:

Algorithm

MATLAB script: MATLAB_testbench_directory

```
%% Create new directory in C:\ drive
PARENTDIR='C:\';
NEWDIR='Examples\PDCCH';
mkdir(PARENTDIR,NEWDIR);
%% Make that your destination directory
DESTDIR=fullfile(PARENTDIR,NEWDIR);
%% Copy 10 necessary files to destination directory
copyfile('Alamouti_DecoderS.m',DESTDIR);
copyfile('Alamouti_EncoderS.m',DESTDIR);
copyfile('fcn_Descrambler.m',DESTDIR);
copyfile('fcn_RateDematcher.m',DESTDIR);
copyfile('fcn_RateMatcher.m',DESTDIR);
copyfile('fcn_Scrambler.m',DESTDIR);
copyfile('MIMOFadingChanS.m',DESTDIR);
copyfile('TransmitDiversityCombinerS.m',DESTDIR);
copyfile('TransmitDiversityEncoderS.m',DESTDIR);
copyfile('zPDCCH_v8.m',DESTDIR);
%% Go to destination directory
cd(DESTDIR);
```

10.12.2   Executing MATLAB Testbench

At this step we execute two scripts: a build script that generates a MEX function for the function *zPDCCH_v8.m* and a calling script, which constitutes our MATLAB testbench. These scripts can be created in the same destination directory as the MATLAB functions are stored in (*C:\Examples\PDCCCH*). Using the first script (*MATLAB_build_version9.m*), we can generate the MEX function of the eighth version of the PDCCH algorithm. The *codegen* command for this build script is as follows:

Algorithm

MATLAB script: MATLAB_build_version9

```
MaxSNR=8;
MaxNumBits=1e7;
MaxNumErrs=MaxNumBits;
fprintf(1,'\nGenerating MEX function for PDCCH algorithm ...\n');
codegen −args { MaxSNR, MaxNumErrs, MaxNumBits} zPDCCH_v8 −o zPDCCH_v9
fprintf(1,'Done.\n\n');
MEX_FCN_NAME='zPDCCH_v9';
fprintf(1,'Output MEX function name:  %s \n',MEX_FCN_NAME);
```

The testbench (*MATLAB_testbench_version9.m*) performs an iterative *Eb/N0* parameter sweep and records the BER values as a function of *Eb/N0*. The testbench uses eight test cases, corresponding to *Eb/N0* values of 0.5–4.0, increasing in steps of 0.5 dB. We compute and record the BER value for each *Eb/N0* value. The stopping criterion for the simulation is a predetermined number of processed bits. This is achieved by setting both the maximum number of errors (*MaxNumErrs*) and the maximum number of bits (*MaxNumBits*) parameters to a single value; for example, 10 million bits. Finally, we record the execution time of the eight test cases by obtaining and then subtracting the system clock values before and after the simulation.

Algorithm

MATLAB testbench: MATLAB_testbench_version9

```
MaxSNR=8;
MaxNumBits=1e7;
MaxNumErrs=1e7;
ber_vector=zeros(MaxSNR,1);
fprintf(1,'\nMATLAB testbench for PDCCH algorithm\n');
fprintf(1,'Maximum number of errors : %9d\n', MaxNumErrs);
fprintf(1,'Maximum number of bits   : %9d\n\n', MaxNumBits);
tic;
for snr=1:MaxSNR
   fprintf(1,'Iteration number %d\r',snr);
   EbNo=snr/2;
   ber= zPDCCH_v9(EbNo, MaxNumErrs, MaxNumBits);
   ber_vector(snr)=ber;
end
```

```
time_8=toc;
fprintf(1,'\nTime to complete %d iterations = %6.4f (sec)\n\n', MaxSNR, time_8);
for snr = 1:MaxSNR
   fprintf(1,'Iteration %2d  EbNo %3.1f  BER  %e\n', snr, snr/2, ber_vector(snr));
end
```

When we execute this testbench, the results shown in Figure 10.32 are displayed in the MATLAB command window. Note that we have obtained reference BER values and simulation times by running the MATLAB testbench. We will compare these values with the results obtained by running the C testbench, which we will create shortly.

```
>> MATLAB_testbench_version9

MATLAB testbench for PDCCH algorithm
Maximum number of errors :   10000000
Maximum number of bits   :   10000000

Iteration number 1
Iteration number 2
Iteration number 3
Iteration number 4
Iteration number 5
Iteration number 6
Iteration number 7
Iteration number 8


Time to complete 8 iterations = 354.8422 (sec)

Iteration   1  EbNo 0.5  BER  5.311296e-03
Iteration   2  EbNo 1.0  BER  1.773132e-03
Iteration   3  EbNo 1.5  BER  5.100804e-04
Iteration   4  EbNo 2.0  BER  1.504942e-04
Iteration   5  EbNo 2.5  BER  3.509865e-05
Iteration   6  EbNo 3.0  BER  4.299835e-06
Iteration   7  EbNo 3.5  BER  8.999654e-07
Iteration   8  EbNo 4.0  BER  0.000000e+00
```

Figure 10.32   MATLAB testbench, furnishing reference values for execution time and output values

10.12.3   Generating C Code

In this step, we generate C code from the function *zPDCCH_v8.m* using the *codegen* command. To generate a static C library we can use either the *codegen* command or the MATLAB Coder Project. When using the *codegen* command line we need only specify *lib* as the configuration option, as illustrated in the following script:

Algorithm

MATLAB script: MATLAB_build_version9

```
MaxSNR=8;
MaxNumBits=2e6;
MaxNumErrs=MaxNumBits;
fprintf(1,'Generating source files (*.c) and header files (*.h) for PDCCH algorithm ...');
codegen −args { MaxSNR, MaxNumErrs, MaxNumBits} zPDCCH_v8 −config:lib -report
fprintf(1,'Done.');
FCN_NAME='zPDCCH_v8';
Location=fullfile(pwd,'codegen','lib',FCN_NAME);
fprintf(1,'All generated files are in the following directory: \n%s\n', Location);
```

Upon completion, the *codegen* command prints a message in the MATLAB command line that includes a hyperlink to the generated C code. When we click on the hyperlink, we open the Code Generation Report (Figure 10.33).

All of the C source files and header files are stored in a unique folder under the Destination directory. In this example, the destination directory is *C:\Examples\PDCCH* and all the source files are in a subdirectory called *codegen\lib\zPDCCH_v8*. Figure 10.34 shows all the source and header files generated, listed by the *ls* command.

10.12.4   Entry-Point Functions in C

Having already generated C code from our MATLAB function, the rest of the development process can be performed completely outside the MATLAB environment. In order to generate a C executable, otherwise known as a C testbench, all we have to do is to write a C main function and call the generated entry-point functions in it.

In our example, the entry-point function in MATLAB is *zPDCCH_v8.m*. The generated C code will therefore have three header files, which define the entry-point C function prototypes for: (i) the main entry-point function, (ii) the initialization function, and (iii) the termination function. These files are *zPDCCH_v8.h*, *zPDCCH_v8_initialize.h*, and *zPDCCH_v8_terminate.h*, respectively.

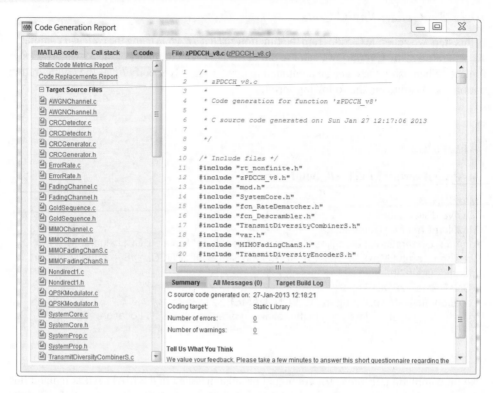

Figure 10.33   Code Generation Report: showing generated code for the *zPDCCH_v8* algorithm

```
>> pwd

ans =

C:\Examples\PDCCH\codegen\lib\zPDCCH_v8

>> ls *.c

AWGNChannel.c          MIMOFadingChanS.c          ViterbiDecoder.c          main.c
CRCDetector.c          Nondirect1.c               diff.c                    mod.c
CRCGenerator.c         QPSKModulator.c            fcn_Descrambler.c         permute.c
ErrorRate.c            SystemCore.c               fcn_RateDematcher.c       rand.c
FadingChannel.c        SystemProp.c               fcn_Scrambler.c           randn.c
GoldSequence.c         TransmitDiversityCombinerS.c   filter.c              repmat.c
MIMOChannel.c          TransmitDiversityEncoderS.c    floor.c               rtGetInf.c

>> ls *.h

AWGNChannel.h          MIMOFadingChanS.h          ViterbiDecoder.h          mod.h
CRCDetector.h          Nondirect1.h               diff.h                    permute.h
CRCGenerator.h         QPSKModulator.h            fcn_Descrambler.h         rand.h
ErrorRate.h            SystemCore.h               fcn_RateDematcher.h       randn.h
FadingChannel.h        SystemProp.h               fcn_Scrambler.h           repmat.h
GoldSequence.h         TransmitDiversityCombinerS.h   filter.h              rtGetInf.h
MIMOChannel.h          TransmitDiversityEncoderS.h    floor.h               rtGetNaN.h
```

Figure 10.34   List of generated C source files and header files

Algorithm

C header file: *zPDCCH_v8.h*

```
/*
 * zPDCCH_v8.h
 *
 * Code generation for function 'zPDCCH_v8'
 *
#ifndef __ZPDCCH_V8_H__
#define __ZPDCCH_V8_H__
/* Include files */
#include <float.h>
#include <math.h>
#include <stddef.h>
#include <stdlib.h>
#include <string.h>
#include "rt_nonfinite.h"
#include "rtwtypes.h"
#include "zPDCCH_v8_types.h"
/* Function Declarations */
extern void zPDCCH_v8(real_T EbNo, real_T maxNumErrs, real_T maxNumBits, real_T
*ber, real_T *bits);
#endif
/* end of code generation (zPDCCH_v8.h) */
```

Algorithm

C header file: *zPDCCH_v8_initialize.h*

```
/*
 * zPDCCH_v8_initialize.h
 *
 * Code generation for function 'zPDCCH_v8_initialize'
 *
 */
#ifndef __ZPDCCH_V8_INITIALIZE_H__
#define __ZPDCCH_V8_INITIALIZE_H__
/* Include files */
#include <float.h>
#include <math.h>
#include <stddef.h>
#include <stdlib.h>
#include <string.h>
#include "rt_nonfinite.h"
#include "rtwtypes.h"
#include "zPDCCH_v8_types.h"
/* Function Declarations */
```

```
extern void zPDCCH_v8_initialize(void);
#endif
/* end of code generation (zPDCCH_v8_initialize.h) */
```

Algorithm

C header file: *zPDCCH_v8_terminate.h*

```
/*
 * zPDCCH_v8_terminate.h
 *
 * Code generation for function 'zPDCCH_v8_terminate'
 *
 */
#ifndef __ZPDCCH_V8_TERMINATE_H__
#define __ZPDCCH_V8_TERMINATE_H__
/* Include files */
#include <float.h>
#include <math.h>
#include <stddef.h>
#include <stdlib.h>
#include <string.h>
#include "rt_nonfinite.h"
#include "rtwtypes.h"
#include "zPDCCH_v8_types.h"
/* Function Declarations */
extern void zPDCCH_v8_terminate(void);
#endif
/* end of code generation (zPDCCH_v8_terminate.h) */
```

These header files must be included in the C main function. It is important to understand how the entry-point functions are called in the main C function. Usually an algorithm needs: (i) an initialization function that sets up data and parameters outside a processing loop, (ii) a main entry-point function that is called inside a processing loop, and (iii) a termination function that cleans up all the resources (data, memory, etc.) that the initialization and entry-point function utilized. The following pseudo-code describes the general structure of the C code inside the main C file and the way it calls the entry-point functions.

Algorithm

```
>> Initialization_function();
>> An iterative processing loop
  >> { that calls Main_entry_point_function many times;}
>> Terminate_function();
```

10.12.5   C Main Function

The following C main function follows exactly the calling structure described in the previous section. Note that the first few lines of the main C file contain the typical variable declarations and reads simulation parameters from the command line. The next portion of the C code is the crux of the simulation. First we record the system clock by calling the clock function in C before simulating our function. Then we call the *zPDCCH_v8_initialize* function to initialize necessary data outside the processing loop. In the processing for loop we iterate through *Eb/N0* values, and by calling the main entry-point function (*zPDCCH_v8*) we obtain the BER measures. Finally, after the processing loop is completed we call the *zPDCCH_v8_terminate* function to release the data we have initialized and once again record the system clock by calling the *clock* function. The elapsed time for the simulation is calculated as the difference between the two recorded clock times. At the end of the function we print the elapsed time and the BER results as a function of the *Eb/N0* values.

Algorithm

C header file: *zPDCCH_v8_terminate.h*

```c
#include <stdio.h>
#include <math.h>
#include <time.h>
#include "rtwtypes.h"
#include "zPDCCH_v8.h"
#define MIN_EBNO 1
#define MAX_EBNO 9
int main( int argc, char *argv[])
{
  int EbNo, maxNumErrs, maxNumBits;
  double snr, elapsed;
  double ber_vector[MAX_EBNO], bits_vector[MAX_EBNO];
  time_t t1,t2;
  printf("\nMain C testbench for PDCCH algorithm\n");
  if ( argc!= 3) {
      printf("Usage : main.exe  Max_Number_of_Errors  Max_Number_of_Bits\n");
      exit(1);
    }
  maxNumBits = atoi(argv[1]);
  maxNumErrs = atoi(argv[2]);
  printf("Maximum Number of Errors :  %d\n", maxNumErrs);
  printf("Maximum Number of Bits   :  %d\n\n", maxNumBits);
  /*****************************************************/
  t1=clock();
  zPDCCH_v8_initialize();
  for (EbNo=MIN_EBNO; EbNo<MAX_EBNO; EbNo++)
  {
    printf("Iteration number %2d\n", EbNo);
    snr = 0.5*((double)EbNo);
    zPDCCH_v8(snr, maxNumErrs, maxNumBits, &ber_vector[EbNo], &bits_vector[EbNo]);
```

```
}
zPDCCH_v8_terminate();
t2=clock();
elapsed = ((double) (t2 - t1)) / CLOCKS_PER_SEC;
/******************************************************/
printf("\nTime to complete %2d iterations = %f (sec) in C\n\n", (MAX_EBNO-MIN_EBNO),
elapsed);
  for (EbNo=MIN_EBNO; EbNo<MAX_EBNO; EbNo++)
    printf("Iteration %2d   EbNo: %3.1f  BER: %e\n", EbNo, 0.5*EbNo ,ber_vector[EbNo]);

  return(0);
} /* end of main() */
```

10.12.6   Compiling and Linking

So far we have added a C main function (*main.c*) to the same directory in which we generated all the files from our MATLAB algorithm. We also need to add a simple Makefile in order to compile and link the source files and create an executable. The Makefile is illustrated in Figure 10.35. It is intended for use on a PC (desktop or laptop) that runs the Microsoft

```
Makefile  ✕

    # Run this makefile with gmake utility


    CC = cl
    CFLAGS = /O2 -I.
    COMPILE = $(CC) $(CFLAGS) -c
    OBJFILES := $(patsubst %.c,%.obj,$(wildcard *.c))


    all: main.exe


    main.exe: $(OBJFILES)
        $(CC) /O2 /Femain.exe $(OBJFILES)


    %.obj: %.c
        $(COMPILE)  $<


    clean:
        del *.obj main.exe
```

Figure 10.35   Simple Makefile for the generation of an executable (Microsoft Windows version)

Windows operating system. The C compiler command is *cl* (the Microsoft Visual C++ compiler command) and a level-2 optimization option is used. With some basic modifications, this Makefile can be used in Linux and other Unix environments that use, for example, *gcc* or other compilers. The Makefile first compiles every source file (**.c*) into an object file (**.obj*) and then links all the object files to generate the output executable file *main.exe*. We use gmake (GNU Makefile utility) to call this Makefile and create the executable.

Figures 10.36 and 10.37 show the steps involved in calling the Makefile utility. First we open the Windows SDK 7.1 Command Prompt, then we navigate to the directory where the generated files, the C main file, and the Makefile are located (*C:\Examples\PDCCH\ codegen\lib\zPDCCH_v8*). Next we call the gmake utility with a clean key in order to remove all the object files and the executable. Finally, by calling the all key, we compile one by one all the source files and link them to create the *main.exe* executable. This executable is the C testbench that we will use to verify the proper operation of code generation for our PDCCH algorithm.

10.12.7 Executing C Testbench

By executing the executable (*main.exe*) as illustrated in Figure 10.38, we are effectively calling the C testbench for our algorithm. In order to make a fair comparison, at the command line we specify the same values for the maximum number of errors and maximum number of bits processed in each iteration as in the MATLAB testbench.

The C testbench performs an iterative *Eb/N0* parameter sweep and records the BER values as a function of *Eb/N0*. It records the total time it takes to complete eight iterations and prints

```
Setting SDK environment relative to C:\Program Files\Microsoft SDKs\Windows\v7.1
\.
Targeting Windows 7 x64 Debug


C:\Program Files\Microsoft SDKs\Windows\v7.1>cd C:\Examples\PDCCH\codegen\lib\zP
DCCH_v8

C:\Examples\PDCCH\codegen\lib\zPDCCH_v8>gmake -f Makefile -k clean
del *.obj main.exe

C:\Examples\PDCCH\codegen\lib\zPDCCH_v8>gmake -f Makefile -k all
cl /O2 -I.  -c  AWGNChannel.c
Microsoft (R) C/C++ Optimizing Compiler Version 16.00.40219.01 for x64
Copyright (C) Microsoft Corporation.  All rights reserved.

AWGNChannel.c
cl /O2 -I.  -c  CRCDetector.c
Microsoft (R) C/C++ Optimizing Compiler Version 16.00.40219.01 for x64
Copyright (C) Microsoft Corporation.  All rights reserved.

CRCDetector.c
cl /O2 -I.  -c  CRCGenerator.c
Microsoft (R) C/C++ Optimizing Compiler Version 16.00.40219.01 for x64
Copyright (C) Microsoft Corporation.  All rights reserved.
```

Figure 10.36 Steps involved in executing a Makefile: compiling each generated C code

```
zPDCCH_v8_rtwutil.c
cl /O2 -I. -c zPDCCH_v8_terminate.c
Microsoft (R) C/C++ Optimizing Compiler Version 16.00.40219.01 for x64
Copyright (C) Microsoft Corporation.  All rights reserved.

zPDCCH_v8_terminate.c
cl /O2 /Femain.exe AWGNChannel.obj CRCDetector.obj CRCGenerator.obj ErrorRate.ob
j FadingChannel.obj GoldSequence.obj MIMOChannel.obj MIMOFadingChanS.obj Nondire
ct1.obj QPSKModulator.obj SystemCore.obj SystemProp.obj TransmitDiversityCombine
rS.obj TransmitDiversityEncoderS.obj ViterbiDecoder.obj diff.obj fcn_Descrambler
.obj fcn_RateDematcher.obj fcn_Scrambler.obj filter.obj floor.obj main.obj mod.o
bj permute.obj rand.obj randn.obj repmat.obj rtGetInf.obj rtGetNaN.obj rt_nonfin
ite.obj setup.obj sum.obj var.obj xor.obj zPDCCH_v8.obj zPDCCH_v8_data.obj zPDCC
H_v8_emxutil.obj zPDCCH_v8_initialize.obj zPDCCH_v8_rtwutil.obj zPDCCH_v8_termin
ate.obj
Microsoft (R) C/C++ Optimizing Compiler Version 16.00.40219.01 for x64
Copyright (C) Microsoft Corporation.  All rights reserved.

Microsoft (R) Incremental Linker Version 10.00.40219.01
Copyright (C) Microsoft Corporation.  All rights reserved.

/out:main.exe
AWGNChannel.obj
CRCDetector.obj
```

Figure 10.37   Steps involved in executing a Makefile: linking the executable

```
C:\Examples\PDCCH\codegen\lib\zPDCCH_v8>main.exe 10000000 10000000

Main C testbench for PDCCH algorithm
Maximum Number of Errors :   10000000
Maximum Number of Bits    :   10000000

Iteration number  1
Iteration number  2
Iteration number  3
Iteration number  4
Iteration number  5
Iteration number  6
Iteration number  7
Iteration number  8

Time to complete  8 iterations = 340.989000 (sec) in C

Iteration  1     EbNo: 0.5    BER: 5.311296e-003
Iteration  2     EbNo: 1.0    BER: 1.773132e-003
Iteration  3     EbNo: 1.5    BER: 5.100804e-004
Iteration  4     EbNo: 2.0    BER: 1.504942e-004
Iteration  5     EbNo: 2.5    BER: 3.509865e-005
Iteration  6     EbNo: 3.0    BER: 4.299835e-006
Iteration  7     EbNo: 3.5    BER: 8.999654e-007
Iteration  8     EbNo: 4.0    BER: 0.000000e+000

C:\Examples\PDCCH\codegen\lib\zPDCCH_v8>
```

Figure 10.38   Screenshot of the C testbench output

Table 10.2 Simulation time for PDCCH processing in MATLAB and C testbenches

Simulation method for PDCCH processing	Simulation time (seconds)
C testbench calling generated C code	339.96
MATLAB testbench calling MEX function	354.84

the values of BER obtained at each. Table 10.2 illustrates the simulation time for the PDCCH processing of 10 million bits iterated over eight Signal-to-Noise Ratio (SNR) values.

The results indicate that the time it takes for the generated C code to process a given amount of data in a native C testbench is very similar to that taken by a MATLAB testbench calling the MEX function for the same algorithm. This is not surprising as any MEX function that is generated by MATLAB Coder essentially translates MATLAB code to C code, compiles it, and then calls it from the MATLAB command line. As such, the performance of the resulting MEX function should be compatible with the results of manually integrating the generated C code within an external C testbench.

10.13   Chapter Summary

In this chapter we examined the process of generating standalone C code from MATLAB code using MATLAB Coder. We first reviewed various use cases for C/C++ code generation, including: (i) accelerating simulation speed, (ii) prototyping a design with a standalone executable, (iii) implementing in an embedded processors, and (iv) integrating with an existing C-based project or software.

We then presented the process of code generation using two distinct methods: (i) calling the *codegen* function from the MATLAB command line and (ii) using the MATLAB Coder Project Application. We elaborated on various language features and constructs supported for code generation and we highlighted the code-generation support given by selected System toolboxes for various data types, including fixed-point data, and for MATLAB programs employing variable-sized data. Finally, we presented the workflow involved in integrating generated code from a MATLAB algorithm with an existing C/C++ testbench.

References

[1] MathWorks Product Releases, http://www.mathworks.com/support/compilers/R2013a/index.html (accessed 16 August 2013).
[2] MathWorks MATLAB Coder, http://www.mathworks.com/help/coder/index.html (accessed 16 August 2013).
[3] 3GPP Evolved Universal Terrestrial Radio Access (E-UTRA); Physical Channels and Modulation, Version 10.0.0 Release 10. TS 36.211.

11

Summary

In this chapter we summarize the topics discussed in the book and provide a framework for future work. We have subdivided the summary into four sections. First, we review our learning objectives regarding the modeling of the LTE (Long Term Evolution) transceiver system. Then we summarize our findings regarding simulation of the system model and how to accelerate it. Third, we relate what we have learned about bridging the gap between modeling and implementation and how to prototype the simulation model as C/C++ software. Finally, we review some of the topics related to the LTE PHY (Physical Layer) that we have not had the chance to study in detail. Considering the level of detail needed to do justice to these topics, we have decided that they cannot be adequately covered in this volume and have left them as subjects of a future work.

11.1 Modeling

As the first learning objective of this book, we provided an overview of the mathematical modeling of the LTE PHY. Our aim was to provide a balanced approach to the discussion in order to foster a deeper understanding. As such, we decided to incorporate three distinct yet complementary conceptual elements: (i) providing an introductory theoretical overview of LTE-enabling technologies such as Orthogonal Frequency Division Multiplexing (OFDM) multicarrier transmission and Multiple Input Multiple Output (MIMO) multi-antenna schemes; (ii) providing an introductory technical overview of LTE specifications, focusing on a more detailed coverage of downlink transmission; and (iii) providing detailed MATLAB® algorithms and testbenches for step-by-step learning and hands-on simulation of the LTE standard. This balanced multitier approach is one of the distinguishing features of this book.

In this section we will summarize what we have presented regarding each of these conceptual elements.

Understanding LTE with MATLAB®: From Mathematical Modeling to Simulation and Prototyping, First Edition.
Houman Zarrinkoub.
© 2014 John Wiley & Sons, Ltd. Published 2014 by John Wiley & Sons, Ltd.

11.1.1   Theoretical Considerations

Throughout the book we have provided discussions regarding the theoretical background of the enabling technologies of the LTE standard. We studied the LTE multicarrier transmission schemes (i.e., OFDM in downlink and its single-carrier counterpart SC-FDM (Single-Carrier Frequency Division Multiplexing) in the uplink), as well as the multi-antenna MIMO transmission schemes.

We presented various aspects of the theoretical underpinnings for the MIMO–OFDM transmission techniques. These revealed how MIMO and OFDM are combined in the standard and helped explain the success of the technology in achieving its goals of high maximum data rates and high throughputs in mobile communications. We also discussed how incorporating the best technologies from previous standards, such as link adaptations through adaptive modulation and coding and efficient turbo coding, contribute to the overall performance of the LTE standard. Examining the theoretical background of the underlying LTE technologies can also be useful in understanding other modern communication systems. OFDM and MIMO technologies also form the fundamental basis of the WiMAX and the new wireless LAN standards.

11.1.2   Standard Specifications

Besides discussing the theoretical foundations, we provided a detailed presentation of PHY signal processing, with a special focus on downlink processing. We reviewed various channels and signals used in the standard. We also provided a more in-depth look at both the Downlink Shared Channel (DLSCH) processing and Physical Downlink Shared Channel (PDSCH) processing.

In particular, we examined in detail the composition of the time–frequency resource grid used in both OFDM and SC-FDM transmission schemes. Understanding the structure of the resource grid shed light on how the LTE standard organizes user data, control information, reference and other signals, and how it performs channel estimation and equalization operations necessary to recover the data at the receiver. It also showed how easily the standard combines the OFDM multicarrier scheme with various MIMO multi-antenna techniques. We highlighted the flexibility of the standard in maintaining a single transmission structure yet accommodating nine different transmission modes for downlink and various uplink transmission modes. We also described how different transmission modes cater to different scheduling conditions and different profiles of mobility and channel quality.

11.1.3   Algorithms in MATLAB

As the distinguishing feature of this book, we presented PHY modeling with a progressive set of algorithms and testbenches in MATLAB and Simulink. Our goal in providing MATLAB algorithms and testbenches was to introduce an initial platform for MATLAB users who are involved in communications system design. Our hope was to offer a starting point that fosters future collaborations among members of this community. Simulating an executable specification of a communications system in MATLAB and Simulink can help take the guesswork out of validating the effects of introducing innovative algorithms in system design.

Starting with MATLAB algorithms characterizing the basic scrambling, modulation, and coding in Chapter 4, we proceeded to include the OFDM multicarrier transmission in

Chapter 5 and various MIMO techniques, including transmit diversity and spatial multiplexing, in Chapter 6. In Chapter 7 we presented MATLAB algorithms that model typical link-adaptation strategies and in Chapter 8 we put together an LTE transceiver covering the first four modes of downlink transmission, then provided various assessments of the quality and performance of the physical-layer simulation model. Finally, in Chapters 9 and 10 we provided MATLAB algorithms that accelerate simulations and generate C code for the prototyping of designs as standalone applications. These topics will be discussed in further detail shortly.

11.1.3.1 Receiver Design

As with most communications standards, the LTE standard only specifies transmitter operations. Since the receiver operations are not explicitly specified, this provides an opportunity to develop innovative receiver algorithms. The innovations, when integrated within the software and hardware implementations by the network equipment and mobile terminal manufacturers, represent the proprietary and value-added contributions of each mobile communications system provider.

MATLAB and its communications system design tools provide an easy-to-use environment for experimenting with the design of various receiver components. In this book we presented various alternatives to different receiver components of the LTE system model. For example, in Chapter 5 we discussed receiver operations related to estimation of the channel-frequency response based on received reference signals. We examined an ideal channel estimator and three different channel-estimation algorithms based on the interpolation of pilot signals. The interpolation functions expand the channel responses computed at the pilots to cover the entire resource grid. As another example, in Chapter 6 we examined various MIMO receiver operations, studying three different approaches to computing best estimates of the transmitted symbols at the receiver. These techniques were based on the Zero Forcing (ZF), Minimum Mean Square Error (MMSE), and Soft-Sphere Decoder (SSD) algorithms. In Chapter 8 we examined the effects of each of these receiver algorithms on the overall system performance. By looking at the algorithm-specific metrics, such as the memory footprint or the computational complexity, as well as the system-level metrics, such as the Bit Error Rate (BER) or the throughput, we can assess the tradeoffs associated with each.

11.1.3.2 Simulation Testbenches

Throughout the book we created and updated MATLAB testbenches (or scripts) to evaluate qualitatively and quantitatively the performance of our LTE transceivers. The testbenches included the transmitter and receiver processing chains and the channel modeling sections needed to represent a transceiver. They also included various qualitative measures, such as spectrum analyzers and constellation diagrams, and quantitative measures, such as BER and throughput computations.

11.1.3.3 Algorithmic Building Blocks

It is important to choose the right granularity for the components of the MATLAB algorithms that model a complex system such as the LTE transceiver. We used a criterion that reflects the system-modeling and simulation mandate of this book. We did not reimplement basic communications building blocks such as modulators, convolutional or turbo encoders, decoders, or space–time block-coding components. For example, in order to implement OFDM transmitter and receiver operations we used MATLAB functions for forward Fast Fourier Transform (FFT) and Inverse Fast Fourier Transform (IFFT). We also used the *dsp.FFT* and *dsp.IFFT* System objects of the DSP System Toolbox as alternative implementations. These System objects can handle fixed-point modeling and block sizes that are not a power of two. We also used modulators, turbo coders, and channel-modeling System objects from the Communications System Toolbox. Leveraging available components from the toolbox, and not spending time redeveloping basic building blocks such as turbo decoders in MATLAB, helped accelerate the pace of the creation of system models for the LTE transceiver.

11.2   Simulation

It is important to develop a full mathematical model for any communications standard. However, to validate a model's accuracy, we must perform a representative number of software simulations. Since many of the performance metrics used in communications systems, such as throughputs and BERs, are measured in a statistical sense, a large amount of data must be processed by a simulation model. Furthermore, in order to verify that a system is robust against occasional outlier degradations, the simulations should be large enough to cover these rare occurrences. These considerations prompted us to look at various ways in which a system model can be optimized for speed. We looked at different methodologies for simulation acceleration and highlighted various tools and techniques that sped up our simulation model in MATLAB and Simulink.

11.2.1   Simulation Acceleration

Acceleration of a software simulation represents a classical tradeoff between an easy-to-understand and readable description of the model on one hand and optimized performance on the other. In our step-by-step approach to developing the components of the LTE model, we took great pains to organize the MATLAB code in a way that is self-contained. To make the code easy to understand, we represented most of the more complicated algorithms as a combination of less complicated subcomponents, without using any shortcuts or code factoring.

As instructive as this approach is, in order to accelerate the simulation speed we sometimes need to take advantage of typical methodologies that factor out repeated operations and fuse processing loops and optimize for various compilers or platform-specific libraries. In Chapter 9, we highlighted many MATLAB programming techniques that rely on these typical acceleration methodologies.

One of the most important and distinguishing features of the acceleration strategies presented in Chapter 9 was their preservation of numerical accuracy. As we went through various code optimizations, we showcased the fact that as successive versions of MATLAB codes execute more rapidly, they still produce the same numerical results. On the other hand, taking a more liberal approach to acceleration can be quite useful in receiver design where no standard specification is available. Many designers use shortcuts or approximations that substantially accelerate the simulation but do not preserve the numerical results. We deliberately took a conservative approach to code optimization and limited its scope to a subset that preserves numerical accuracy in order to make the process of validation easier and more direct.

11.2.2   Acceleration Methods

In Chapter 9, we showcased various techniques used to accelerate simulations of our LTE system model in MATLAB and Simulink. We presented a series of six types of optimization applied to control-channel processing. The techniques either provide ways of optimizing MATLAB programs or gain performance improvements through the use of additional computing power or by retargeting the design to compiled C code. We started with a baseline algorithm and through successive profiling and code updates introduced the following optimizations:

- Better MATLAB serial programming techniques (vectorization, preallocation)
- Use of System objects
- MATLAB-to-C code generation (MATLAB Executable, MEX)
- Parallel computing (parfor, spmd)
- GPU (Graphics Processing Unit)-optimized System objects
- Rapid accelerator mode for simulation in Simulink.

We also showed how to further accelerate simulations by combining two or more of these techniques. To take advantage of some of their benefits, specialized product capabilities beyond what is offered in application-specific toolboxes must be used. For example, MATLAB parallel-computing products provide computing techniques that take advantage of multicore processors, computer clusters, and GPUs. MATLAB Coder provides the ability to automatically convert a MATLAB code to C code, which can be compiled to provide faster simulations.

11.2.3   Implementation

Besides discussions regarding modeling and simulation, in Chapter 10 we went through the first steps involved in implementing the LTE-standard model. In order to bridge the gap between modeling and implementation, we used the MATLAB Coder to generate a prototype of the model as C code. We showed how the ANSI/ISO C source code generated by MATLAB Coder can be integrated with existing C/C++ testbenches and applications.

11.3   Directions for Future Work

There is a lot more to be done before we can adequately specify every detail of the PHY model of the LTE standard in MATLAB. In this book, our approach has mostly been pedagogic and educational. We focused on the LTE-enabling technologies, aiming to shed light on user-plane signal processing. We also covered as much detail as needed regarding various physical signals and channels, the organization of data in the OFDM resource grid, and the handling of multi-antenna techniques. These discussions clarified the underlying approach to transmission and explained the feasibility of achieving high data rates and improved system throughputs, as mandated by the standard specifications.

The next level of modeling is to provide a software solution that can be used as a reference to verify conformity to the LTE-standard requirements. If our objective is to ensure standard compliance, we must incorporate much more detail in our simulation model. The resulting LTE simulation model in MATLAB needs to incorporate all standard tests and cover all transmission modes and scenarios.

Next we will go through a list of modeling components that need to be added in order to evolve our baseline simulation model to the next level. With these upgrades, we can ultimately turn the LTE system model into a simulation platform for LTE-standard compliance testing. We will present these details in three sections: user-plane modeling, control-plane modeling, and system-access modules.

11.3.1   User-Plane Details

In order to update the LTE simulation model developed in this book, we need first to cover all aspects of user-plane modeling. These include the inclusion of both FDD (Frequency Division Duplex) and TDD (Time Division Duplex) duplexing for time framing, a complete treatment of both downlink and uplink shared-channel processing, and the inclusion of the LTE-Advanced features. These items are discussed in this section.

11.3.1.1   FDD and TDD Duplexing

As we saw in this book, two types of frame structure are specified in the LTE standard. Type 1 frames are used in FDD mode and type 2 frames in TDD mode. We have provided details relating to the FDD and type 1 frames. With minor modifications, we can present MATLAB functions that represent the time framing applicable to the TDD duplexing modes. Similarly, throughout the book we used normal cyclic prefix lengths, and again with minor modifications of the MATLAB code we can also accommodate extended cyclic prefixes in OFDM and SC-FDM transmissions.

11.3.1.2   Uplink Processing (PUSCH)

We have focused entirely on downlink transmission details in this book. The future work should contain the signal processing chain of the Physical Uplink Shared Channel (PUSCH). Many of the MATLAB components developed for downlink transmission can be used for uplink modeling almost without modification. However, there are some differences specifically related to the reference signals that are based on Zadoff–Chu sequences in the uplink specifications.

11.3.1.3   Complete Downlink Transmission Modes

We examined in detail the first four downlink-transmission modes. A complete model should include all of the modes, including the Downlink Enhanced MIMO modes (modes 7, 8, and 9), UE (User Equipment)-specific beamforming modes, and single-layer spatial-multiplexing modes. The modeling should include the generation and placement of various types of reference signal, including the Channel State Information Reference Signal (CSI-RS) and the Demodulation Reference Signal (DM-RS).

11.3.1.4   LTE-Advanced Features

LTE-Advanced features should also be included in the LTE MATLAB receiver model. These include in particular an uplink MIMO transmission and carrier aggregation. A multi-user uplink MIMO example populates and transmits PUSCH subframes in such a way that multiple UEs can share resources in a transmission. This technique is quite effective in boosting the uplink throughput. Carrier aggregation is another LTE-Advanced feature that enables downlink transmission to cover multiple carriers. By leveraging up to five contiguous carriers, carrier aggregation is the main technique responsible for achieving the maximum data rate of 1 Gbps provisioned within the LTE-Advanced standard. Functions that handle these two features must be part of a standard compliant MATLAB model for LTE PHY. As each of the processing chains in each of the carrier-aggregation bandwidths is independent, parallel processing can provide an obvious boost to the processing time needed for implementation. As such, the techniques we learned in Chapter 9 are directly applicable here.

11.3.2   Control-Plane Processing

As one of the features of this book, we focused on user-plane shared-channel processing. We did not study in any depth the control information needed to make the user-plane transmission possible. The collection of Downlink Control Information (DCI) and Uplink Control Information (UCI) must be part of a comprehensive LTE system model in MATLAB.

11.3.3   Hybrid Automatic Repeat Request

In the LTE standard, a Hybrid Automatic Repeat Request (HARQ) protocol is specified to ensure the reliability of data packet transmission and to manage occasional retransmissions. With a positive acknowledgment of a received packet, new data is transmitted. However, a negative acknowledgment initiates the retransmission of a previously sent packet. In order to provide a continuous supply of data packets at the receiver and minimize the waiting time for new data, we can send different data packets on different HARQ process numbers. In the LTE downlink specification, the DCI format contains explicit signaling related to the HARQ process. This includes an incremental-redundancy version and a new data indicator. In this book we have not presented the MATLAB functions necessary to implement the HARQ process. As an area of future work, inclusion of these routines will help contain the system delay resulting from excessive retransmissions and will update the way DLSCH handles channel coding with the inclusion of HARQ information.

11.3.4   System-Access Modules

In this book we focused on developing routines and functions that enable communications between UE and eNodeB (enhanced Node Base station) once initial access has been established. The LTE standard provides many components, signals, and capabilities for the initial phase of system access, cell search, and handoff procedures. A comprehensive system model in MATLAB should include these types of functionality. Two particular examples are described in further detail in this section.

11.3.4.1   Cell Search and Frame Timing

Encoded within the resource grid in the downlink transmitted signals are blocks of information that are essential to system access, cell search, and frame timing by a mobile unit. As we saw earlier, some of the initial system information is conveyed in the Master Information Block (MIB) and encoded and represented in the grid with a fixed modulation and coding scheme. The MIB contains information regarding system bandwidth, System Frame Number (SFN), and Physical Hybrid ARQ Indicator Channel (PHICH) configuration. We studied the Primary Synchronization Signal (PSS) and Secondary Synchronization Signal (SSS) and the Physical Broadcast Channel (PBCH) (containing the MIB) in Chapter 5. However, we did not present the MATLAB algorithms and functions that encode and transmit this information or the receiver operations that use it to obtain the initial system bandwidth and other critical information.

11.3.4.2   Random Access

In order to initiate access to the network, the UE uses the Physical Random Access Channel (PRACH) to transmit a preamble. Since this corresponds to the first communication from the UE to the eNodeB, the system does not know the type or specifications of the UE device. Various transmission modes, such as Cyclic Delay Diversity (CDD) and Precoding Vector Switching (PVS), provide a transparent way of decoding the preamble information. As we have not presented the uplink transmission details, we have not presented the MATLAB algorithms and functions needed for initial system access.

11.4   Concluding Remarks

In this chapter we summarized the learning objectives of this book and provided directions for further study. We subdivided the topics covered into two main categories: modeling and simulation. Within the modeling context, we elaborated on our stated goal of providing a balanced approach in presenting three distinct aspects related to understanding the LTE standard. We covered theoretical and mathematical descriptions of various enabling technologies, presented standard specifications as needed, and provided MATLAB programs and testbenches that enable hands-on experimentation with concepts through simulation. In the section on simulation, we highlighted the necessity of an adequate simulation speed for effective use of software that models a complex system like LTE. We reviewed various simulation acceleration techniques and prototyping mechanisms presented in this book. Finally, we presented a list of additional topics that need to be covered in a future work in order to provide a complete treatment of LTE-standard PHY modeling.

Depending on the interest in the LTE and MATLAB communities, the completion of our work in producing a fully standard-compliant LTE model in MATLAB may require another book. Having laid the foundation here by focusing on the enabling technologies and principles, the next volume would focus on standard compliance and full coverage of standard specifications.

Index

Understanding LTE with MATLAB®: From Mathematical Modeling to Simulation and Prototyping, First Edition.
Houman Zarrinkoub.
© 2014 John Wiley & Sons, Ltd. Published 2014 by John Wiley & Sons, Ltd.